INTRODUCTION TO MODEL SPACES
AND THEIR OPERATORS

The study of model spaces, the closed invariant subspaces of the backward shift operator, is a vast area of research with connections to complex analysis, operator theory, and functional analysis. This self-contained text is the ideal introduction for newcomers to the field. It sets out the basic ideas and quickly takes the reader through the history of the subject before ending up at the frontier of mathematical analysis. Open questions point to potential areas of future research, offering plenty of inspiration to graduate students wishing to advance further.

Stephan Ramon Garcia is an Associate Professor at Pomona College, California. He has earned multiple NSF research grants and five teaching awards from three different institutions. He has also authored over 60 research articles in operator theory, complex analysis, matrix analysis, and number theory.

Javad Mashreghi is a Professor of Mathematics at Université Laval, Québec, where he has been selected Star Professor of the Year seven times for excellence in teaching. His main fields of interest are complex analysis, operator theory and harmonic analysis. He is the author of several mathematical textbooks, monographs, and research articles. He won the G. de B. Robinson Award, the publication prize of the Canadian Mathematical Society, in 2004.

William T. Ross is the Roger Francis and Mary Saunders Richardson chair in mathematics at the University of Richmond, Virginia. He is the author of over 40 research papers in function theory and operator theory and also four books.

Already published

Introduction to Model Spaces and their Operators

STEPHAN RAMON GARCIA
Pomona College, California

JAVAD MASHREGHI
Université Laval, Québec

WILLIAM T. ROSS
University of Richmond, Virginia

CAMBRIDGE
UNIVERSITY PRESS

Shaftesbury Road, Cambridge CB2 8EA, United Kingdom

One Liberty Plaza, 20th Floor, New York, NY 10006, USA

477 Williamstown Road, Port Melbourne, VIC 3207, Australia

314–321, 3rd Floor, Plot 3, Splendor Forum, Jasola District Centre, New Delhi – 110025, India

103 Penang Road, #05–06/07, Visioncrest Commercial, Singapore 238467

Cambridge University Press is part of Cambridge University Press & Assessment,
a department of the University of Cambridge.

We share the University's mission to contribute to society through the pursuit of
education, learning and research at the highest international levels of excellence.

www.cambridge.org
Information on this title: www.cambridge.org/9781107108745

First published 2016

A catalogue record for this publication is available from the British Library

Library of Congress Cataloging-in-Publication data
Garcia, Stephan Ramon.
Introduction to model spaces and their operators / Stephan Ramon Garcia,
Pomona College, California, Javad Mashreghi, Université Laval, Québec, William T. Ross,
University of Richmond, Virginia.
pages   cm. – (Cambridge studies in advanced mathematics)
Includes bibliographical references and index.
ISBN 978-1-107-10874-5
1. Hardy spaces.   2. Operator theory.   I. Mashreghi, Javad.   II. Ross, William T.,
1964–   III. Title.
QA320.G325   2015
515´.73–dc23
2015012064

ISBN   978-1-107-10874-5   Hardback

To our families:
>Gizem; Reyhan, and Altay
>Shahzad; Dorsa, Parisa, and Golsa
>Fiona

Contents

Preface

This is an introductory text on model spaces that is aimed towards both graduate students and active researchers who wish to enter this evolving subject at a comfortable and digestible pace.

Model spaces have been studied, in one form or another, for the past 40 years, making connections to many areas of complex analysis (boundary behavior of analytic functions, analytic continuation, zero sets), operator theory (spectral properties, cyclic vectors, invariant subspaces, model operators for contractions, commutant lifting theorems, Hankel operators), engineering (the Darlington synthesis problem, control theory), and, more recently, to mathematical physics (completeness problems for Schrödinger and Sturm–Liouville operators). The purpose of this book is to present some of the basics of the subject in order to whet the appetite and to provide the newcomer with a solid foundation.

Many of the topics in this book were inspired by several series of lectures given by us at workshops in Montréal, Helsinki, Lens, Rennes, and Kashan where, during the course of these lectures, we became convinced of the need for students and researchers from adjacent fields to have a friendly introduction to model spaces and their operators.

This book is largely self-contained, although the reader is expected to be familiar with the basics of complex analysis, measure theory, and functional analysis. We briefly review these topics to establish our notation and, if necessary, to re-heat some possibly forgotten foundational topics. We develop and prove almost everything else, including a thorough treatment of the fundamentals of Hardy space theory, which is essential to the study of model spaces.

Since the list of topics we plan to cover is readily available in the table of contents, we would like to spend a few moments making the case for model spaces. Why should you keep on reading? Historically, model spaces began

with the desire to characterize the cyclic vectors and invariant subspaces of the
backward shift operator S^* on the Hardy space H^2, Beurling's 1949 theorem
completely characterized the non-trivial invariant subspaces of the unilateral
shift $Sf = zf$ on H^2 as uH^2, where u is an inner function on the open unit
disk $\mathbb{D}$. By taking orthogonal complements, we see that the proper invariant
subspaces of S^* are the so-called model spaces $\mathcal{K}_u = (uH^2)^\perp$. The elements
of uH^2 are easy to describe (multiples of the inner function u) while the elements of $\mathcal{K}_u$ are hidden behind annihilators and hence are more difficult to
characterize. Indeed, which functions actually belong to $\mathcal{K}_u$?

In 1970, Douglas, Shapiro, and Shields linked membership to a model space
with certain continuation (analytic and pseudocontinuation) properties of these
functions. Around the same time, Ahern and Clark explored the close relationship between the boundary behavior of functions in $\mathcal{K}_u$ and the existence
of angular derivatives, building upon earlier work of Carathéodory, Frostman,
and Riesz.

Some of the most important theorems in operator theory involve model-
ing a class of abstract operators by concrete operators on familiar spaces.
For example, there is the spectral theorem which models normal operators as
multiplication operators on Lebesgue spaces. There are other representation
theorems for subnormal operators and n-isometries in terms of multiplication
operators on certain Hilbert spaces of analytic functions. Pushing this even further, there are the theorems of Sz.-Nagy and Foiaş which model certain types
of contractions as the compression of the unilateral shift to a model space. This
program was highly successful and gave reasons to broaden the study of model
spaces from the scalar-valued case discussed above to the vector-valued case.
The discussion was broadened even further with the discovery of the close
cousins of the model spaces, the de Branges–Rovnyak spaces.

Sarason, in 1967, identified the commutant of the compressed shift. This
result was greatly generalized by Sz.-Nagy and Foiaş to a wider class of operators and became known as the commutant lifting theorem – now regarded as
one of the crowning gems of operator theory.

In 1972, Clark discovered a fascinating family of unitary operators whose
associated spectral measures are ubiquitous in operator theory, complex analysis, and mathematical physics. These ideas were investigated further by
Aleksandrov. Aleksandrov–Clark measures, as they have come to be known,
have been used to study composition operators and are proving relevant in
harmonic analysis. They also make connections to completeness problems for
solutions to Schrödinger and Sturm–Liouville operators.

A seminal article of Sarason from 2008 initiated the study of truncated
Toeplitz operators, close relatives of Toeplitz operators whose domains are

model spaces. We discuss some of the foundational results of this evolving field, providing detailed proofs of the key results.

The topics mentioned above, as well as some others, are covered in this book along with all the necessary background material and historical references. Since this is an introduction to model spaces, we cannot cover everything. Although certain topics are missing, the topics that we do cover, we cover in great detail. We do not skimp on the explanations or examples. First and foremost, this book is meant to help the reader learn about model spaces and to become fluent with the fundamental ideas.

Finally, writing good books depends on valuable feedback from your colleagues. It this regard, we would like to thank John B. Conway, John E. McCarthy, Dan Timotin, and Dragan Vukotic for their comments on an earlier draft of this book. We also would like to thank Zachary Glassman for the wonderful drawings and Elizabeth Sarapata for the careful editing.

Notation

$\mathbb{C}$	complex numbers
$\widehat{\mathbb{C}} = \mathbb{C} \cup \{\infty\}$	Riemann sphere
$\mathbb{C}_+$	upper half plane
$\mathbb{C}_-$	lower half plane
$\mathbb{N} = \{1, 2, \cdots\}$	natural numbers
$\mathbb{D} = \{z \in \mathbb{C} : |z| < 1\}$	open unit disk
$\mathbb{D}_e = \{z \in \mathbb{C} : |z| > 1\} \cup \{\infty\}$	extended exterior disk
$\mathbb{T} = \{z \in \mathbb{C} : |z| = 1\}$	unit circle
A^-	the closure of A
$M(\mathbb{T})$	finite Borel measures on $\mathbb{T}$ (p. 1)
$M_+(\mathbb{T})$	positive finite Borel measures on $\mathbb{T}$ (p. 1)
m	normalized Lebesgue measure on $\mathbb{T}$ (p. 1)
$D\mu$	symmetric derivative of a measure (p. 2)
$L^2 = L^2(\mathbb{T}, m)$	standard Lebesuge space (p. 7)
$\widehat{f}(n)$	Fourier coefficient of f (p. 8)
$\ell^2(\mathbb{Z}), \ell^2(\mathbb{N})$	square summable sequences (p. 8)
$P_z(\zeta)$	Poisson kernel (p. 9)
$\mathscr{P}\mu$	Poisson integral of a measure $\mu \in M(\mathbb{T})$ (p. 9)
$\widehat{\mu}(n)$	Fourier coefficient of a measure (p. 10)
$\angle \lim_{z \to \zeta} f(z)$	non-tangential limit of f at ζ (p. 14)
$\mathcal{B}(\mathcal{H})$	bounded operators on a Hilbert space $\mathcal{H}$ (p. 24)
$\sigma(A)$	spectrum of an operator A (p. 25)
$\sigma_p(A)$	point spectrum of an operator A (p. 25)
$\mathbf{x} \otimes \mathbf{y}$	a rank one operator (p. 27)
$\sigma_e(A)$	essential spectrum of an operator A (p. 31)
$\tau_{\zeta,a}$	disk automorphism (p. 32)
H^∞	bounded analytic functions on $\mathbb{D}$ (p. 33)
H^2	Hardy space (p. 58)

$c_\lambda(z) = (1 - \bar{\lambda}z)^{-1}$	reproducing kernel for H^2 (p. 59)
P	Riesz projection onto H^2 (p. 66)
S	unilateral shift on H^2 (p. 83)
T_φ	Toeplitz operator on H^2 (p. 90)
S^*	backward shift on H^2 (p. 99)
$\mathcal{K}_u = (uH^2)^\perp$	model space (p. 104)
Q_λ	difference quotient operator (p. 101)
$k_\lambda(z) = \frac{1 - \overline{u(\lambda)}u(z)}{1 - \bar{\lambda}z}$	reproducing kernel for $\mathcal{K}_u$ (p. 111)
P_u	projection onto $\mathcal{K}_u$ (p. 111)
$\mathcal{O}(\mathbb{D})$	analytic functions on $\mathbb{D}$ (p. 126)
C_φ	composition operator (p. 126)
$\sigma(u)$	spectrum of an inner u (p. 152)
S_u	compressed shift (p. 189)
$\kappa_\lambda(z)$	normalized reproducing kernel for $\mathcal{K}_u$ (p. 208)
σ_α	Clark measure (p. 232)
U_α	Clark unitary operator (p. 236)
A_φ^u	truncated Toeplitz operator (p. 282)
$\mathcal{T}_u$	space of truncated Toeplitz operators on $\mathcal{K}_u$ (p. 283)

1

Preliminaries

1.1 Measure and integral

1.1.1 Borel sets and measures

Most of the "measuring" in this book will take place on the unit circle $\mathbb{T} = \{z \in \mathbb{C} : |z| = 1\}$. Since we assume that the reader has a background in graduate analysis, we quickly review the standard definitions without much fanfare.

We let $m := d\theta/2\pi$ denote *Lebesgue measure* on $\mathbb{T}$, normalized so that $m(\mathbb{T}) = 1$. A subset of $\mathbb{T}$ is called a *Borel set* if it is contained in the *Borel σ-algebra*, the smallest σ-algebra of subsets of $\mathbb{T}$ that contains all of the open arcs of $\mathbb{T}$. A *Borel measure* on $\mathbb{T}$ is a countably additive function that assigns a complex number to each Borel subset of $\mathbb{T}$. Unless otherwise stated, our measures will always be finite. A Borel measure is *positive* if it assigns a non-negative number to each Borel set. We let $M(\mathbb{T})$ denote the set of all complex Borel measures on $\mathbb{T}$ and we let $M_+(\mathbb{T})$ denote the set of all positive Borel measures on $\mathbb{T}$. A function $f : \mathbb{T} \to \widehat{\mathbb{C}}$ (where $\widehat{\mathbb{C}}$ denotes the Riemann sphere $\mathbb{C} \cup \{\infty\}$) satisfying the condition that $f^{-1}(U)$ is a Borel set for any open set $U \subset \widehat{\mathbb{C}}$ is called a *Borel function*.

We often need to distinguish between the "support" and a "carrier" of a measure. For $\mu \in M_+(\mathbb{T})$, consider the union $\mathcal{U}$ of all the open subsets $U \subset \mathbb{T}$ for which $\mu(U) = 0$. The complement $\mathbb{T} \setminus \mathcal{U}$ is called the *support* of μ. On the other hand, a Borel set $E \subset \mathbb{T}$ for which

$$\mu(E \cap A) = \mu(A) \tag{1.1}$$

for all Borel subsets $A \subset \mathbb{T}$ is called a *carrier of* μ. The support of μ is certainly a carrier, but a carrier need not be the support. Indeed, a carrier of a measure might not even be closed. For example, if $f \geqslant 0$ is continuous and $d\mu = f\,dm$, then a carrier of μ is $\mathbb{T} \setminus f^{-1}(\{0\})$ (which is open) while the support of μ is the closure of this set. The support of a measure is unique while a carrier is not.

The *Hahn–Jordan Decomposition Theorem* says that each $\mu \in M(\mathbb{T})$ can be written uniquely as

$$\mu = (\mu_1 - \mu_2) + i(\mu_3 - \mu_4), \qquad \mu_j \in M_+(\mathbb{T}), \tag{1.2}$$

in which μ_1, μ_2 and μ_3, μ_4, respectively, are carried on disjoint sets.

Since $\mathbb{T}$ is a compact Hausdorff space, each Borel measure μ on $\mathbb{T}$ is *regular* in the sense that each positive measure μ_j in the Hahn–Jordan Decomposition of μ satisfies

$$\inf\{\mu_j(U) : U \supset E, U \text{ open}\} = \sup\{\mu_j(F) : F \subset E, F \text{ closed}\} \tag{1.3}$$

for each Borel set $E \subset \mathbb{T}$ [158, p. 48]. Moreover, the quantity above is equal to $\mu_j(E)$.

Recall that $\mu \in M_+(\mathbb{T})$ is *absolutely continuous* with respect to m (written $\mu \ll m$) if $\mu(A) = 0$ whenever $m(A) = 0$. We say that μ is *singular* with respect to m (written $\mu \perp m$) if there are disjoint Borel sets A and B such that $\mathbb{T} = A \cup B$ and $\mu(A) = m(B) = 0$. Also recall that the Radon–Nikodym Theorem says that $\mu \ll m$ if and only if $d\mu = f dm$, where f is a Lebesgue integrable function on $\mathbb{T}$ (that is, $\int |f| dm < \infty$). By this we mean that $\mu(A) = \int_A f dm$ for each Borel set $A \subset \mathbb{T}$. The function f is unique up to a set of Lebesgue measure zero and is called the *Radon–Nikodym derivative* of μ (with respect to m). It is denoted by $d\mu/dm$. One can also obtain $d\mu/dm$ as a "derivative" as follows.

Definition 1.1 For $\mu \in M(\mathbb{T})$, the *symmetric derivative* $(D\mu)(w)$ of μ at $w \in \mathbb{T}$ is defined to be

$$(D\mu)(w) := \lim_{t \to 0^+} \frac{\mu((e^{-it}w, e^{it}w))}{m((e^{-it}w, e^{it}w))}, \tag{1.4}$$

whenever this limit exists. Here $(e^{-it}w, e^{it}w)$ denotes the arc of $\mathbb{T}$ subtended by the points $e^{-it}w$ and $e^{it}w$.

Theorem 1.2 *For each $\mu \in M(\mathbb{T})$, we have:*

(i) $(D\mu)(w)$ exists for m-almost every $w \in \mathbb{T}$ and

$$D\mu = \frac{d\mu}{dm}$$

m-almost everywhere;

(ii) $\mu \perp m$ if and only if $D\mu = 0$ m-almost everywhere;

(iii) If $\mu \in M_+(\mathbb{T})$ and $\mu \perp m$, then $D\mu = \infty$ μ-almost everywhere. Moreover, μ is carried by the set $\{\zeta : (D\mu)(\zeta) = \infty\}$.

The Lebesgue Decomposition Theorem says that every $\mu \in M(\mathbb{T})$ can be decomposed uniquely as

$$\mu = \mu_a + \mu_s, \tag{1.5}$$

where $\mu_a \ll m$ and $\mu_s \perp m$. The measure μ_a is called the *absolutely continuous part* of μ while μ_s is called the *singular part* of μ. Furthermore, the singular part μ_s can be decomposed as $\mu_s = \nu_d + \nu_c$, where

$$\nu_d = \sum_{n \geqslant 1} c_n \delta_{\zeta_n}$$

is a measure with distinct atoms at $\zeta_n \in \mathbb{T}$ (that is to say, for each Borel set $E \subset \mathbb{T}$, $\delta_{\zeta_n}(E) = 1$ if $\zeta_n \in E$ and zero otherwise) and weights $c_n = \nu_d(\{\zeta_n\})$, and where ν_c is a singular measure with no atoms (that is, $\nu_c(\{\zeta\}) = 0$ for all $\zeta \in \mathbb{T}$). The measure ν_d is called the *discrete part* of μ_s while ν_c is called the *singular continuous part* of μ_s. See below for a more classical approach to measures using functions of bounded variation.

We now review the *weak-∗ topology* on $M(\mathbb{T})$. Let $C(\mathbb{T})$ denote the algebra of complex-valued continuous functions on $\mathbb{T}$ endowed with the sup-norm

$$\|f\|_\infty := \sup_{\zeta \in \mathbb{T}} |f(\zeta)|.$$

Note that $C(\mathbb{T})$ is complete with respect to this norm and hence a Banach space. For each $\mu \in M(\mathbb{T})$, the linear functional

$$\ell_\mu : C(\mathbb{T}) \to \mathbb{C}, \qquad \ell_\mu(f) := \int_{\mathbb{T}} f \, d\mu$$

is bounded. The norm of ℓ_μ is defined by

$$\|\ell_\mu\| := \sup \{ |\ell_\mu(f)| : f \in C(\mathbb{T}) : \|f\|_\infty \leqslant 1 \}$$

and is equal to

$$\|\mu\| := |\mu|(\mathbb{T}), \tag{1.6}$$

where $|\mu|(\mathbb{T})$ is the supremum of $\sum_{n \geqslant 1} |\mu(E_n)|$ as $\{E_n\}_{n \geqslant 1}$ runs over all finite partitions of $\mathbb{T}$ into disjoint Borel subsets. The quantity $\|\mu\|$ is called the *total variation norm* of μ. In terms of the Hahn decomposition (1.2) of μ, it satisfies

$$\frac{1}{\sqrt{2}} \sum_{1 \leqslant j \leqslant 4} \mu_j(\mathbb{T}) \leqslant \|\mu\| \leqslant \sum_{1 \leqslant j \leqslant 4} \mu_j(\mathbb{T}).$$

Theorem 1.3 (Riesz Representation Theorem) *If ℓ is a bounded linear functional on $C(\mathbb{T})$, then $\ell = \ell_\mu$ for some unique $\mu \in M(\mathbb{T})$.*

This allows us to define the *weak-∗ topology* on $M(\mathbb{T})$. A sequence $\{\mu_n\}_{n\geqslant 1} \subset M(\mathbb{T})$ converges *weak-∗* to μ if

$$\lim_{n\to\infty} \int_{\mathbb{T}} f \, d\mu_n = \int_{\mathbb{T}} f \, d\mu \qquad \forall f \in C(\mathbb{T}). \qquad (1.7)$$

The following tells us that closed balls in $M(\mathbb{T})$ are weak-∗ compact.

Theorem 1.4 (Banach–Alaoglu) *If $\{\mu_n\}_{n\geqslant 1}$ is a sequence in $M(\mathbb{T})$ for which*

$$\sup_{n\geqslant 1} \|\mu_n\| < \infty,$$

then there is a measure $\mu \in M(\mathbb{T})$ and a subsequence μ_{n_k} that converges to μ in the weak-∗ topology.

We will also need a version of the Hahn–Banach Theorem for $C(\mathbb{T})$ and $M(\mathbb{T})$.

Theorem 1.5 *Suppose that $\mathcal{M}$ is a linear manifold in $C(\mathbb{T})$ whose annihilator*

$$\left\{\mu \in M(\mathbb{T}) : \int_{\mathbb{T}} f \, d\mu = 0 \ \ \forall f \in \mathcal{M}\right\}$$

is zero. Then $\mathcal{M}$ is dense in $C(\mathbb{T})$. Furthermore, suppose that $\mathcal{N}$ is a linear manifold in $M(\mathbb{T})$ whose pre-annihilator

$$\left\{f \in C(\mathbb{T}) : \int_{\mathbb{T}} f \, d\mu = 0 \ \ \forall \mu \in \mathcal{N}\right\}$$

is zero. Then $\mathcal{N}$ is weak-∗ dense in $M(\mathbb{T})$.

1.1.2 Classical approach to measures

The following classical approach to measure theory requires a discussion of functions of bounded variation and the Lebesgue–Stieltjes integral. We cover this material not only for students to reconnect with the classical roots of analysis, but to also help out with several proofs and examples later on.

Definition 1.6 A function $F : [0, 2\pi] \to \mathbb{C}$ is of *bounded variation* if

$$\|F\|_{BV} := \sup_P \sum_{0\leqslant j\leqslant n_P-1} |F(x_{j+1}) - F(x_j)| < \infty, \qquad (1.8)$$

where the supremum is taken over all partitions $P = \{x_0, x_1, \ldots, x_{n_P}\}$ of $[0, 2\pi]$, where $0 = x_0 < x_1 < x_2 \cdots < x_{n_P} = 2\pi$.

The expression (1.8) defines a semi-norm $\|F\|_{BV}$, the total variation (semi)-norm, on the set BV of all functions of bounded variation. Notice that $\|F\|_{BV}$ is not a true norm since $\|F\|_{BV} = 0$ if F is a constant function. We gather together some important facts about BV.

Proposition 1.7 *If $F \in BV$, then:*

(i) $F'(x)$ exists for m-almost every $x \in [0, 2\pi]$;
(ii) The one-sided limits

$$F(x^+) := \lim_{t \to x^+} F(t), \quad F(x^-) := \lim_{t \to x^-} F(t)$$

exist for every $x \in (0, 2\pi)$. Moreover, $F(0^+)$ and $F(2\pi^-)$ exist;
(iii) F has at most a countable number of discontinuities;
(iv) $F = (F_1 - F_2) + i(F_3 - F_4)$, where each F_j is increasing.

For $F \in BV$ and right continuous (that is, $F(x) = \lim_{t \to x^+} F(t)$ for all x), define μ_F on the set of half-open intervals

$$\{[a, b) : a, b \in [0, 2\pi], a \leqslant b\}$$

by

$$\mu_F([a, b)) := F(b) - F(a).$$

By the Carathéodory Extension Theorem, μ_F extends to a unique Borel measure on $[0, 2\pi]$. Moreover

$$\|\mu_F\| = \|F\|_{BV},$$

that is to say, the total variation norm of μ_F defined in (1.6) equals the bounded variation norm of F defined in (1.8). The integral

$$\int_{[0, 2\pi]} f \, dF = \int_{[0, 2\pi]} f \, d\mu_F, \tag{1.9}$$

defined for every $f \in C[0, 2\pi]$ (continuous functions on $[0, 2\pi]$), is called the *Lebesgue–Stieltjes integral* of f with respect to F.

In this classical setting, the Lebesgue Decomposition Theorem says that every $F \in BV$ can be written as

$$F = F_a + F_s,$$

where F_a is *absolutely continuous* (that is, f is the anti-derivative of a Lebesgue integrable function) on $[0, 2\pi]$ and F_s is *singular* (that is, $F_s' = 0$ almost everywhere with respect to Lebesgue measure). Note that $\mu_F = \mu_{F_a} + \mu_{F_s}$ is the

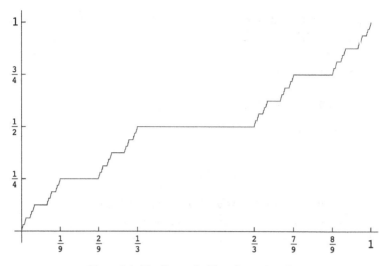

Figure 1.1 The Cantor devil's staircase function

decomposition of the measure μ_F into its absolutely continuous and singular parts from (1.5). Furthermore, F_s can be decomposed as

$$F_s = F_d + F_c,$$

where

$$F_d(x) = \sum_{y \leqslant x} (F(y^+) - F(y^-))$$

is a jump function and F_c is continuous with $F_c' = 0$ almost everywhere with respect to Lebesgue measure on $[0, 2\pi]$. This gives us the decomposition $\mu_{F_s} = \mu_{F_d} + \mu_{F_c}$ of μ_{F_s} into its discrete part μ_{F_d} and its continuous part μ_{F_c}.

For example, to produce a singular measure with no atoms one could take F to be the Cantor "devil's staircase" function (Figure 1.1). Note that F is continuous and $F' = 0$ almost everywhere with respect to Lebesgue measure on $[0, 2\pi]$. Thus $F = F_c$. The desired singular continuous measure is then μ_F.

The Riesz Representation Theorem tells us that every continuous linear functional on $C[0, 2\pi]$ takes the form

$$f \mapsto \int_{[0,2\pi]} f \, dF$$

for some unique (up to an additive constant) $F \in BV$. Moreover, the norm of this linear functional is $\|F\|_{BV}$.

The Banach–Alaoglu Theorem (Theorem 1.4) now takes the form of the *Helly Selection Theorem*: if $\{F_n\}_{n \geqslant 1} \subset BV$ and $\sup_{n \geqslant 1} \|F_n\|_{BV} < \infty$, then there is an $F \in BV$ and a subsequence F_{n_k} such that

$$\lim_{k \to \infty} \int_{[0,2\pi]} f \, dF_{n_k} = \int_{[0,2\pi]} f \, dF \qquad \forall f \in C[0, 2\pi].$$

1.1.3 Lebesgue spaces

Let $L^2 := L^2(\mathbb{T}, m)$ denote the space of m-measurable (that is, Lebesgue measurable) functions $f : \mathbb{T} \to \widehat{\mathbb{C}}$ such that

$$\|f\| := \left(\int_{\mathbb{T}} |f|^2 dm \right)^{\frac{1}{2}} < \infty. \tag{1.10}$$

Retaining tradition, we equate two measurable functions which are equal almost everywhere. With this norm, L^2 is a Hilbert space endowed with the inner product

$$\langle f, g \rangle := \int_{\mathbb{T}} f \overline{g} \, dm.$$

From time to time we will need the spaces L^p $(0 < p < \infty)$ of measurable functions for which

$$\|f\|_p := \left(\int_{\mathbb{T}} |f|^p dm \right)^{\frac{1}{p}} < \infty.$$

We also require the space $L^\infty := L^\infty(\mathbb{T})$ of all essentially bounded measurable functions on $\mathbb{T}$, equipped with the essential supremum norm

$$\|f\|_\infty := \text{ess-sup}_{\zeta \in \mathbb{T}} |f(\zeta)|.$$

Here,

$$\text{ess-sup}_{\zeta \in \mathbb{T}} |f(\zeta)| := \sup \{a \geqslant 0 : m(\{\zeta \in \mathbb{T} : |f(\zeta)| > a\}) > 0\}$$

is the *essential supremum* of $|f|$.

For $\mu \in M_+(\mathbb{T})$, we will also need the corresponding $L^p(\mu)$ $(0 < p < \infty)$ spaces of Borel measurable functions f on $\mathbb{T}$ such that

$$\|f\|_{L^p(\mu)} := \left(\int_{\mathbb{T}} |f|^p d\mu \right)^{\frac{1}{p}} < \infty.$$

The identity

$$\int_{\mathbb{T}} \zeta^n \, dm(\zeta) = \int_{[0,2\pi]} e^{in\theta} \frac{d\theta}{2\pi} = \begin{cases} 1 & \text{if } n = 0, \\ 0 & \text{otherwise,} \end{cases} \tag{1.11}$$

shows that the family of functions $\{\zeta \mapsto \zeta^n : n \in \mathbb{Z}\}$ is an orthonormal set in L^2. The coefficients

$$\widehat{f}(n) := \langle f, \zeta^n \rangle = \int_{\mathbb{T}} f(\zeta) \overline{\zeta}^n \, dm(\zeta)$$

of an $f \in L^2$ with respect to this orthonormal set are called the (complex) *Fourier coefficients* of f.

Theorem 1.8 (Parseval's Theorem)   *For each $f \in L^2$,*

$$\|f\|^2 = \sum_{n \in \mathbb{Z}} |\widehat{f}(n)|^2.$$

Furthermore,

$$\lim_{N \to \infty} \left\| f - \sum_{-N \leqslant n \leqslant N} \widehat{f}(n) \zeta^n \right\| = 0.$$

The previous theorem tells us several things. First, $\{\zeta^n : n \in \mathbb{Z}\}$ is an orthonormal basis for L^2. Second, the *Fourier series*

$$\sum_{n \in \mathbb{Z}} \widehat{f}(n) \zeta^n$$

converges to f in the norm of L^2. In general, the Fourier series of an L^2 function need not converge pointwise. However, a deep theorem of Carleson says that it converges pointwise m-a.e. to f [17]. Although we will not use this fact, the reader should be aware that such delicate matters exist. Finally, Theorem 1.8 tells us that the L^2 norm of f coincides with the norm of the sequence $\{\widehat{f}(n) : n \in \mathbb{Z}\}$ of Fourier coefficients in the Hilbert space

$$\ell^2(\mathbb{Z}) := \left\{ \mathbf{a} = \{a_n\}_{n \in \mathbb{Z}} : \|\mathbf{a}\|_{\ell^2(\mathbb{Z})} := \left(\sum_{n \in \mathbb{Z}} |a_n|^2 \right)^{\frac{1}{2}} < \infty \right\}$$

of all square-summable sequences of complex numbers, endowed with the inner product

$$\langle \mathbf{a}, \mathbf{b} \rangle = \sum_{n \in \mathbb{Z}} a_n \overline{b_n}.$$

We therefore identify the Hilbert spaces L^2 and $\ell^2(\mathbb{Z})$ via the correspondence

$$f \leftrightarrow \{\widehat{f}(n) : n \in \mathbb{Z}\}.$$

1.2 Poisson integrals

The function

$$P_z(\zeta) := \frac{1 - |z|^2}{|\zeta - z|^2}, \qquad \zeta \in \mathbb{T}, \ z \in \mathbb{D},$$

is called the *Poisson kernel* of the unit disk $\mathbb{D} = \{z : |z| < 1\}$. Note that

$$P_z(\zeta) > 0.$$

A computation verifies that

$$P_z(\zeta) = \operatorname{Re}\left(\frac{\zeta + z}{\zeta - z}\right) \tag{1.12}$$

and a computation with geometric series yields

$$P_{rw}(\zeta) = \sum_{n \in \mathbb{Z}} r^{|n|} w^n \overline{\zeta}^n, \quad w \in \mathbb{T}, \, r \in (0, 1). \tag{1.13}$$

With $w = e^{i\theta}$ and $\zeta = e^{it}$, one can also establish the formula

$$P_{re^{i\theta}}(e^{it}) = \frac{1 - r^2}{1 - 2r\cos(\theta - t) + r^2}. \tag{1.14}$$

For fixed $\zeta \in \mathbb{T}$, $P_z(\zeta)$ is the real part of the analytic function

$$z \mapsto \frac{\zeta + z}{\zeta - z},$$

which makes $z \mapsto P_z(\zeta)$ a harmonic function on $\mathbb{D}$. Integrating the series in (1.13) term by term and using the orthogonality relations (1.11), we see that

$$\int_{\mathbb{T}} P_z(\zeta)\, dm(\zeta) = 1, \qquad z \in \mathbb{D}. \tag{1.15}$$

An important property of the Poisson kernel is that for fixed $\delta > 0$,

$$\lim_{r \to 1^-} \left(\sup_{\delta \leqslant |t| \leqslant \pi} P_r(e^{it}) \right) = 0. \tag{1.16}$$

This is illustrated in Figure 1.2. One can also see this from the estimate

$$P_r(e^{it}) \leqslant \frac{1 - r^2}{1 - 2\cos\delta + r^2}, \quad \delta \leqslant |t| \leqslant \pi.$$

For $\mu \in M(\mathbb{T})$, define the *Poisson integral of μ* by

$$\mathscr{P}(\mu)(z) := \int_{\mathbb{T}} P_z(\zeta) d\mu(\zeta), \quad z \in \mathbb{D}.$$

By differentiating under the integral sign, we see that $\mathscr{P}(\mu)$ is harmonic on $\mathbb{D}$. Furthermore,

$$\mathscr{P}(\mu)(rw) = \sum_{n \in \mathbb{Z}} \widehat{\mu}(n) r^{|n|} w^n, \quad w \in \mathbb{T}, \, r \in (0, 1), \tag{1.17}$$

where

$$\widehat{\mu}(n) := \int_{\mathbb{T}} \overline{\zeta}^n d\mu(\zeta), \quad n \in \mathbb{Z},$$

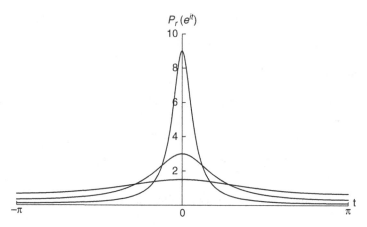

Figure 1.2 The graphs of $P_r(e^{it})$ for $r = 0.2, 0.5, 0.8$. Notice that $P_r(e^{it}) > 0$. Furthermore, notice how P_r peaks higher and higher near $t = 0$, for increasing values of r, while decaying rapidly away from the origin.

are the *Fourier coefficients* of the measure μ. We often write $\mathscr{P}(f)$ in place of the more cumbersome $\mathscr{P}(f\,dm)$ for $f \in L^1$.

For $f \in C(\mathbb{T})$, we have the Poisson Integral Formula for the solution of the Dirichlet problem on $\mathbb{D}^-$. The classical Dirichlet problem for a planar domain Ω is: given a continuous function f on the boundary $\partial\Omega$ of Ω, find a function u which is continuous on Ω^- that is harmonic on Ω and agrees with f on $\partial\Omega$.

Theorem 1.9 (Poisson Integral Formula) *If $f \in C(\mathbb{T})$, then $\mathscr{P}(f)$ is harmonic on $\mathbb{D}$ and extends continuously to $\mathbb{D}^-$. Furthermore,*

$$\lim_{z \to w} \mathscr{P}(f)(z) = f(w)$$

for every $w \in \mathbb{T}$.

Proof Let $u := \mathscr{P}(f)$ and recall from our earlier discussion that u is harmonic on $\mathbb{D}$. To complete the proof, we will show that for every fixed $w \in \mathbb{T}$,

$$\lim_{z \to w} u(z) = f(w). \tag{1.18}$$

Let $\varepsilon > 0$ be given. Use the continuity of f at w to produce a $\delta > 0$ so that whenever $\zeta \in \mathbb{T}$ and $|w - \zeta| < \delta$ we have $|f(w) - f(\zeta)| < \varepsilon$. From here we get

$$|u(z) - f(w)| = \left| \int_{\mathbb{T}} f(\zeta) P_z(\zeta)\,dm(\zeta) - f(w) \int_{\mathbb{T}} P_z(\zeta)\,dm(\zeta) \right| \quad \text{(by (1.15))}$$

$$\leqslant \int_{\mathbb{T}} |f(\zeta) - f(w)| P_z(\zeta)\,dm(\zeta)$$

$$= \int\limits_{\mathbb{T}\cap\{|\zeta-w|<\delta\}} + \int\limits_{\mathbb{T}\cap\{|\zeta-w|\geqslant\delta\}}$$

$$= \varepsilon \int\limits_{\mathbb{T}\cap\{|\zeta-w|<\delta\}} P_z(\zeta)dm(\zeta) + 2\|f\|_\infty \int\limits_{\mathbb{T}\cap\{|\zeta-w|\geqslant\delta\}} P_z(\zeta)dm(\zeta)$$

$$\leqslant \varepsilon + 2\|f\|_\infty \int\limits_{\mathbb{T}\cap\{|\zeta-w|\geqslant\delta\}} P_z(\zeta)dm(\zeta). \qquad \text{(by (1.15))}.$$

Now use (1.16) to see that

$$\varlimsup_{z\to w} \int\limits_{\mathbb{T}\cap\{|\zeta-w|\geqslant\delta\}} P_z(\zeta)dm(\zeta) = 0$$

and so

$$\varlimsup_{z\to w} |u(z) - f(w)| \leqslant \varepsilon$$

for all $\varepsilon > 0$ from which (1.18) follows.   □

1.2.1 Fatou's Theorem

An important result of Fatou is the following generalization of the Poisson Integral Formula. Since this result plays an important role in understanding the boundary behavior of certain classes of analytic functions on $\mathbb{D}$, we include a proof which follows [63, 112].

Theorem 1.10 (Fatou)   *Let $\mu \in M(\mathbb{T})$ and $w \in \mathbb{T}$. If $(D\mu)(w)$ exists, then*

$$\lim_{r\to 1^-} \mathscr{P}(\mu)(rw) = (D\mu)(w).$$

Thus $\mathscr{P}(\mu)$ has a finite radial limit m-almost everywhere on $\mathbb{T}$.

Proof   Without loss of generality assume that $w = 1$. Let us rephrase the problem in terms of Lebesgue–Stieltjes integrals (see (1.9)) and show that if $F \in BV$ and

$$D := \lim_{t\to 0^+} \frac{F(t) - F(-t)}{2t}$$

exists, then

$$\lim_{r\to 1^-} \frac{1}{2\pi} \int\limits_{[-\pi,\pi]} P_r(e^{it})\, dF(t) = D.$$

Use the fact that

$$\frac{1}{2\pi} \int\limits_{[-\pi,\pi]} P_r(e^{it})\, dt = 1, \qquad r \in (0,1),$$

(see (1.15)) to get

$$\frac{1}{2\pi} \int\limits_{[-\pi,\pi]} P_r(e^{it})\, dF(t) - D = \frac{1}{2\pi} \int\limits_{[-\pi,\pi]} P_r(e^{it})\, (dF(t) - D\, dt).$$

Integrating by parts, the right-hand side of the previous equation equals

$$\frac{1}{2\pi} \left(P_r(e^{it})[F(t) - Dt] \right)\Big|_{t=-\pi}^{t=\pi} - \frac{1}{2\pi} \int\limits_{[-\pi,\pi]} (F(t) - Dt)\frac{\partial}{\partial t}(P_r(e^{it}))\, dt. \qquad (1.19)$$

The first term in (1.19) evaluates to

$$\frac{1}{2\pi} \frac{1-r^2}{(1+r)^2}(F(\pi) - F(-\pi) - D\pi - D\pi),$$

which goes to zero as $r \to 1^-$. For fixed $\delta \in (0,\pi)$ (to be optimized in a moment), we write the second term in (1.19) as the sum of two integrals

$$\int\limits_{|t|>\delta} + \int\limits_{|t|\leqslant\delta} \qquad (1.20)$$

each having the integrand $-\frac{1}{2\pi}(F(t) - Dt)\frac{\partial}{\partial t}(P_r(e^{it}))$.

To estimate these two integrals, recall from (1.14) that

$$P_r(e^{it}) = \frac{1-r^2}{1 - 2r\cos t + r^2}$$

and so

$$\frac{\partial}{\partial t}P_r(e^{it}) = -\frac{2r(1-r^2)\sin t}{(1 - 2r\cos t + r^2)^2}. \qquad (1.21)$$

This means that if $|t| > \delta$ then

$$\left| \frac{\partial}{\partial t}P_r(e^{it}) \right| \leqslant \frac{2r(1-r^2)}{(1 - 2r\cos\delta + r^2)^2} \leqslant c(1-r^2),$$

for some $c > 0$ independent of t and r, and so

$$\int\limits_{|t|>\delta} (F(t) - Dt)\frac{\partial}{\partial t}(P_r(e^{it}))dt \to 0 \quad \text{as } r \to 1^-.$$

Thus the first integral in (1.20) tends to zero as $r \to 1^-$.

Now write the second integral $\int\limits_{|t|\leqslant\delta}$ in (1.20) as

$$\int\limits_{|t|\leqslant\delta} = \frac{1}{2\pi}\int\limits_{|t|\leqslant\delta}(F(t) - Dt)\frac{\partial}{\partial t}(P_r(e^{it}))\,dt$$

$$= \frac{1}{\pi}\int\limits_{0\leqslant t\leqslant\delta}\left(\frac{F(t) - F(-t)}{2t} - D\right)t\left(-\frac{\partial}{\partial t}P_r(e^{it})\right)dt.$$

Given $\varepsilon > 0$, choose $\delta \in (0,\pi)$ so that

$$\left|\frac{F(t) - F(-t)}{2t} - D\right| < \varepsilon, \qquad 0 < t < \delta.$$

Now use the fact, from (1.21), that $-t\frac{\partial}{\partial t}P_r(e^{it}) \geqslant 0$ for $0 < t < \delta$ to get

$$\left|\int\limits_{|t|\leqslant\delta}\right| \leqslant \frac{1}{\pi}\int\limits_{0\leqslant t\leqslant\delta}\left|\frac{F(t) - F(-t)}{2t} - D\right|t\left(-\frac{\partial}{\partial t}P_r(e^{it})\right)dt$$

$$\leqslant \frac{\varepsilon}{\pi}\int\limits_{0\leqslant t\leqslant\delta}t\left(-\frac{\partial}{\partial t}P_r(e^{it})\right)dt$$

$$\leqslant \frac{\varepsilon}{\pi}\int\limits_{[-\pi,\pi]}t\left(-\frac{\partial}{\partial t}P_r(e^{it})\right)dt$$

$$= \frac{\varepsilon}{\pi}\left(-tP_r(e^{it})\Big|_{t=-\pi}^{t=\pi} + \int\limits_{[-\pi,\pi]}P_r(e^{it})\,dt\right)$$

$$= \frac{\varepsilon}{\pi}(-2\pi P_r(-1) + 2\pi)$$

$$= \frac{\varepsilon}{\pi}\left(-2\pi\frac{1-r^2}{1-2r\cos(-1)+r^2} + 2\pi\right).$$

It follows that

$$\overline{\lim_{r\to 1^-}}\left|\frac{1}{2\pi}\int\limits_{[-\pi,\pi]}P_r(e^{it})\,dF(t) - D\right| < \eta$$

for every $\eta > 0$, from which the result follows. $\square$

Remark 1.11 With a little more work (see [158, Thm. 11.10]), one can also show that for $\mu \in M_+(\mathbb{T})$ and $w \in \mathbb{T}$

$$\lim_{r\to 1^-}\mathscr{P}(\mu)(rw) \leqslant (\underline{D}\mu)(w) \leqslant (\overline{D}\mu)(w) \leqslant \overline{\lim_{r\to 1^-}}\,\mathscr{P}(\mu)(rw),$$

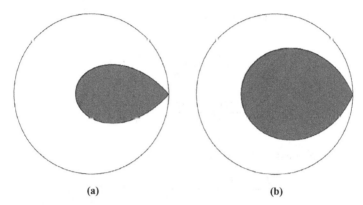

(a)   **(b)**

Figure 1.3 Two Stolz domains $\Gamma_\alpha(\zeta)$ (shaded) anchored at $\zeta = 1$. The one on the left is $\Gamma_{3/2}(1)$ while the one on the right is $\Gamma_{5/2}(1)$.

where

$$(\underline{D}\mu)(w) = \lim_{t \to 0^+} \frac{\mu((e^{-it}w, e^{it}w))}{m((e^{-it}w, e^{it}w))}, \qquad (\overline{D}\mu)(w) = \overline{\lim_{t \to 0^+}} \frac{\mu((e^{-it}w, e^{it}w))}{m((e^{-it}w, e^{it}w))}$$

are the lower and upper symmetric derivatives of μ at w.

Fatou's theorem (Theorem 1.10) can be extended in the following way. For m-almost every $\zeta \in \mathbb{T}$, the Poisson integral $\mathscr{P}(\mu)(z)$ approaches the finite limit $(D\mu)(\zeta)$, whenever this exists, as $z \to \zeta$ in any *Stolz domain*

$$\Gamma_\alpha(\zeta) := \{z \in \mathbb{D} : |z - \zeta| < \alpha(1 - |z|)\}, \qquad \alpha > 1$$

anchored at ζ; see Figure 1.3. One also has

$$\Gamma_\alpha(\zeta) \subset \Gamma_\beta(\zeta), \qquad \alpha < \beta.$$

Definition 1.12 For a function $f : \mathbb{D} \to \mathbb{C}$ and $\zeta \in \mathbb{T}$, we say that $f(z)$ approaches $L \in \mathbb{C}$ *non-tangentially*, denoted

$$L = \angle \lim_{z \to \zeta} f(z), \qquad (1.22)$$

if $f(z) \to L$ holds whenever $z \to \zeta$ in every fixed Stolz domain.

Observe that the Poisson Integral Formula (Theorem 1.9) follows from Fatou's Theorem since, in this case, $d\mu = f\,dm$. Thus for *every* $w \in \mathbb{T}$ we have

$$D\mu(w) = \lim_{t \to 0^+} \frac{1}{m(e^{-it}w, e^{it}w)} \int_{[e^{-it}w, e^{it}w]} f(\zeta)\,dm(\zeta) = f(w)$$

by the Fundamental Theorem of Calculus.

Remark 1.13 Theorem 1.10 is known as Fatou's *Theorem*. We will also use Fatou's *Lemma*: if $\{f_n\}_{n \geqslant 1}$ is a sequence of non-negative Borel functions on $\mathbb{T}$, then

$$\int_{\mathbb{T}} \varliminf_{n \to \infty} f_n \, d\mu \leqslant \varliminf_{n \to \infty} \int_{\mathbb{T}} f_n \, d\mu.$$

We need several other facts about Poisson integrals and harmonic functions.

Theorem 1.14 (Mean Value Property) *If u is harmonic on $\mathbb{D}$, then*

$$\int_{\mathbb{T}} u(r\zeta) dm(\zeta) = u(0)$$

for each $r \in (0, 1)$.

Proof Note that $u = \operatorname{Re} f$ for some analytic function f on $\mathbb{D}$. By the Cauchy Integral Formula,

$$f(0) = \frac{1}{2\pi i} \int_{r\mathbb{T}} \frac{f(z)}{z} dz.$$

Writing $z = re^{i\theta}$ for $0 \leqslant \theta \leqslant 2\pi$, we write the integral above as

$$f(0) = \int_0^{2\pi} f(re^{i\theta}) \frac{d\theta}{2\pi}.$$

Now take real parts of both sides of the previous equation. □

Theorem 1.15 (Maximum Modulus Theorem for Poisson integrals) *If $f \in L^\infty$, then*

$$|\mathscr{P}(f)(z)| \leqslant \|f\|_\infty, \qquad z \in \mathbb{D}.$$

Proof For each $z \in \mathbb{D}$ we have

$$\begin{aligned}
|\mathscr{P}(f)(z)| &= \left| \int_{\mathbb{T}} P_z(\zeta) f(\zeta) \, dm(\zeta) \right| \\
&\leqslant \int_{\mathbb{T}} P_z(\zeta) |f(\zeta)| \, dm(\zeta) && (P_z(\zeta) \geqslant 0) \\
&\leqslant \|f\|_\infty \int_{\mathbb{T}} P_z(\zeta) \, dm(\zeta) \\
&= \|f\|_\infty. && \text{(by 1.15)} \qquad \square
\end{aligned}$$

Here are two gems that often come in handy.

Proposition 1.16 (Uniqueness of Fourier Coefficients) *If $\mu \in M(\mathbb{T})$ and $\widehat{\mu}(n) = 0$ for all $n \in \mathbb{Z}$, then μ is the zero measure.*

Proof   If $\mu \in M(\mathbb{T})$, the dual of $C(\mathbb{T})$, and

$$\hat{\mu}(n) = \int_{\mathbb{T}} \overline{\zeta}^n d\mu(\zeta) = 0, \qquad n \in \mathbb{Z},$$

then by linearity,

$$\int_{\mathbb{T}} p \, d\mu = 0$$

for every trigonometric polynomial

$$p(\zeta) = \sum_{-N \leqslant n \leqslant N} a_n \zeta^n, \quad \zeta \in \mathbb{T}.$$

By the Stone–Weierstrass Theorem, such polynomials are dense in $C(\mathbb{T})$ and so

$$\int_{\mathbb{T}} f d\mu = 0$$

for all $f \in C(\mathbb{T})$. By Theorem 1.3, μ is the zero measure. $\square$

For each fixed $z \in \mathbb{D}$, the function $\zeta \mapsto P_z(\zeta)$ is continuous on $\mathbb{T}$. In fact. linear combinations of these functions form a dense subset of $C(\mathbb{T})$.

Proposition 1.17   *The span of $\{P_z : z \in \mathbb{D}\}$ is dense in $C(\mathbb{T})$.*

Proof   If $\mu \in M(\mathbb{T})$ and

$$\int_{\mathbb{T}} P_z(\zeta) d\mu(\zeta) = 0 \qquad \forall z \in \mathbb{D},$$

(that is, μ annihilates the span of $\{P_z : z \in \mathbb{D}\}$), then the Poisson integral $\mathscr{P}(\mu)$ is identically zero on $\mathbb{D}$. Using the series expansion of $\mathscr{P}(\mu)(r\zeta)$ from (1.17) we see that

$$0 = \int_{\mathbb{T}} \overline{\zeta}^k \mathscr{P}(\mu)(r\zeta) \, d\mu(\zeta) = r^{|k|} \hat{\mu}(k), \qquad k \in \mathbb{Z}. \tag{1.23}$$

Thus μ is a measure whose Fourier coefficients vanish. By the previous proposition, μ is the zero measure. By Theorem 1.5, the span of $\{P_z : z \in \mathbb{D}\}$ is dense in $C(\mathbb{T})$. $\square$

1.2.2 Herglotz's theorem

Since $P_z(\zeta) > 0$ for all $z \in \mathbb{D}$ and $\zeta \in \mathbb{T}$, we see that $\mathscr{P}(\mu)$ is a positive harmonic function on $\mathbb{D}$ whenever $\mu \in M_+(\mathbb{T})$. The following theorem of Herglotz says that the converse is true. We follow [63].

Theorem 1.18 (Herglotz)   *If u is a positive harmonic function on $\mathbb{D}$, then there is a unique $\mu \in M_+(\mathbb{T})$ such that $u = \mathscr{P}(\mu)$.*

Proof   The set $\mathscr{U} := \{u_r\, dm : 0 < r < 1\}$, where

$$u_r(\zeta) = u(r\zeta), \quad \zeta \in \mathbb{T},$$

is a collection of positive measures satisfying

$$\int_{\mathbb{T}} u_r\, dm = u(0)$$

(Theorem 1.14). Since $\mathscr{U}$ is uniformly bounded in total variation norm, we can apply the Banach–Alaoglu Theorem (Theorem 1.4) to conclude that $\mathscr{U}$ has a weak-$*$ accumulation point μ. Thus there is a sequence $r_n \uparrow 1$ such that the measures $u_{r_n} dm$ converge weak-$*$ to μ. It follows from the Poisson Integral Formula (Theorem 1.9) and the definition (1.7) of weak-$*$ convergence that $\mu \in M_+(\mathbb{T})$ and

$$\begin{aligned}
\mathscr{P}(\mu)(z) &= \int P_z(\zeta) d\mu(\zeta) \\
&= \lim_{r_n \to 1^-} \int_{\mathbb{T}} P_z(\zeta) u(r_n\zeta) dm(\zeta) \\
&= \lim_{r_n \to 1^-} u(r_n z) \\
&= u(z)
\end{aligned}$$

for each $z \in \mathbb{D}$.

To prove that μ is unique, suppose that $u = \mathscr{P}(\mu_1) = \mathscr{P}(\mu_2)$ for some $\mu_1, \mu_2 \in M_+(\mathbb{T})$. Then the measure $\nu = \mu_1 - \mu_2$ satisfies $\mathscr{P}(\nu)(r\zeta) = 0$ for every $0 < r < 1$ and $\zeta \in \mathbb{T}$. However,

$$0 = \mathscr{P}(\nu)(r\zeta) = \sum_{n \in \mathbb{Z}} r^{|n|} \zeta^n \widehat{\nu}(n) = \widehat{\nu}(0) + \sum_{k \geqslant 1} r^k (\zeta^k \widehat{\nu}(k) + \overline{\zeta}^k \widehat{\nu}(-k)).$$

Since the last expression is a power series in the variable r we must have (by the uniqueness of power series coefficients)

$$\widehat{\nu}(0) = 0,$$

and, for each fixed $k \geqslant 1$,

$$\zeta^k \widehat{\nu}(k) + \overline{\zeta}^k \widehat{\nu}(-k) = 0, \quad \forall \zeta \in \mathbb{T}.$$

By the uniqueness of the coefficients of a trigonometric polynomial (ultimately coming from the orthogonality relations from (1.11)), we see that $\widehat{\nu}(k) = \widehat{\nu}(-k) = 0$. It now follows that

$$\widehat{\nu}(n) = 0 \quad \forall n \in \mathbb{Z},$$

and so, by the uniqueness of Fourier coefficients (Proposition 1.16), we see that ν is the zero measure.   □

A slight variation of the preceding proof says a bit more:

Corollary 1.19   *If u is harmonic on $\mathbb{D}$ and*

$$\sup_{0<r<1} \int_{\mathbb{T}} |u(r\zeta)|\, dm(\zeta) < \infty,$$

then there is a unique $\mu \in M(\mathbb{T})$ such that $u = \mathscr{P}(\mu)$.

Proof   As in the proof of Herglotz's Theorem, observe that the family of measures

$$\mathscr{U} = \{u_r\, dm : 0 < r < 1\}$$

is uniformly bounded in total variation norm since

$$\sup_{0<r<1} \|u_r\, dm\| = \sup_{0<r<1} \int_{\mathbb{T}} |u_r|\, dm < \infty.$$

By the Banach–Alaoglu theorem, $\mathscr{U}$ has a weak-$*$ accumulation point μ. Now follow the rest of the proof of Herglotz's Theorem.   □

Corollary 1.20 (Herglotz)   *If f is analytic on $\mathbb{D}$ and $\operatorname{Re} f > 0$ on $\mathbb{D}$, then there is a unique $\mu \in M_+(\mathbb{T})$ such that*

$$f(z) = \int_{\mathbb{T}} \frac{\zeta + z}{\zeta - z}\, d\mu(\zeta) + i \operatorname{Im} f(0).$$

Proof   First observe that $\operatorname{Re} f$ is a positive harmonic function on $\mathbb{D}$. By Herglotz's Theorem,

$$\operatorname{Re} f(z) = \int_{\mathbb{T}} P_z(\zeta)\, d\mu(\zeta)$$

for some unique $\mu \in M_+(\mathbb{T})$. Since

$$P_z(\zeta) = \operatorname{Re}\left(\frac{\zeta + z}{\zeta - z}\right),$$

we conclude that

$$f(z) = \int_{\mathbb{T}} \frac{\zeta + z}{\zeta - z}\, d\mu(\zeta) + i \operatorname{Im} f(0)$$

since any two harmonic conjugates of $\operatorname{Re} f$ differ by a real constant.   □

1.2.3 The F. and M. Riesz Theorem

The following result of the brothers Riesz explores measures with vanishing negative Fourier coefficients. We follow the proof from [136].

Theorem 1.21 (F. and M. Riesz) *Suppose that $\mu \in M(\mathbb{T})$ and $\widehat{\mu}(n) = 0$ for all $n \leqslant -1$. Then $\mu \ll m$.*

The proof depends on a construction of Fatou. Let E be a closed subset of $\mathbb{T}$ with $m(E) = 0$. Fatou constructed a function F on $\mathbb{D}^-$ satisfying the following properties:

(i) F is continuous on $\mathbb{D}^-$;
(ii) F is analytic on $\mathbb{D}$;
(iii) $|F(z)| < 1$ for all $z \in \mathbb{D}$;
(iv) $|F(\zeta)| < 1$ for all $\zeta \in \mathbb{T} \setminus E$;
(iv) $F(\zeta) = 1$ for all $\zeta \in E$.

Such a function is called a *Fatou function* for E. To construct such a function F, we may assume, without loss of generality, that $1 \in E$. Since E is closed, we can write the open set $\mathbb{T} \setminus E$ as

$$\mathbb{T} \setminus E = \bigcup_{n \geqslant 1} I_n,$$

a finite or countable disjoint union of open arcs

$$I_n = \{e^{i\theta} : 0 < \alpha_n < \theta < \beta_n < 2\pi\}.$$

The condition $m(E) = 0$ is equivalent to

$$\sum_{n \geqslant 1} (\beta_n - \alpha_n) = 2\pi.$$

Therefore, one can find a sequence $c_n > 0$ such that $c_n \to \infty$ and

$$\sum_{n \geqslant 1} c_n (\beta_n - \alpha_n) < \infty.$$

Define

$$u(e^{it}) = \begin{cases} +\infty & \text{if } e^{it} \in E, \\ \dfrac{c_n \ell_n}{\sqrt{\ell_n^2 - (t - \gamma_n)^2}} & \text{if } e^{it} \in I_n, \end{cases}$$

where

$$\ell_n = \frac{\beta_n - \alpha_n}{2}, \quad \gamma_n = \frac{\alpha_n + \beta_n}{2}.$$

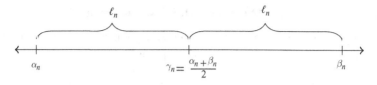

Figure 1.4

See Figure 1.4.

First we prove that $u \in L^1$. To see this we note that

$$\int_E u(e^{it})\,dt = 0$$

due to the fact that $m(E) = 0$ (even though $u = +\infty$ on E [158, Ch. 1]). Thus we have

$$
\begin{aligned}
\int_{\mathbb{T}} |u(e^{it})|\,dt &= \int_{\mathbb{T}} u(e^{it})\,dt \\
&= \int_E u(e^{it})\,dt + \sum_{n \geq 1} \int_{I_n} u(e^{it})\,dt \\
&= \sum_{n \geq 1} \int_{\alpha_n}^{\beta_n} \frac{c_n\,\ell_n}{\sqrt{\ell_n^2 - (t - \gamma_n)^2}}\,dt \\
&= \frac{\pi}{2} \sum_{n \geq 1} c_n (\beta_n - \alpha_n) < \infty.
\end{aligned}
$$

Second, observe that u is a continuous function from $\mathbb{T}$ to $[0, +\infty]$. As a matter of fact, u is infinitely differentiable on each I_n and tends to infinity as t approaches α_n or β_n from within I_n. Moreover, $u \geq c_n$ on I_n and c_n tends to infinity as $n \to \infty$. Hence,

$$\lim_{t \to t_0} u(e^{it}) = +\infty \tag{1.24}$$

for all $e^{it_0} \in E$. Define

$$
U(re^{i\theta}) =
\begin{cases}
\dfrac{1}{2\pi} \displaystyle\int_{-\pi}^{\pi} P_{re^{i\theta}}(e^{it})\,u(e^{it})\,dt & \text{if } 0 \leq r < 1, \\[2ex]
u(e^{i\theta}) & \text{if } r = 1.
\end{cases}
$$

Then $U : \mathbb{D}^- \to [0, +\infty]$ is continuous and harmonic on $\mathbb{D}$, where we mean that if $u(\zeta) = +\infty$, then $U(z)$ approaches $+\infty$ whenever z approaches ζ from within $\mathbb{D}$. In particular, for each $e^{it_0} \in E$,

$$\lim_{z \to e^{it_0}} U(z) = +\infty. \tag{1.25}$$

Since u is continuously differentiable on each I_n, the harmonic conjugate

$$V(re^{i\theta}) = \frac{1}{2\pi} \int_{-\pi}^{\pi} \frac{2r \sin(\theta - t)}{1 + r^2 - 2r\cos(\theta - t)} u(e^{it}) \, dt$$

has a continuous extension up to I_n [136, Sect. 5.4] and its boundary values on I_n are given by $v(e^{i\theta})$, the boundary function of V for $e^{i\theta} \in I_n$. Note that we did not define v on E and moreover, for the rest of the discussion, we do not need to verify whether or not $v(e^{i\theta})$ exists when $e^{i\theta} \in E$. Finally, let $F : \mathbb{D}^- \to \mathbb{C}$ be defined by

$$F(z) = \begin{cases} \dfrac{U(z) + iV(z)}{1 + U(z) + iV(z)} & \text{if } z \in \mathbb{D}, \\[3mm] \dfrac{u(z) + iv(z)}{1 + u(z) + iv(z)} & \text{if } z \in \mathbb{T} \setminus E, \\[3mm] 1 & \text{if } z \in E. \end{cases}$$

Clearly, F is analytic on $\mathbb{D}$ and continuous at all points of $\mathbb{T} \setminus E$. Since $u > 0$ and $U > 0$ on $\mathbb{D}$, we also have $|F(z)| < 1$ for all $z \in \mathbb{D}$ and all $z \in \mathbb{T} \setminus E$. Moreover, by (1.24) and (1.25), F is continuous at each point of E. Hence, F is continuous on $\mathbb{D}^-$.

Now that we have constructed the Fatou function for a closed subset of $\mathbb{T}$, we are ready to prove the F. and M. Riesz Theorem.

Proof of Theorem 1.21 Let $dv = \zeta d\mu$. Hence $d\mu = \bar{\zeta} dv$, and thus it is enough to show that $v \ll m$. The advantage of v is that

$$\widehat{v}(n) = 0, \qquad n \leqslant 0.$$

Thus, for any analytic function G on $\mathbb{D}$ whose power series is

$$G(z) = \sum_{n \geqslant 0} a_n z^n,$$

we have, for each $r \in (0, 1)$,

$$\begin{aligned} \int_{\mathbb{T}} G(r\zeta) \, dv(\zeta) &= \int_{\mathbb{T}} \left(\sum_{n \geqslant 0} a_n r^n \zeta^n \right) dv(\zeta) \\ &= \sum_{n \geqslant 0} a_n r^n \int_{\mathbb{T}} \zeta^n \, dv(\zeta) \\ &= \sum_{n \geqslant 0} a_n r^n \widehat{v}(-n) = 0. \end{aligned}$$

Suppose $E \subset \mathbb{T}$ with $m(E) = 0$. Let K be a compact subset of E and let F be a Fatou function for K. According to the preceding observation,

$$\int_{\mathbb{T}} (F(r\zeta))^\ell \, d\nu(\zeta) = 0, \qquad r \in (0, 1),$$

for all integers $\ell \geqslant 1$. Since F is bounded on $\mathbb{D}^-$, the Bounded Convergence Theorem tells us that

$$\int_{\mathbb{T}} (F(\zeta))^\ell \, d\nu(\zeta) = \lim_{r \to 1^-} \int_{\mathbb{T}} (F(r\zeta))^\ell d\nu(\zeta) = 0,$$

which, since $F \equiv 1$ on K, is equivalent to

$$\nu(K) + \int_{\mathbb{T} \setminus K} (F(\zeta))^\ell d\nu(\zeta) = 0.$$

Since $|F(\zeta)| < 1$ for all $\zeta \in \mathbb{T} \setminus K$, we see, letting $\ell \to \infty$ in the equation above, that $\nu(K) = 0$ for every compact subset K of E. Thus $\nu(E) = 0$ follows by regularity of the measure ν (see (1.3)). Since ν places no mass on any set of Lebesgue measure zero, it follows that $\nu \ll m$. $\qquad \square$

Remark 1.22 One can go further and prove that if $\widehat{\mu}(n) = 0$ for all $n \leqslant -1$, then $d\mu = f dm$ for f belonging to a certain class of functions on $\mathbb{T}$ (the Hardy space H^1). There are other proofs of the F. and M. Riesz Theorem (perhaps shorter) but they depend on more advanced knowledge of boundary values of analytic functions. See [158, p. 335] for more on this.

1.3 Hilbert spaces and their operators

1.3.1 Norm and weak topologies

Let $\mathcal{H}$ be a separable complex Hilbert space with inner product $\langle \mathbf{x}, \mathbf{y} \rangle$ and corresponding norm $\|\mathbf{x}\| = \sqrt{\langle \mathbf{x}, \mathbf{x} \rangle}$. Since $\mathcal{H}$ is separable, it has an *orthonormal basis* $\{\mathbf{e}_j\}_{1 \leqslant j \leqslant N}$, where we mean that

$$\langle \mathbf{e}_j, \mathbf{e}_k \rangle = \delta_{j,k}$$

and every $\mathbf{x} \in \mathcal{H}$ can be written uniquely as

$$\mathbf{x} = \sum_{1 \leqslant j \leqslant N} a_j \mathbf{e}_j,$$

where the sum above converges in the norm of $\mathcal{H}$. Furthermore, $a_j = \langle \mathbf{x}, \mathbf{e}_j \rangle$. The number $N \in \mathbb{N} \cup \{\infty\}$ is called the *dimension* of $\mathcal{H}$. By the term *subspace* we mean a closed linear manifold. Note that every subspace of a Hilbert space is itself a Hilbert space.

Theorem 1.23 (Cauchy–Schwarz Inequality)   *For $x, y \in \mathcal{H}$,*

$$|\langle x, y \rangle| \leqslant \|x\| \, \|y\|$$

with equality if and only if x and y are linearly dependent.

There are two natural topologies on a Hilbert space that will appear in this book. The first is the *norm topology*, where a sequence $\mathbf{x}_n$ converges in norm to $\mathbf{x}$ if $\|\mathbf{x}_n - \mathbf{x}\| \to 0$ as $n \to \infty$. The next is the *weak topology*, where a sequence $\mathbf{x}_n$ converges weakly to $\mathbf{x}$ if $\langle \mathbf{x}_n, \mathbf{y} \rangle \to \langle \mathbf{x}, \mathbf{y} \rangle$ for every $\mathbf{y} \in \mathcal{H}$. By the Cauchy–Schwarz Inequality, norm convergence implies weak convergence, although the converse does not hold (any orthonormal basis in an infinite-dimensional Hilbert space converges weakly to zero). However, weakly convergent sequences are bounded.

Theorem 1.24 (Principle of Uniform Boundedness)   *If $x_n \to x$ weakly in $\mathcal{H}$ then $\|x_n\|$ is bounded.*

Unless the ambient Hilbert space is finite dimensional, a closed and bounded set need not be compact. For example, the closed unit ball in an infinite-dimensional Hilbert space is not compact since it contains an infinite orthonormal set. However, it is *weakly compact*.

Theorem 1.25 (Banach–Alaoglu)   *If $\|x_n\|$ is bounded, then there is a subsequence x_{n_k} converging weakly to some $x \in \mathcal{H}$. That is to say, closed balls in a Hilbert space are weakly compact.*

1.3.2 Linear functionals and annihilators

For each fixed $\mathbf{y} \in \mathcal{H}$, define the *linear functional*

$$\ell : \mathcal{H} \to \mathbb{C}, \quad \ell(\mathbf{x}) = \langle \mathbf{x}, \mathbf{y} \rangle.$$

The norm of this linear functional is defined to be

$$\|\ell\| := \sup_{\|\mathbf{x}\|=1} |\ell(\mathbf{x})| = \sup_{\|\mathbf{x}\|=1} |\langle \mathbf{x}, \mathbf{y} \rangle|.$$

An application of the Cauchy–Schwarz Inequality says that

$$|\ell(\mathbf{x})| = |\langle \mathbf{x}, \mathbf{y} \rangle| \leqslant \|\mathbf{x}\| \, \|\mathbf{y}\| \qquad \forall \mathbf{x} \in \mathcal{H},$$

which means that $\|\ell\| \leqslant \|\mathbf{y}\|$. Plugging the unit vector $\mathbf{x} = \mathbf{y}/\|\mathbf{y}\|$ into the formula for ℓ shows us that $\|\ell\| = \|\mathbf{y}\|$. A theorem of F. Riesz classifies all of the bounded linear functionals on a Hilbert space.

Theorem 1.26 (Riesz Representation Theorem) *If $\ell : \mathcal{H} \to \mathbb{C}$ is a continuous linear functional, then there is a unique $\mathbf{y} \in \mathcal{H}$ such that*

$$\ell(\mathbf{x}) = \langle \mathbf{x}, \mathbf{y} \rangle \qquad \forall \mathbf{x} \in \mathcal{H}.$$

Furthermore, $\|\ell\| = \|\mathbf{y}\|$.

For a set $X \subset \mathcal{H}$, the *span* of X, denoted by span X, is the set of all finite linear combinations of elements of X. We let $\bigvee X$ denote the norm closure of span X, called the *closed span of X*.

For $X \subset \mathcal{H}$, the set

$$X^{\perp} = \{\mathbf{y} \in \mathcal{H} : \langle \mathbf{x}, \mathbf{y} \rangle = 0 \ \forall \mathbf{x} \in X\}$$

is a subspace of $\mathcal{H}$ called the *orthogonal complement* of X. Moreover, for any non-empty set $X \subset \mathcal{H}$, the double orthogonal complement $(X^{\perp})^{\perp}$ is equal to the closed span of X. This implies the following:

Theorem 1.27 *A set $X \subset \mathcal{H}$ has dense span if and only if $X^{\perp} = \{\mathbf{0}\}$.*

1.3.3 Operators

For Hilbert spaces $\mathcal{H}_1$ (with norm $\|\cdot\|_1$) and $\mathcal{H}_2$ (with norm $\|\cdot\|_2$), we say that a linear transformation $A : \mathcal{H}_1 \to \mathcal{H}_2$ is *bounded* if

$$\|A\| := \sup \{\|A\mathbf{x}\|_2 : \|\mathbf{x}\|_1 \leqslant 1\} < \infty.$$

The quantity $\|A\|$ is called the *operator norm* of A. By linearity, one can show that an operator is bounded if and only if it is continuous. We use $\mathcal{B}(\mathcal{H}_1, \mathcal{H}_2)$ to denote the set of bounded linear operators from $\mathcal{H}_1$ to $\mathcal{H}_2$. When $\mathcal{H} = \mathcal{H}_1 = \mathcal{H}_2$, we let $\mathcal{B}(\mathcal{H}) = \mathcal{B}(\mathcal{H}, \mathcal{H})$. Note that if $A, B \in \mathcal{B}(\mathcal{H})$, then $AB \in \mathcal{B}(\mathcal{H})$ and

$$\|AB\| \leqslant \|A\| \|B\|. \tag{1.26}$$

When dealing with operators on Hilbert spaces of analytic functions, the following often comes in handy when trying to determine if a linear transformation is bounded.

Theorem 1.28 (Closed Graph Theorem) *Suppose that A is a linear transformation on $\mathcal{H}$ whose graph $\{(\mathbf{x}, A\mathbf{x}) : \mathbf{x} \in \mathcal{H}\}$ is closed in $\mathcal{H} \oplus \mathcal{H}$. Then A is bounded.*

For $A \in \mathcal{B}(\mathcal{H}_1, \mathcal{H}_2)$, the *adjoint* of A is the unique $A^* \in \mathcal{B}(\mathcal{H}_2, \mathcal{H}_1)$ satisfying

$$\langle A\mathbf{x}, \mathbf{y} \rangle_2 = \langle \mathbf{x}, A^* \mathbf{y} \rangle_1 \qquad \forall \mathbf{x} \in \mathcal{H}_1, \mathbf{y} \in \mathcal{H}_2.$$

Moreover, we also have $\|A\| = \|A^*\|$ and $\ker A = (\operatorname{ran} A^*)^\perp$.

Proposition 1.29 $\mathcal{B}(\mathcal{H})$ *is an algebra that is complete with respect to the operator norm. Moreover, for* $A, B \in \mathcal{B}(\mathcal{H})$,

(i) $A^{**} = A$;
(ii) $(zA + B)^* = \bar{z}A^* + B^*$ *for all* $z \in \mathbb{C}$;
(iii) $(AB)^* = B^*A^*$;
(iv) $\|A^*A\| = \|A\|^2$.

We say that $A \in \mathcal{B}(\mathcal{H})$ is *invertible* if there exists a $B \in \mathcal{B}(\mathcal{H})$ such that $AB = BA = I$, where I is the identity operator on $\mathcal{H}$. In finite dimensions, $AB = I$ if and only if $BA = I$. In infinite dimensions, both conditions must be checked.

An operator $A \in \mathcal{B}(\mathcal{H}_1, \mathcal{H}_2)$ is *isometric* if $\|A\mathbf{x}\|_2 = \|\mathbf{x}\|_1$ for all $\mathbf{x} \in \mathcal{H}_1$ (equivalently $A^*A = I_{\mathcal{H}_1}$) and *contractive* if $\|A\| \leqslant 1$. If $A \in \mathcal{B}(\mathcal{H}_1, \mathcal{H}_2)$ is a surjective isometry, then we say that A is *unitary*. In this case we have $AA^* = I_{\mathcal{H}_2}$ and $A^*A = I_{\mathcal{H}_1}$. Two operators $A \in \mathcal{B}(\mathcal{H}_1)$ and $B \in \mathcal{B}(\mathcal{H}_2)$ are said to be *unitarily equivalent* if there is a unitary operator $U \in \mathcal{B}(\mathcal{H}_1, \mathcal{H}_2)$ such that $UA = BU$. Unitary equivalence is an equivalence relation on bounded operators that we denote by $\cong$.

The *spectrum* $\sigma(A)$ of $A \in \mathcal{B}(\mathcal{H})$ is the set of all $\lambda \in \mathbb{C}$ such that $A - \lambda I$ is not invertible in $\mathcal{B}(\mathcal{H})$. The *point spectrum* $\sigma_p(A)$ of $A \in \mathcal{B}(\mathcal{H})$ is the set of eigenvalues of A. In other words,

$$\sigma_p(A) = \{\lambda \in \mathbb{C} : \ker(A - \lambda I) \neq \{\mathbf{0}\}\}.$$

Note that when $\mathcal{H}$ is infinite dimensional, $\sigma_p(A)$ might be the empty set. We gather some important facts about $\sigma(A)$.

Proposition 1.30 *For* $A \in \mathcal{B}(\mathcal{H})$,

(i) $\sigma(A)$ *is a non-empty compact subset of* $\mathbb{C}$;
(ii) $\sigma_p(A) \subset \sigma(A) \subset \{\lambda : |\lambda| \leqslant \|A\|\}$.

To every subspace $\mathcal{M}$ of $\mathcal{H}$ there is an associated *orthogonal projection* $P_{\mathcal{M}} \in \mathcal{B}(\mathcal{H})$ defined by

$$P_{\mathcal{M}}\mathbf{x} = \begin{cases} \mathbf{x} & \text{if } \mathbf{x} \in \mathcal{M}, \\ \mathbf{0} & \text{if } \mathbf{x} \in \mathcal{M}^\perp. \end{cases}$$

Note that $P_{\mathcal{M}}^2 = P_{\mathcal{M}}$, $P_{\mathcal{M}}^* = P_{\mathcal{M}}$, and $I - P_{\mathcal{M}}$ is the orthogonal projection onto $\mathcal{M}^\perp$.

An operator $A \in \mathcal{B}(\mathcal{H})$ is *cyclic* if there is an $\mathbf{x} \in \mathcal{H}$ such that

$$\bigvee \{A^n \mathbf{x} : n \geqslant 0\} = \mathcal{H},$$

where $A^0 := I$. Such a vector $\mathbf{x}$ is called a *cyclic vector* for A. We now recall several forms of the Spectral Theorem.

Theorem 1.31 (Spectral Theorem)   *Let $A \in \mathcal{B}(\mathcal{H})$ be cyclic. Then:*

(i) *If A is normal ($A^*A = AA^*$), then there exists a $\mu \in M_+(\sigma(A))$ such that A is unitarily equivalent to the operator $f \mapsto zf$ on $L^2(\mu)$;*

(ii) *If A is self-adjoint ($A = A^*$), then $\sigma(A) \subset \mathbb{R}$ and there exists a $\nu \in M_+(\sigma(A))$ so that A is unitarily equivalent to the operator $f \mapsto xf$ on $L^2(\nu)$;*

(iii) *If A is unitary ($A^{-1} = A^*$), then $\sigma(A) \subset \mathbb{T}$ and there exists an $\eta \in M_+(\sigma(A))$ so that A is unitarily equivalent to the operator $f \mapsto \zeta f$ on $L^2(\eta)$.*

1.3.4 Three topologies

There are three natural topologies on $\mathcal{B}(\mathcal{H})$. The first is the norm topology, where we say a sequence $A_n \to A$ in *norm* if $\|A_n - A\| \to 0$. A weaker topology on $\mathcal{B}(\mathcal{H})$ is the *strong operator topology* (SOT), which says that a sequence $A_n \to A$ (SOT) if $A_n \mathbf{x} \to A\mathbf{x}$ for each $\mathbf{x} \in \mathcal{H}$. The weakest topology on $\mathcal{B}(\mathcal{H})$ we will encounter is the *weak operator topology* (WOT), where $A_n \to A$ (WOT) if $\langle A_n \mathbf{x}, \mathbf{y} \rangle \to \langle A\mathbf{x}, \mathbf{y} \rangle$ for every $\mathbf{x}, \mathbf{y} \in \mathcal{H}$. It is straightforward to see that

$$A_n \to A \;(\text{norm}) \quad \Longrightarrow \quad A_n \to A \;(\text{SOT}) \quad \Longrightarrow \quad A_n \to A \;(\text{WOT}).$$

In general, none of the implications above can be reversed.

An operator A belongs to the WOT closure of a set $\mathcal{X} \subset \mathcal{B}(\mathcal{H})$ if there is a net $\{A_\alpha\}_{\alpha \in I} \subset \mathcal{X}$ such that $A_\alpha \to A$ (WOT). We refer the reader to [117] for a precise treatment of nets. By the Principle of Uniform Boundedness, every (WOT) convergent sequence A_n is norm bounded. Using the definition of the weak operator topology, one can show that for $S \in \mathcal{B}(\mathcal{H})$, the *commutant*

$$\{S\}' := \{A \in \mathcal{B}(\mathcal{H}) : AS = SA\}$$

is WOT closed.

1.3.5 Rank-one operators

The rank of an operator $A \in \mathcal{B}(\mathcal{H})$ is equal to the dimension of the closure of its range. For $\mathbf{x}, \mathbf{y} \in \mathcal{H} \setminus \{\mathbf{0}\}$, the operator

$$\mathbf{x} \otimes \mathbf{y} : \mathcal{H} \to \mathcal{H}, \qquad (\mathbf{x} \otimes \mathbf{y})\mathbf{z} := \langle \mathbf{z}, \mathbf{y} \rangle \mathbf{x}$$

has rank one since the range of $\mathbf{x} \otimes \mathbf{y}$ is the one-dimensional subspace $\mathbb{C}\mathbf{x}$. We leave it to the reader to verify the following facts.

Proposition 1.32

(i) If $A \in \mathcal{B}(\mathcal{H})$ has rank one, then $A = \mathbf{x} \otimes \mathbf{y}$ for some $\mathbf{x}, \mathbf{y} \in \mathcal{H} \setminus \{\mathbf{0}\}$;

(ii) $\|\mathbf{x} \otimes \mathbf{y}\| = \|\mathbf{x}\| \|\mathbf{y}\|$;

(iii) $(\mathbf{x} \otimes \mathbf{y})^* = \mathbf{y} \otimes \mathbf{x}$;

(iv) If $A, B \in \mathcal{B}(\mathcal{H})$, then $A(\mathbf{x} \otimes \mathbf{y})B = (A\mathbf{x}) \otimes (B^*\mathbf{y})$.

1.3.6 Contractions and partial isometries

Recall that an operator $T \in \mathcal{B}(\mathcal{H})$ is called a contraction if $\|T\| \leqslant 1$. Certainly any unitary operator ($T^*T = TT^* = I$) is a contraction, as is any isometry ($T^*T = I$), or co-isometry ($TT^* = I$). Another type of contraction which will appear in this book is a partial isometry.

Definition 1.33 An operator $T \in \mathcal{B}(\mathcal{H})$ is called a *partial isometry* if the restriction $T|_{(\ker T)^\perp}$ is an isometry. The space $(\ker T)^\perp$ is called the *initial space* of T while the range of T is called the *final space* of T.

Proposition 1.34 *For $T \in \mathcal{B}(\mathcal{H})$ the following are equivalent:*

(i) *T is a partial isometry;*

(ii) *T^* is a partial isometry;*

(iii) *$T = TT^*T$;*

(iv) *$T^* = T^*TT^*$;*

(v) *TT^* is an orthogonal projection;*

(vi) *T^*T is an orthogonal projection.*

Moreover, T^*T is the orthogonal projection onto the initial space of T while TT^* is the orthogonal projection onto the final space of T.

For finite-dimensional Hilbert spaces, partial isometries are easy to spot [71].

Proposition 1.35 *Let A be an $n \times n$ matrix with complex entries. Then A defines a partial isometry on $\mathbb{C}^n$ if and only if there are orthonormal vectors $\{\mathbf{u}_1, \mathbf{u}_2, \ldots, \mathbf{u}_r\}$, $r \leqslant n$, in $\mathbb{C}^n$ and an $n \times n$ unitary matrix Q such that*

$$A = Q[\mathbf{u}_1|\mathbf{u}_2|\cdots|\mathbf{u}_r|\mathbf{0}|\mathbf{0}|\cdots|\mathbf{0}]Q^*.$$

In the proposition above, $[\mathbf{u}_1|\mathbf{u}_2|\cdots|\mathbf{u}_r|\mathbf{0}|\mathbf{0}|\cdots|\mathbf{0}]$ denotes the $n \times n$ matrix whose first r columns consists of the (column) vectors $\mathbf{u}_1, \ldots, \mathbf{u}_r$ and the last $n - r$ columns are zero vectors.

Let $T \in \mathcal{B}(\mathcal{H})$ be a contraction. As such, we know that $\|T\mathbf{x}\| \leqslant \|\mathbf{x}\|$ for every $\mathbf{x} \in \mathcal{H}$ and so

$$\langle (I - T^*T)\mathbf{x}, \mathbf{x} \rangle = \langle \mathbf{x}, \mathbf{x} \rangle - \langle T^*T\mathbf{x}, \mathbf{x} \rangle$$
$$= \|\mathbf{x}\|^2 - \langle T\mathbf{x}, T\mathbf{x} \rangle$$
$$= \|\mathbf{x}\|^2 - \|T\mathbf{x}\|^2 \geqslant 0.$$

Thus $I - T^*T$ is a positive operator and so, by the Spectral Theorem, $I - T^*T$ has a unique positive square root $(I - T^*T)^{1/2}$. This positive operator

$$D_T := (I - T^*T)^{1/2}$$

is called the *defect operator*. The subspace

$$\mathscr{D}_T := (D_T\mathcal{H})^-,$$

where the closure is taken in $\mathcal{H}$, is called the *defect space* and

$$\mathfrak{d}_T := \dim \mathscr{D}_T$$

is called the *defect index*. Since $\|T^*\| = \|T\| \leqslant 1$, T^* is also a contraction and one can define D_{T^*}, as well as $\mathscr{D}_{T^*}$ and $\mathfrak{d}_{T^*}$.

The defect operators and defect indices of T and T^* are tools to measure how far the contraction T is from being a unitary operator. In fact, $D_T = 0$ if and only if T is isometric and $D_T = D_{T^*} = 0$ if and only if T is unitary. For example, if T is a partial isometry recall from Proposition 1.34 that

$$D_T = P_{\ker T}, \quad \mathscr{D}_T = \ker T, \tag{1.27}$$

$$D_{T^*} = P_{(\operatorname{ran} T)^\perp}, \quad \text{and} \quad \mathscr{D}_{T^*} = (\operatorname{ran} T)^\perp. \tag{1.28}$$

From here one can more clearly see that a partial isometry T is unitary if and only if $D_T = D_{T^*} = 0$.

In Chapter 11 we will consider unitary extensions of a certain partial isometry. To place that discussion in a broader operator theory context, we consider the following situation. Suppose T is a partial isometry with defect indices

$$\mathfrak{d}_T = \mathfrak{d}_{T^*} = 1.$$

By Proposition 1.34,

$$T^*T = I - P_{\ker T}, \quad TT^* = I - P_{\ker T^*}.$$

However, since both defect indices are equal to one, we have

$$\ker T = \mathbb{C}\mathbf{x}, \quad \ker T^* = \mathbb{C}\mathbf{y}, \quad \|\mathbf{x}\| = \|\mathbf{y}\| = 1,$$

and so

$$\mathbf{x} \otimes \mathbf{x} = P_{\ker T}, \quad \mathbf{y} \otimes \mathbf{y} = P_{\ker T^*}.$$

Thus

$$T^*T = I - \mathbf{x} \otimes \mathbf{x}, \quad TT^* = I - \mathbf{y} \otimes \mathbf{y}.$$

We leave it to the reader to use the facts above to prove the following.

Theorem 1.36 *If $T \in \mathcal{B}(\mathcal{H})$ is a partial isometry with $\eth_T = \eth_{T^*} = 1$, then for each $\alpha \in \mathbb{T}$, the operator*

$$U_\alpha = T + \alpha(\mathbf{y} \otimes \mathbf{x})$$

is unitary on $\mathcal{H}$ with

$$U_\alpha|_{(\ker T)^\perp} = T|_{(\ker T)^\perp}.$$

Furthermore, if U is any unitary operator on $\mathcal{H}$ satisfying

$$U|_{(\ker T)^\perp} = T|_{(\ker T)^\perp},$$

then $U = U_\alpha$ for some $\alpha \in \mathbb{T}$.

1.3.7 The Calkin algebra

An operator $A \in \mathcal{B}(\mathcal{H})$ is *compact* if the image of every bounded set under A has compact closure in $\mathcal{H}$. For example, every operator whose range is a finite-dimensional subspace (such operators are called *finite rank*) is compact by the Heine–Borel Theorem.

Proposition 1.37 *An operator $A \in \mathcal{B}(\mathcal{H})$ is compact if and only if $\|A\mathbf{x}_n\| \to 0$ whenever $\mathbf{x}_n \to \mathbf{0}$ weakly.*

Here are some well-known facts about the set $\mathcal{K} := \mathcal{K}(\mathcal{H})$ of compact operators on $\mathcal{H}$.

Proposition 1.38

(i) $A \in \mathcal{K} \implies A^* \in \mathcal{K}$;

(ii) $\mathcal{K}$ *is a norm closed ideal in $\mathcal{B}(\mathcal{H})$;*

(iii) The finite-rank operators belong to $\mathcal{K}$;

(iv) If $A \in \mathcal{K}$, then A is the operator norm limit of a sequence of finite-rank operators.

$\mathcal{B}(\mathcal{H})$ is complete with respect to the operator norm and $\mathcal{K}$ is a norm closed, $*$-closed ideal of $\mathcal{B}(\mathcal{H})$. This allows us to form the quotient space $\mathcal{B}(\mathcal{H})/\mathcal{K}$, that is, the collection of cosets

$$\{A + \mathcal{K} : A \in \mathcal{B}(\mathcal{H})\}.$$

Note that $\mathcal{B}(\mathcal{H})/\mathcal{K}$ is a vector space with addition

$$(A + \mathcal{K}) + (B + \mathcal{K}) := (A + B) + \mathcal{K},$$

and scalar multiplication

$$z(A + \mathcal{K}) := zA + \mathcal{K}.$$

Since $\mathcal{K}$ is an ideal, we also have a well-defined multiplication

$$(A + \mathcal{K}) \cdot (B + \mathcal{K}) := AB + \mathcal{K}.$$

There is an involution $*$ defined on $\mathcal{B}(\mathcal{H})/\mathcal{K}$ by

$$(A + \mathcal{K})^* := A^* + \mathcal{K},$$

along with a quotient norm on $\mathcal{B}(\mathcal{H})/\mathcal{K}$ defined by

$$\|A + \mathcal{K}\| := \inf_{K \in \mathcal{K}} \|A - K\|.$$

For $A \in \mathcal{B}(\mathcal{H})$ we define

$$\|A\|_e := \|A + \mathcal{K}\|,$$

the *essential norm* (the distance to the compact operators) and note that $\|A\|_e \leqslant \|A\|$. It is also possible to show that

$$\|A^*A\|_e = \|A\|_e^2,$$

and, most importantly, $\mathcal{B}(\mathcal{H})$ is complete with respect to this semi-norm whose kernel is $\mathcal{K}$. Putting this all together, we see that $\mathcal{B}(\mathcal{H})/\mathcal{K}$ is a C^*-algebra called the *Calkin algebra*.

When is a coset in $\mathcal{B}(\mathcal{H})/\mathcal{K}$ invertible? The answer is contained in the following theorem:

Theorem 1.39 (Atkinson)   *For $A \in \mathcal{B}(\mathcal{H})$ the following are equivalent:*

(i) $A + \mathcal{K}$ is invertible in $\mathcal{B}(\mathcal{H})/\mathcal{K}$;

(ii) A has closed range and both $\ker A$ and $\ker A^*$ are finite dimensional.

The operators satisfying condition (ii) are called *Fredholm operators*.

The *essential spectrum* of an operator A, denoted by $\sigma_e(A)$, is the set of $\lambda \in \mathbb{C}$ for which $A - \lambda I$ is not Fredholm. Equivalently, $\lambda \in \sigma_e(A)$ if $(A - \lambda I) + \mathcal{K}$ is not an invertible element in $\mathcal{B}(\mathcal{H})/\mathcal{K}$.

Corollary 1.40 *For $A \in \mathcal{B}(\mathcal{H})$ and $\lambda \in \mathbb{C}$, the following are equivalent:*

(i) $\lambda \notin \sigma_e(A)$;
(ii) $A - \lambda I$ has closed range and both $\ker(A - \lambda I)$ and $\ker(A^* - \overline{\lambda} I)$ are finite dimensional.

Proposition 1.41 *For $A \in \mathcal{B}(\mathcal{H})$, we have:*

(i) $\sigma_e(A)$ is a non-empty compact subset of $\mathbb{C}$;
(ii) $\sigma_e(A) \subset \sigma(A)$.

1.4 Notes

1.4.1 Further references

The reader needing a further review of measure theory (measures, derivatives, decomposition theorems, Lebesgue spaces, functions of bounded variation, etc.) should consult [157, 158]. More on Poisson integrals (in particular Fatou's Theorem) is found in [63, 112, 158] while the basics of operator theory are presented in [51, 159].

2

Inner functions

Model spaces are intimately connected with the function-theoretic properties of inner functions. In this chapter, we will cover the basics of inner functions, sometimes following the exposition and proofs from the classic texts [50, 63, 97, 158].

2.1 Disk automorphisms

For fixed $a \in \mathbb{D}$ and $\zeta \in \mathbb{T}$, consider the Möbius transformation

$$\tau_{\zeta,a}(z) := \zeta \frac{a - z}{1 - \overline{a}z}, \qquad z \in \mathbb{D}. \tag{2.1}$$

For each $w \in \mathbb{T}$ we have

$$|\tau_{\zeta,a}(w)| = \frac{|a - w|}{|1 - \overline{a}w|} = \frac{|a - w|}{|w(\overline{w} - \overline{a})|} = \frac{|a - w|}{|\overline{w} - \overline{a}|} = 1; \tag{2.2}$$

so $\tau_{\zeta,a}$ maps the unit circle onto itself. Since $\tau_{\zeta,a}(0) = \zeta a \in \mathbb{D}$, it follows from properties of Möbius transformations that $\tau_{\zeta,a}$ maps $\mathbb{D}$ onto itself. Another computation shows that

$$\tau_{\zeta,a}^{-1} = \tau_{\overline{\zeta},\zeta a};$$

and so $\tau_{\zeta,a}$ is an automorphism (bijective analytic map) of $\mathbb{D}$. It turns out that these are all of the disk automorphisms.

Proposition 2.1 *Suppose $\varphi : \mathbb{D} \to \mathbb{D}$ is an analytic bijective map. Then $\varphi = \tau_{\zeta,a}$ for some $\zeta \in \mathbb{T}$ and $a \in \mathbb{D}$.*

The proof requires the Schwarz Lemma.

Lemma 2.2 (Schwarz Lemma) *Suppose $f : \mathbb{D} \to \mathbb{D}$ is analytic with $f(0) = 0$. Then $|f(z)| \leqslant |z|$ for all $z \in \mathbb{D}$ and $|f'(0)| \leqslant 1$. Moreover, if $|f(z_0)| = |z_0|$ for*

*some $z_0 \in \mathbb{D} \setminus \{0\}$ or if $|f'(0)| = 1$, then there exists a $\zeta \in \mathbb{T}$ such that $f(z) = \zeta z$
for all $z \in \mathbb{D}$.*

Proof   Define the function g on $\mathbb{D}$ by

$$g(z) = \begin{cases} \dfrac{f(z)}{z} & \text{if } z \neq 0, \\ f'(0) & \text{if } z = 0, \end{cases}$$

and note that g is analytic on $\mathbb{D}$. For $r \in [0, 1)$, apply the Maximum Modulus
Principle to $r\mathbb{D}$ to produce a $\zeta \in \mathbb{T}$ so that

$$|g(rz)| \leqslant |g(r\zeta)| = \frac{|f(r\zeta)|}{|r\zeta|} \leqslant \frac{1}{r}, \quad \forall z \in \mathbb{D}.$$

Letting $r \to 1^-$ proves the first part of the Schwarz Lemma.

For the proof of the second part, suppose that $|f(z_0)| = |z_0|$ for some $z_0 \in \mathbb{D} \setminus \{0\}$, or $|f'(0)| = 1$. Then $|g(z_0)| = 1$ for some $z_0 \in \mathbb{D}$. By the Maximum
Modulus Principle (applied to $\mathbb{D}$), g is equal to a constant $\zeta \in \mathbb{T}$. Therefore,
$f(z) = \zeta z$ for all $z \in \mathbb{D}$. $\square$

Proof of Proposition 2.1   Suppose that φ is a disk automorphism satisfying
$\varphi(a) = 0$. Then the automorphism $\psi = \varphi \circ \tau_{1,a}^{-1}$ satisfies $\psi(0) = 0$. By the
Schwarz Lemma, $|\psi(z)| \leqslant |z|$ for all $z \in \mathbb{D}$. Since ψ^{-1} is also a disk automor-
phism satisfying $\psi^{-1}(0) = 0$, we can apply the Schwarz Lemma again to see
that $|\psi^{-1}(z)| \leqslant |z|$ on $\mathbb{D}$. These two inequalities yield $|\psi(z)| = |z|$ on $\mathbb{D}$ and so,
by the second part of the Schwarz Lemma, we get $\psi(z) = \zeta z$. From here we get
$\varphi \circ \tau_{1,a}^{-1} = \zeta z$ from which it follows that $\varphi = \zeta \tau_{1,a} = \tau_{\zeta,a}$. $\square$

2.2 Bounded analytic functions

Let H^∞ denote the set of bounded analytic functions on $\mathbb{D}$. When H^∞ is
endowed with the norm

$$\|f\|_\infty := \sup_{z \in \mathbb{D}} |f(z)|,$$

it becomes a Banach space. It turns out that functions in H^∞ possess boundary
values in the following sense.

Theorem 2.3 (Fatou)   *For each $f \in H^\infty$, the radial limit*

$$\lim_{r \to 1^-} f(r\zeta)$$

exists for almost every $\zeta \in \mathbb{T}$.

Proof   Each $f \in H^\infty$ is harmonic on $\mathbb{D}$ and satisfies

$$\sup_{0<r<1} \int_{\mathbb{T}} |f(r\zeta)| \, dm(\zeta) \leqslant \|f\|_\infty < \infty.$$

By Corollary 1.19, $f = \mathscr{P}(\mu)$, the Poisson integral of a measure $\mu \in M(\mathbb{T})$. Using Fatou's Theorem (Theorem 1.10), we conclude that f has radial limits m-almost everywhere.   □

From the comments following Theorem 1.10, bounded analytic functions have non-tangential limits almost everywhere. In fact, a theorem of Lindelöf [50, p. 19] says that if a bounded analytic function has a radial limit at ζ, then it has a non-tangential limit at ζ. To be more precise, if $f \in H^\infty$ and $f(z) \to A$ as z approaches ζ along a continuous path $L \subset \mathbb{D}$ terminating at $\zeta \in \mathbb{T}$, then $f(z) \to A$ as z approaches ζ in any Stolz domain $\Gamma_\alpha(\zeta)$. We will see Lindelöf's Theorem again in Chapter 6.

Theorem 2.4 (Maximum Modulus Theorem)   *If $f \in H^\infty$, then*

$$\sup_{z \in \mathbb{D}} |f(z)| = \||f|_{\mathbb{T}}\|_\infty,$$

where

$$f(\zeta) = \lim_{r \to 1^-} f(r\zeta)$$

is the almost m-everywhere defined boundary function on $\mathbb{T}$.

Proof   If f is analytic in an open neighborhood of $\mathbb{D}^-$, then this is just the classical Maximum Modulus Theorem from elementary complex analysis [158]. The tricky part comes when dealing with the (almost everywhere defined) radial limit function. For $f = \sum_{n \geqslant 0} a_n z^n \in H^\infty$, the orthogonality of the functions $\{\zeta^n : n \in \mathbb{Z}\}$ with respect to the L^2 inner product (see (1.11)) implies that

$$\int_{\mathbb{T}} |f(r\zeta)|^2 dm(\zeta) = \int_{\mathbb{T}} f(r\zeta)\overline{f(r\zeta)} \, dm(\zeta)$$
$$= \sum_{m,n \geqslant 0} a_n \overline{a_m} r^{n+m} \int_{\mathbb{T}} \zeta^{n-m} dm(\zeta)$$
$$= \sum_{n \geqslant 0} |a_n|^2 r^{2n} \tag{2.3}$$

for all $0 < r < 1$, and so

$$r \mapsto \int_{\mathbb{T}} |f(r\zeta)|^2 dm(\zeta) \tag{2.4}$$

is an increasing function of r. From here we see that

$$\int_{\mathbb{T}} |f(r\zeta)|^2 dm(\zeta) \leqslant \sup_{0<r<1} \int_{\mathbb{T}} |f(r\zeta)|^2 dm(\zeta)$$

$$= \lim_{r\to 1^-} \int_{\mathbb{T}} |f(r\zeta)|^2 dm(\zeta) \qquad \text{(by (2.4))}$$

$$= \int_{\mathbb{T}} \lim_{r\to 1^-} |f(r\zeta)|^2 dm(\zeta) \qquad \text{(Bounded Conv. Thm.)}$$

$$= \int_{\mathbb{T}} |f(\zeta)|^2 dm(\zeta). \qquad (2.5)$$

Thus for each $r \in (0, 1)$ we have

$$|f(0)| = \left| \int_{\mathbb{T}} f(r\zeta) dm(\zeta) \right| \qquad \text{(Theorem 1.14)}$$

$$\leqslant \left(\int_{\mathbb{T}} |f(r\zeta)|^2 dm(\zeta) \right)^{1/2} \qquad \text{(Cauchy–Schwarz)}$$

$$\leqslant \left(\int_{\mathbb{T}} |f(\zeta)|^2 dm(\zeta) \right)^{1/2} \qquad \text{(by (2.5))}$$

$$\leqslant \|f|_{\mathbb{T}}\|_\infty.$$

For each $a \in \mathbb{D}$ recall from (2.1) that the map

$$\tau_{1,a}(z) = \frac{a - z}{1 - \bar{a}z}$$

maps $\mathbb{D}$ onto itself and $\mathbb{T}$ onto itself. Noting that $\tau_{1,a}(0) = a$, apply the argument above to the function $f \circ \tau_{1,a}$ (which belongs to H^∞) to see that

$$|f(a)| \leqslant \|f|_{\mathbb{T}}\|_\infty \qquad (2.6)$$

for each $a \in \mathbb{D}$. Thus

$$\|f\|_\infty \leqslant \|f|_{\mathbb{T}}\|_\infty. \qquad (2.7)$$

For the reverse inequality in (2.7), given $\varepsilon > 0$, one can choose a $\zeta_\varepsilon \in \mathbb{T}$ so that the radial limit of f at ζ_ε exists and satisfies

$$\|f|_{\mathbb{T}}\|_\infty - \epsilon < |f(\zeta_\varepsilon)| \leqslant \|f|_{\mathbb{T}}\|_\infty.$$

For r close enough to 1, we use the definition of a radial limit to get

$$\|f|_{\mathbb{T}}\|_\infty - \varepsilon < |f(r\zeta_\varepsilon)| \leqslant \|f|_{\mathbb{T}}\|_\infty.$$

Now use (2.7) to get

$$\|f|_{\mathbb{T}}\|_\infty - \varepsilon < \|f\|_\infty \leqslant \|f|_{\mathbb{T}}\|_\infty,$$

which proves the result. $\qquad\qquad\qquad\qquad\qquad\qquad\qquad\qquad\qquad\qquad\square$

2.3 Inner functions

Recall from Theorem 2.3 (Fatou's Theorem) that every $f \in H^\infty$ has a well-defined radial boundary value $f(\zeta)$ for a.e. $\zeta \in \mathbb{T}$.

Definition 2.5   A function $u \in H^\infty$ is said to be *inner* if $|u(\zeta)| = 1$ a.e. on $\mathbb{T}$.

By the Maximum Modulus Theorem (Theorem 2.4), every inner function u satisfies $|u(z)| \leqslant 1$ on $\mathbb{D}$. Here are some examples of inner functions.

Example 2.6   From (2.1), the function $u = \tau_{\zeta,a}$ is inner for every $\zeta \in \mathbb{T}$ and $a \in \mathbb{D}$. It follows that if $a_1, a_2, \ldots, a_n \in \mathbb{D}$, then $u = \tau_{1,a_1} \tau_{1,a_2} \cdots \tau_{1,a_n}$ is also inner. Under certain circumstances, one can also consider infinite products of such functions (see Theorem 2.10).

Example 2.7   Another simple example of an inner function is

$$u(z) = \exp\left(\frac{z+1}{z-1}\right). \tag{2.8}$$

From the identity $|e^w| = e^{\text{Re}\, w}$, we see that

$$\left|\exp\left(\frac{z+1}{z-1}\right)\right| = \exp\left(\text{Re}\left(\frac{z+1}{z-1}\right)\right) = \exp\left(-\frac{1-|z|^2}{|1-z|^2}\right) < 1$$

for all $z \in \mathbb{D}$. Thus u is a bounded (non-vanishing) analytic function on $\mathbb{D}$. Furthermore, $|u(e^{i\theta})| = 1$ for $\theta \in (0, 2\pi)$, and hence u is an inner function. If $\alpha_1, \alpha_2, \ldots, \alpha_n$ are positive numbers and $\theta_1, \theta_2, \ldots, \theta_n \in [0, 2\pi]$, one can also check that

$$u(z) = \exp\left(\alpha_1 \frac{z + e^{i\theta_1}}{z - e^{i\theta_1}} + \cdots + \alpha_n \frac{z + e^{i\theta_n}}{z - e^{i\theta_n}}\right)$$

is an inner function. More examples of this type will be given in Theorem 2.12.

Remark 2.8   Consider the function

$$f(z) = \exp\left(\frac{1+z}{1-z}\right).$$

Since this function is the reciprocal of the function u from (2.8), it follows that $|f(e^{i\theta})| = 1$ for $\theta \in (0, 2\pi)$. However, for $r \in (0, 1)$,

$$|f(r)| = \exp\left(\frac{1+r}{1-r}\right) \to \infty, \quad r \to 1^-.$$

Although f has unimodular boundary values almost everywhere on $\mathbb{T}$, it is unbounded on $\mathbb{D}$ and hence is not an inner function. Thus when checking to see whether or not an analytic function is inner, one must be careful to first check that the function is actually *bounded* on $\mathbb{D}$.

We now classify all inner functions and show that they are built from two basic types. The first type is described in Theorem 2.10 below.

Definition 2.9 (Blaschke condition)   A sequence $\{a_n\}_{n \geqslant 1} \subset \mathbb{D}$ satisfies the *Blaschke condition* if

$$\sum_{n \geqslant 1} (1 - |a_n|) < \infty. \tag{2.9}$$

A sequence that satisfies the Blaschke condition is called a *Blaschke sequence*.

Theorem 2.10 (Blaschke, F. Riesz)   *If $\{a_n\}_{n \geqslant 1} \subset \mathbb{D} \setminus \{0\}$ is a Blaschke sequence, then the infinite product*

$$B(z) := \prod_{n \geqslant 1} \frac{\overline{a_n}}{|a_n|} \frac{a_n - z}{1 - \overline{a_n}z}, \qquad z \in \mathbb{D},$$

converges uniformly on compact subsets of $\mathbb{D}$. Moreover, B is an inner function.

Definition 2.11   For $\xi \in \mathbb{T}$, $N \in \mathbb{N} \cup \{0\}$, and a Blaschke sequence $\{a_n\}_{n \geqslant 1} \subset \mathbb{D} \setminus \{0\}$, we call the function

$$B(z) = \xi z^N \prod_{n \geqslant 1} \frac{\overline{a_n}}{|a_n|} \frac{a_n - z}{1 - \overline{a_n}z}$$

a *Blaschke product*.

Proof of Theorem 2.10   Our proof follows [63, 112]. We assume that $a_n \neq 0$. First let us show that for each $r \in (0, 1)$, the product converges uniformly on $r\mathbb{D}$. To do this, let

$$b_{a_n}(z) = \frac{\overline{a_n}}{|a_n|} \frac{a_n - z}{1 - \overline{a_n}z}$$

(note that $b_{a_n} = \tau_{\overline{a_n}/|a_n|, a_n}$ is a disk automorphism from (2.1)) and recall from basic complex analysis [158, Ch. 15] that if

$$\sum_{n \geqslant 1} |1 - b_{a_n}(z)|$$

converges uniformly on $r\mathbb{D}$, then the infinite product

$$\prod_{n \geqslant 1} b_{a_n}(z)$$

converges uniformly on $r\mathbb{D}$.

For $z \in r\mathbb{D}$, notice that

$$|1 - b_{a_n}(z)| = \left|1 - \frac{\overline{a_n}}{|a_n|}\frac{a_n - z}{1 - \overline{a_n}z}\right| \tag{2.10}$$

$$= \left|\left(1 - \frac{1}{|a_n|}\right) + \left(\frac{1}{|a_n|} - \frac{\overline{a_n}}{|a_n|}\frac{a_n - z}{1 - \overline{a_n}z}\right)\right|$$

$$- \left|\frac{1}{|a_n|}\left(1 - \frac{|a_n|^2 - \overline{a_n}z}{1 - \overline{a_n}z}\right) + \left(1 - \frac{1}{|a_n|}\right)\right|$$

$$= \left|\frac{1}{|a_n|}\left(\frac{1 - \overline{a_n}z}{1 - \overline{a_n}z} - \frac{|a_n|^2 - \overline{a_n}z}{1 - \overline{a_n}z}\right) + \left(1 - \frac{1}{|a_n|}\right)\right|$$

$$= \left|\frac{1}{|a_n|}\frac{1 - |a_n|^2}{1 - \overline{a_n}z} + \left(1 - \frac{1}{|a_n|}\right)\right|$$

$$= \left|\frac{1}{|a_n|}\frac{1 - |a_n|^2}{1 - \overline{a_n}z} + \frac{|a_n| - 1}{|a_n|}\right|$$

$$= \left|\frac{1 - |a_n|}{|a_n|}\left(\frac{1 + |a_n|}{1 - \overline{a_n}z} - 1\right)\right|$$

$$\leqslant \frac{1 - |a_n|}{|a_n|}\left(\left|\frac{1 + |a_n|}{1 - \overline{a_n}z}\right| + 1\right)$$

$$\leqslant \frac{1 - |a_n|}{|a_n|}\left(\frac{2}{|1 - \overline{a_n}z|} + 1\right) \tag{2.11}$$

$$\leqslant \frac{1 - |a_n|}{|a_n|}\left(\frac{2}{1 - r} + 1\right).$$

Since we are assuming that $\sum_{n\geqslant 1}(1 - |a_n|) < \infty$, it follows that $|a_n|$ is bounded away from zero. Thus the series

$$\sum_{n\geqslant 1}|1 - b_{a_n}(z)|,$$

and hence the product

$$\prod_{n\geqslant 1}b_{a_n}(z),$$

converge uniformly on $r\mathbb{D}$.

Since the Blaschke product B converges uniformly on compact subsets of $\mathbb{D}$, it defines an analytic function on $\mathbb{D}$. Furthermore, since each factor b_{a_n} is inner (Example 2.6), it follows that $|B(z)| \leqslant 1$ for all $z \in \mathbb{D}$. Since B belongs to H^∞, we can apply Theorem 2.3 to see that $B(\zeta)$ (the radial limit of B at ζ) exists and satisfies $|B(\zeta)| \leqslant 1$ for almost every $\zeta \in \mathbb{T}$. To finish showing that B is inner, we need to prove that $|B(\zeta)| = 1$ almost everywhere.

To do this, note that if

$$B_N(z) = \prod_{1 \leqslant n \leqslant N} b_{a_n}(z)$$

denotes the Nth partial product of B, then $B/B_N \in H^\infty$. For $r \in (0, 1)$ we have

$$\int_{\mathbb{T}} \left| \frac{B}{B_N}(r\zeta) \right|^2 dm(\zeta) \leqslant \int_{\mathbb{T}} \frac{|B(\zeta)|^2}{|B_N(\zeta)|^2} dm(\zeta) \qquad \text{(by (2.5))}$$

$$= \int_{\mathbb{T}} |B(\zeta)|^2 dm(\zeta).$$

Now use the fact that $B_N \to B$ uniformly on $r\mathbb{T}$ to see that

$$1 = \lim_{N \to \infty} \int_{\mathbb{T}} \left| \frac{B}{B_N}(r\zeta) \right|^2 dm(\zeta) \leqslant \int_{\mathbb{T}} |B(\zeta)|^2 dm(\zeta) \leqslant 1.$$

Thus $|B| = 1$ almost everywhere on $\mathbb{T}$. $\qquad\square$

Blaschke products form one distinguished class of inner functions. Another variety of inner function is furnished by the following construction.

Theorem 2.12 *For any $\mu \in M_+(\mathbb{T})$ with $\mu \perp m$, the function*

$$s_\mu(z) - \exp\left(-\int_{\mathbb{T}} \frac{\xi + z}{\xi - z} d\mu(\xi)\right), \qquad z \in \mathbb{D}, \tag{2.12}$$

is inner.

Proof The Poisson integral $\mathscr{P}(\mu)(z)$ is non-negative since $\mu \in M_+(\mathbb{T})$. For any $z \in \mathbb{D}$,

$$|s_\mu(z)| = \exp\left[\mathrm{Re}\left(-\int_{\mathbb{T}} \frac{\xi + z}{\xi - z} d\mu(\xi)\right)\right]$$

$$= \exp\left(-\int_{\mathbb{T}} P_z(\xi) d\mu(\xi)\right) \qquad \text{(by (1.12))}$$

$$= \exp\left(-\mathscr{P}(\mu)(z)\right)$$

$$\leqslant 1,$$

so that $s_\mu \in H^\infty$. Since μ is singular, it follows that $D\mu = 0$ m-almost everywhere (Theorem 1.2) and so, by Fatou's Theorem (Theorem 1.10),

$$\lim_{r \to 1^-} \mathscr{P}(\mu)(r\zeta) = (D\mu)(\zeta) = 0$$

for m-almost every $\zeta \in \mathbb{T}$. This shows that $|s_\mu(\zeta)| = 1$ a.e. on $\mathbb{T}$ and thus s_μ is an inner function. $\qquad\square$

Definition 2.13 A function of the form ξs_μ, where $\xi \in \mathbb{T}$, is called a *singular inner function*.

Blaschke products and singular inner functions form the building blocks of all non-constant inner functions.

Theorem 2.14 (Nevanlinna, F. Riesz) *Any inner function u can be factored as*

$$u(z) = z^k B(z) s_\mu(z),$$

where $k \geqslant 0$, B is a Blaschke product, and s_μ is a singular inner function. This factorization is unique up to a unimodular constant.

Proof We follow [112]. Let $\{a_n\}_{n\geqslant 1}$ be the zeros of u in $\mathbb{D}$, repeated according to multiplicity. We can assume that $u(0) \neq 0$; since otherwise we may replace u by u/z^k for some appropriate k. Let

$$B_N(z) = \prod_{1 \leqslant n \leqslant N} \frac{\overline{a_n}}{|a_n|} \frac{a_n - z}{1 - \overline{a_n} z}$$

be the Nth partial Blaschke product formed from these zeros. Since $|B_N(\zeta)| = 1$ for every $\zeta \in \mathbb{T}$, we see that $u/B_N \in H^\infty$. Hence, by the Maximum Modulus Theorem (Theorem 2.4),

$$\left| \frac{u}{B_N}(z) \right| \leqslant \text{ess-sup}_{\zeta \in \mathbb{T}} \frac{|u(\zeta)|}{|B_N(\zeta)|} = \text{ess-sup}_{\zeta \in \mathbb{T}} |u(\zeta)| = 1$$

for all $z \in \mathbb{D}$. Thus $|u| \leqslant |B_N|$ on $\mathbb{D}$. In particular, this means that

$$0 < |u(0)| \leqslant |B_N(0)| = \prod_{1 \leqslant n \leqslant N} |a_n|.$$

Since $|a_n| < 1$ and the partial products are bounded away from zero, we see that the infinite product $\prod_{n \geqslant 1} |a_n|$ converges. Thus $\sum_{n \geqslant 1}(1 - |a_n|) < \infty$ [158, Ch. 15].

By Theorem 2.10, the partial products B_N converge uniformly on compact subsets of $\mathbb{D}$ to the Blaschke product B with zeros $\{a_n\}_{n \geqslant 1}$. Thus u/B_N converges uniformly on compact subsets of $\mathbb{D}$ to u/B, which consequently must belong to H^∞. Furthermore, since both u and B have unimodular boundary values almost everywhere, u/B is an inner function s.

To finish we need to show that s is a *singular* inner function. Note that since s has no zeros in $\mathbb{D}$ (they were divided out when we constructed B), $-\log|s|$ is a positive harmonic function on $\mathbb{D}$ and so, by Herglotz's Theorem (Theorem 1.18),

$$-\log|s(z)| = \mathscr{P}(\mu)(z)$$

for some unique $\mu \in M_+(\mathbb{T})$. By analytic completion, as in the proof of Corollary 1.20,

$$s(z) = \exp\left(-\int_{\mathbb{T}} \frac{\zeta + z}{\zeta - z} d\mu(\zeta) + i\gamma\right)$$

for some $\gamma \in [0, 2\pi)$. Since s is inner, we can apply the argument in the proof of Theorem 2.12 to see that $\mu \perp m$. $\square$

Example 2.15   One can show that the only inner polynomials

$$p(z) = a_0 + a_1 z + a_2 z^2 + \cdots + a_n z^n$$

are ξz^n, $n = 0, 1, 2, \ldots$ and $\xi \in \mathbb{T}$.

To see why this works, note that if p is an inner function, we have

$$1 = |p(z)|^2 = p(z)\overline{p(z)}, \quad |z| = 1.$$

Thus

$$p(z) = \frac{1}{\overline{p(z)}} = \frac{1}{\overline{a_0} + \overline{a_1}\overline{z} + \cdots + \overline{a_n}\overline{z}^n} = \frac{z^n}{z^n \overline{a_0} + \cdots + z\overline{a_{n-1}} + \overline{a_n}}.$$

The functions $p(z)$ and

$$\frac{z^n}{z^n \overline{a_0} + \cdots + z\overline{a_{n-1}} + \overline{a_n}}$$

are rational functions which agree on $\mathbb{T}$ and thus everywhere. Moreover, since p is a polynomial, $z^n \overline{a_0} + \cdots + z\overline{a_{n-1}} + \overline{a_n}$ divides z^n. However, since $a_n \neq 0$, it must be the case that $a_0 = a_1 = \cdots = a_{n-1} = 0$. Thus $p(z) = a_n z^n$ and $a_n \in \mathbb{T}$ since p is inner.

There is a notion of "divisibility" of inner functions which mirrors that of divisibility of integers.

Definition 2.16   Let u_1, u_2 be inner functions.

(i) We say that u_1 *divides* u_2, written $u_1 | u_2$, if u_2/u_1 is inner.
(ii) We say that u_1 and u_2 are *relatively prime* if the only common inner divisors of u_1 and u_2 are constant functions of unit modulus.

For example, if u_1 and u_2 are Blaschke products with simple zeros, then u_1 divides u_2 if and only if the zero set of u_1 is contained in the zero set of u_2. Moreover, u_1 is relatively prime to u_2 if and only if u_1 and u_2 have no common zeros. If $u_1 = s_{\mu_1}, u_2 = s_{\mu_2}$ are singular inner functions, then $u_1 | u_2$ if and only if $\mu_2 - \mu_1$ is a positive measure.

2.4 Unimodular boundary limits

For an inner function u and fixed $\zeta \subset \mathbb{T}$, when does the non tangential limit of u exist and have unit modulus at ζ? The answer is contained in the following theorem.

Theorem 2.17   *An inner function $u = Bs_\mu$ and all of its inner divisors have non-tangential limits of modulus one at $\zeta \in \mathbb{T}$ if and only if*

$$\sum_{n \geqslant 1} \frac{1 - |a_n|}{|\zeta - a_n|} + \int_{\mathbb{T}} \frac{d\mu(\xi)}{|\xi - \zeta|} < \infty.$$

Here $\{a_n\}_{n \geqslant 1}$ denotes the zeros, repeated according to their multiplicity, of the Blaschke product B and μ denotes the singular measure corresponding to the singular inner function s_μ.

Since this theorem is not integral to our treatment of model spaces, we will only sketch a few ideas for one direction of the proof. When u is a Blaschke product, the theorem is one of Frostman [75] (see also [50]). The general case is due to Ahern and Clark [4, 5].

Suppose that $\zeta = 1$ and $\{a_n\}_{n \geqslant 1}$ satisfy

$$\sum_{n \geqslant 1} \frac{1 - |a_n|}{|1 - a_n|} < \infty. \tag{2.13}$$

For each $r \in (0, 1)$, we have

$$|1 - \overline{a_n}r| > 1 - r \quad \text{and} \quad |1 - \overline{a_n}r| > \tfrac{1}{2}|1 - a_n|.$$

To see the first inequality, notice that

$$|1 - \overline{a_n}r| \geqslant 1 - |a_n|r > 1 - r.$$

To see the second inequality, observe from the first inequality that

$$\begin{aligned}
|1 - \overline{a_n}| &\leqslant |1 - \overline{a_n}r + \overline{a_n}r - \overline{a_n}| \\
&\leqslant |1 - \overline{a_n}r| + |a_n|(1 - r) \\
&\leqslant |1 - \overline{a_n}r| + (1 - r) \\
&< |1 - \overline{a_n}r| + |1 - \overline{a_n}r| \\
&= 2|1 - \overline{a_n}r|. \tag{2.14}
\end{aligned}$$

Apply these inequalities to see that

$$\frac{(1 - r^2)(1 - |a_n|^2)}{|1 - \overline{a_n}r|^2} = \frac{(1 - r^2)(1 - |a_n|^2)}{|1 - \overline{a_n}r||1 - \overline{a_n}r|}$$

$$\leqslant 2\frac{(1 - r^2)(1 - |a_n|^2)}{(1 - r)|1 - a_n|}$$

$$\leqslant 8\frac{1 - |a_n|}{|1 - a_n|}.$$

This estimate shows that the series

$$\sum_{n\geqslant 1} \frac{(1 - r^2)(1 - |a_n|^2)}{|1 - \overline{a}_n r|^2}$$

is bounded above by

$$8\sum_{n\geqslant 1} \frac{1 - |a_n|}{|1 - a_n|} < \infty$$

(see (2.13)) and thus converges uniformly in r. Observe the estimate

$$|B(r)|^2 = \prod_{n\geqslant 1} \frac{|a_n - r|^2}{|1 - \overline{a}_n r|^2}$$

$$= \prod_{n\geqslant 1} \frac{|a_n|^2 - 2\,\mathrm{Re}(a_n r) + r^2}{|1 - \overline{a}_n r|^2}$$

$$= \prod_{n\geqslant 1} \frac{(1 - 2\mathrm{Re}(a_n r) + |a_n|^2 r^2) - (1 - r^2 - |a_n|^2 + |a_n|^2 r^2)}{|1 - \overline{a}_n r|^2}$$

$$= \prod_{n\geqslant 1} \frac{|1 - \overline{a}_n r|^2 - (1 - r^2)(1 - |a_n|^2)}{|1 - \overline{a}_n r|^2}$$

$$= \prod_{n\geqslant 1} \left(1 - \frac{(1 - r^2)(1 - |a_n|^2)}{|1 - \overline{a}_n r|^2}\right)$$

$$\geqslant 1 - \sum_{n\geqslant 1} \frac{(1 - r^2)(1 - |a_n|^2)}{|1 - \overline{a}_n r|^2}.$$

Notice in the last step the use of the fact that for a sequence $\{x_j\}_{j\geqslant 1} \subset (0, 1)$ and $n \geqslant 1$,

$$(1 - x_1)(1 - x_2)\cdots(1 - x_n) \geqslant 1 - \sum_{1\leqslant j\leqslant n} x_j.$$

One proves this by induction. The Dominated Convergence Theorem ensures that

$$\lim_{r\to 1^-} |B(r)|^2 \geqslant 1 - \lim_{r\to 1^-} \sum_{n\geqslant 1} \frac{(1 - r^2)(1 - |a_n|^2)}{|1 - \overline{a}_n r|^2}$$

$$= 1 - \sum_{n\geqslant 1} \lim_{r\to 1^-} \left(\frac{(1 - r^2)(1 - |a_n|^2)}{|1 - \overline{a}_n r|^2}\right)$$

$$= 1 - 0 = 1.$$

Hence

$$\lim_{r \to 1^-} |B(r)| = 1.$$

Now we need to focus on arg $B(r)$. The infinite product defining the quantity $B(r)$ converges and thus

$$\arg B(r) = \sum_{n \geq 1} \arg b_{a_n}(r), \tag{2.15}$$

where, as always,

$$b_a(z) = \frac{\bar{a}}{|a|} \frac{a - z}{1 - \bar{a}z}, \quad a \in \mathbb{D}.$$

We then get

$$
\begin{aligned}
|1 - b_{a_n}(r)| &\leq c(1 - |a_n|) + 2c\frac{1 - |a_n|}{|1 - \bar{a}_n r|} && \text{(by (2.11))} \\
&\leq c(1 - |a_n|) + 4c\frac{1 - |a_n|}{|1 - a_n|} && \text{(by (2.14))} \\
&= c\frac{1 - |a_n|}{|1 - a_n|}|1 - a_n| + 4c\frac{1 - |a_n|}{|1 - a_n|} \\
&\leq 2c\frac{1 - |a_n|}{|1 - a_n|} + 4c\frac{1 - |a_n|}{|1 - a_n|} \\
&= 6c\frac{1 - |a_n|}{|1 - a_n|}. && (2.16)
\end{aligned}
$$

Now observe that

$$
\begin{aligned}
|1 - Re^{i\theta}|^2 &= 1 + R^2 - 2R\cos\theta \\
&= (1 - R)^2 + 4R\sin^2(\theta/2) \\
&\geq 4R\sin^2(\theta/2) \geq (\theta/\pi)^2, \qquad \frac{1}{4} < R < 1.
\end{aligned}
$$

This implies that

$$|\theta| \leq \pi|1 - Re^{i\theta}|.$$

Combining this with (2.16) gives us the inequality

$$|\arg b_{a_n}(r)| \leq 6\pi\frac{1 - |a_n|}{|1 - a_n|},$$

which says that the series on the right-hand side of (2.15) converges absolutely and uniformly. This proves that $\lim_{r \to 1^-} \arg B(r)$ exists. But we have already shown that

$$\lim_{r \to 1^-} |B(r)| = 1.$$

Thus

$$\lim_{r \to 1^-} B(r) \in \mathbb{T}.$$

Remark 2.18 If B has a unimodular radial limit at ζ, then no Stolz domain anchored at ζ can contain infinitely many zeros of B. Otherwise, by Lindelöf's Theorem [50, p. 19], the non-tangential limit would also exist and have unit modulus. In other words, if B has a unimodular limit at ζ, then the zeros of B must approach ζ tangentially.

For a singular inner function

$$s_\mu(z) = \exp\left(-\int_\mathbb{T} \frac{\xi + z}{\xi - z} d\mu(\xi)\right)$$

for which (we still maintain the convention that $\zeta = 1$)

$$\int_\mathbb{T} \frac{1}{|1 - \xi|} d\mu(\xi) < \infty, \tag{2.17}$$

note that

$$s_\mu(r) = \exp\left(-\int_\mathbb{T} \frac{\xi + r}{\xi - r} d\mu(\xi)\right)$$

$$= \exp\left(-\int_\mathbb{T} \mathrm{Re}\left(\frac{\xi + r}{\xi - r}\right) d\mu(\xi) - i \int_\mathbb{T} \mathrm{Im}\left(\frac{\xi + r}{\xi - r}\right) d\mu(\xi)\right)$$

$$= \exp\left(-\int_\mathbb{T} \frac{1 - r^2}{|\xi - r|^2} d\mu(\xi)\right) \exp\left(-i \int_\mathbb{T} \frac{2\,\mathrm{Im}(r\bar{\xi})}{|\xi - r|^2} d\mu(\xi)\right).$$

The integrability condition in (2.17) shows that $\mu(\{1\}) = 0$ and so

$$\lim_{r \to 1^-} \frac{1 - r^2}{|\xi - r|^2} = 0$$

for μ-almost every $\xi \in \mathbb{T}$. Furthermore,

$$\frac{1 - r^2}{|\xi - r|^2} \leqslant \frac{2}{|1 - \xi|}, \qquad r \in (0, 1);$$

and, by hypothesis, the function on the right-hand side of the equation above belongs to $L^1(\mu)$. Thus, by the Dominated Convergence Theorem, we see that

$$\lim_{r \to 1^-} \int_\mathbb{T} \frac{1 - r^2}{|\xi - r|^2} d\mu(\xi) = 0$$

and so $|s_\mu(r)| \to 1$. In a similar way,

$$\lim_{r \to 1^-} \frac{\mathrm{Im}(r\bar{\xi})}{|\xi - r|^2} = \frac{\mathrm{Im}(\bar{\xi})}{|1 - \xi|^2}$$

for μ-almost every $\xi \in \mathbb{T}$. Another estimate shows that

$$\frac{|\operatorname{Im}(r\bar{\xi})|}{|\xi - r|^2} \leq \frac{M}{|1 - \xi|}, \qquad r \in (0, 1)$$

for some positive constant M. By the integrability assumption in (2.17), the function on the right-hand side of the equation above belongs to $L^1(\mu)$. By the Dominated Convergence Theorem, we see that

$$\lim_{r \to 1^-} \int_{\mathbb{T}} \frac{\operatorname{Im}(r\bar{\xi})}{|\xi - r|^2} d\mu(\xi) = \int_{\mathbb{T}} \frac{\operatorname{Im}(\bar{\xi})}{|\xi - 1|^2} d\mu(\xi).$$

From here we conclude that

$$\lim_{r \to 1^-} s_\mu(r) = \exp\left(-i \int_{\mathbb{T}} \frac{2\operatorname{Im}(\bar{\xi})}{|\xi - 1|^2} d\mu(\xi)\right)$$

exists.

2.5 Angular derivatives

The following result reveals the connection between the values of $f'(z)$ and the difference quotient

$$\frac{f(z) - f(\zeta_0)}{z - \zeta_0}$$

as z approaches the boundary point $\zeta_0 \in \mathbb{T}$ from within a Stolz domain (see Definition 1.12).

Theorem 2.19 *For an analytic function f on $\mathbb{D}$ and $\zeta_0 \in \mathbb{T}$, the following are equivalent:*

(i) *The non-tangential limits*

$$f(\zeta_0) = \angle \lim_{z \to \zeta_0} f(z) \quad and \quad \angle \lim_{z \to \zeta_0} \frac{f(z) - f(\zeta_0)}{z - \zeta_0}$$

exist;

(ii) *There is a complex number λ such that*

$$\angle \lim_{z \to \zeta_0} \frac{f(z) - \lambda}{z - \zeta_0}$$

exists;

(iii) *The function f' has a non-tangential limit at ζ_0.*

Moreover, under the equivalent conditions above, we have

$$\angle \lim_{z \to \zeta_0} \frac{f(z) - f(\zeta_0)}{z - \zeta_0} = \angle \lim_{z \to \zeta_0} f'(z).$$

Proof (i) $\implies$ (ii): Automatic.

(ii) $\implies$ (iii): For simplicity, let

$$L = \angle \lim_{z \to \zeta_0} \frac{f(z) - \lambda}{z - \zeta_0}$$

and consider the auxiliary function

$$g(z) = \frac{f(z) - \lambda}{z - \zeta_0} - L, \qquad z \in \mathbb{D}.$$

Fix a Stolz domain $\Gamma_\alpha(\zeta_0)$. Note that our assumption (applied to the wider Stolz domain $\Gamma_{2\alpha+1}(\zeta_0)$) says that, given any $\varepsilon > 0$, there is $\delta > 0$ such that

$$|g(w)| < \varepsilon \tag{2.18}$$

if we choose w inside $\Gamma_{2\alpha+1}(\zeta_0)$ but close to the boundary, that is, $|w - \zeta_0| < \delta$.

For a point $z \in \Gamma_\alpha(\zeta_0)$, let C_z denote the circle with center z and radius $(1 - |z|)/2$. By the Cauchy Integral Formula,

$$f'(z) = \frac{1}{2\pi i} \int_{C_z} \frac{f(w)}{(w - z)^2} \, dw$$

$$= L + \frac{1}{2\pi i} \int_{C_z} \frac{g(w) (w - \zeta_0)}{(w - z)^2} \, dw.$$

Since z will eventually tend to a boundary point, we may further assume that

$$1 - |z| < \frac{2\delta}{1 + 2\alpha}.$$

This restriction leads to the required conclusion. To see this, first note that for each point w on the circle C_z we have

$$|w - \zeta_0| \leqslant |w - z| + |z - \zeta_0|$$

$$\leqslant \frac{1 - |z|}{2} + \alpha(1 - |z|) \qquad \text{(since } z \in \Gamma_\alpha(\zeta_0))$$

$$= \frac{2\alpha + 1}{2}(1 - |z|).$$

Hence, $|w - \zeta_0| < \delta$. Second, we have

$$1 - |z| = 2|w - z| \geqslant 2|w| - 2|z|,$$

or equivalently, $1 - |z| \leqslant 2(1 - |w|)$. This implies

$$|w - \zeta_0| \leqslant (2\alpha + 1)(1 - |w|).$$

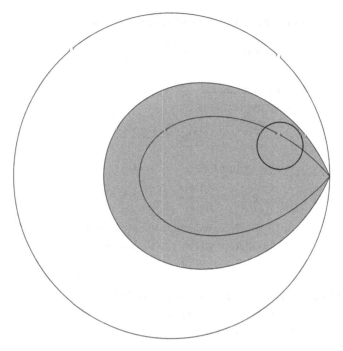

Figure 2.1 The domains $\Gamma_\alpha(\zeta_0) \subset \Gamma_{2\alpha+1}(\zeta_0)$ and the circle C_z.

In other words,

$$w \in C_z \implies |w - \zeta_0| \leqslant (2\alpha + 1)(1 - |w|);$$

that is, C_z entirely resides in Stolz domain $\Gamma_{2\alpha+1}(\zeta_0)$ and, moreover, its points are close enough to the boundary so that we can use (2.18) whenever $w \in C_z$ (see Figure 2.1).

Appealing to

$$1 - |z| = 2|w - z| \quad \text{and} \quad |w - \zeta_0| \leqslant \frac{2\alpha + 1}{2}(1 - |z|),$$

we have

$$\left| \frac{g(w)\,(w - \zeta_0)}{(w - z)^2} \right| \leqslant \frac{2\varepsilon\,(2\alpha + 1)}{1 - |z|}, \qquad w \in C_z.$$

Since the length of the circle C_z is $\pi(1 - |z|)$, the Cauchy Integral Formula yields

$$|f'(z) - L| \leqslant \varepsilon\,(2\alpha + 1)$$

whenever $z \in \Gamma_\alpha(\zeta_0)$ with $1 - |z| < 2\delta/(2\alpha + 1)$. This is precisely what we want to prove; in other words,

$$\angle \lim_{z \to \zeta_0} f'(z) = L.$$

(iii) $\implies$ (i): Since f' has a non-tangential limit at ζ_0, the integral

$$\int_{[0,\zeta_0]} f'(w)\, dw$$

is well-defined. We use this fact to define

$$\lambda = f(0) + \int_{[0,\zeta_0]} f'(w)\, dw.$$

For any $z \in \mathbb{D}$, we can apply the Fundamental Theorem of Calculus to see that

$$\lambda = f(z) + \int_{[z,\zeta_0]} f'(w)\, dw. \tag{2.19}$$

For z in the Stolz domain $\Gamma_\alpha(\zeta_0)$, we have

$$\left| \int_{[z,\zeta_0]} f'(w)\, dw \right| \leqslant \left(\sup_{w \in \Gamma_\alpha(\zeta_0)} |f'(w)| \right) |z - \zeta_0|.$$

Therefore, the quantity λ is precisely the non-tangential limit of f:

$$f(\zeta_0) := \angle \lim_{z \to \zeta_0} f(z) = \lambda.$$

If we rewrite (2.19) as

$$\frac{f(z) - f(\zeta_0)}{z - \zeta_0} = -\frac{1}{z - \zeta_0} \int_{[z,\zeta_0]} f'(w)\, dw$$

$$= f'(\zeta_0) + \frac{1}{z - \zeta_0} \int_{[z,\zeta_0]} (f'(\zeta_0) - f'(w))\, dw,$$

then, since in each Stolz domain the last integral tends to zero as $z \to \zeta_0$, we deduce that

$$\angle \lim_{z \to \zeta_0} \frac{f(z) - f(\zeta_0)}{z - \zeta_0} = f'(\zeta_0). \qquad \square$$

Definition 2.20 Assuming that the conditions of Theorem 2.19 are fulfilled, the quantity

$$f'(\zeta_0) = \angle \lim_{z \to \zeta_0} \frac{f(z) - f(\zeta_0)}{z - \zeta_0} = \angle \lim_{z \to \zeta_0} f'(z)$$

is called the *angular derivative* of f at ζ_0. Up to now, we put no restriction on the range of f. Consider the collection of analytic functions which map $\mathbb{D}$ into itself. We say that f has an *angular derivative in the sense of Carathéodory* at $\zeta_0 \in \mathbb{T}$ if two conditions hold:

(i) f has an angular derivative at ζ_0;
(ii) $f(\zeta_0) \in \mathbb{T}$.

There is a beautiful theorem of Julia and Carathéodory that gives an alternate method of testing whether or not a function has an angular derivative at ζ_0. We will present this result in Chapter 7 when we work with angular derivatives in the broader context of non-tangential limits of functions in model spaces.

When u is inner, there is a more tangible condition related to Theorem 2.17 that gives explicit conditions for the existence of an angular derivative in the sense of Carathéodory.

Theorem 2.21 *An inner function $u = Bs_\mu$ has a finite angular derivative in the sense of Carathéodory at $\zeta \in \mathbb{T}$ if and only if*

$$\sum_{n \geqslant 1} \frac{1 - |a_n|}{|\zeta - a_n|^2} + \int_{\mathbb{T}} \frac{d\mu(\xi)}{|\xi - \zeta|^2} < \infty.$$

Here $\{a_n\}_{n \geqslant 1}$ denotes the zeros, repeated according to their multiplicity, of the Blaschke product B and μ denotes the singular measure corresponding to the singular inner function s_μ.

Compare this theorem with Theorem 2.17 where the condition is weaker due to the lack of a square in the denominator of the two terms. When u is a Blaschke product, Theorem 2.21 was established by Frostman [75]. The general case is due to Ahern and Clark [4, 5]. To give the reader some idea of why this result is true, we start with a sketch of one direction of Frostman's original proof. Assume that $\zeta = 1$ and that B is a Blaschke product whose zeros $\{a_n\}_{n \geqslant 1}$ satisfy

$$\sum_{n \geqslant 1} \frac{1 - |a_n|^2}{|1 - a_n|^2} < \infty. \tag{2.20}$$

Writing

$$B(z) = \prod_{n \geqslant 1} b_{a_n}(z),$$

where

$$b_{a_n}(z) = \frac{\overline{a_n}}{|a_n|} \frac{a_n - z}{1 - \overline{a_n}z},$$

and using formal logarithmic differentiation, we get

$$\frac{B'(z)}{B(z)} = \sum_{n \geqslant 1} \frac{b'_{a_n}(z)}{b_{a_n}(z)} = \sum_{n \geqslant 1} \frac{1 - |a_n|^2}{(z - a_n)(1 - \overline{a_n}z)}.$$

This yields the formula

$$B'(z) = B(z) \sum_{n \geqslant 1} \frac{1 - |a_n|^2}{(z - a_n)(1 - \overline{a_n}z)}; \qquad (2.21)$$

which can be written as

$$B'(z) = \sum_{n \geqslant 1} \beta_n(z) \frac{1 - |a_n|^2}{(1 - \overline{a_n}z)^2}, \qquad (2.22)$$

where

$$\beta_n(z) = \frac{B(z)}{\frac{z - a_n}{1 - \overline{a_n}z}}.$$

For any $a \in \mathbb{D}$ and $0 < r < 1$, we can use (2.14) to get

$$\frac{1 - |a|^2}{|1 - \overline{a}r|^2} \leqslant 4 \frac{1 - |a|^2}{|1 - a|^2}.$$

This estimate, along with the fact that the modulus of $B_n(r)$ is at most 1, allows us to use the series representation in (2.22) to see that the series in (2.21) converges uniformly in r. The summability condition in (2.20) implies the summability condition in (2.13), which implies that $\lim_{r \to 1^-} B(r)$ exists and is unimodular. Thus

$$\lim_{r \to 1^-} B'(r) = B(1) \sum_{n \geqslant 1} \frac{1 - |a_n|^2}{|1 - a_n|^2}.$$

When $u = s_\mu$ is a singular inner function and

$$\int_{\mathbb{T}} \frac{1}{|1 - \xi|^2} d\mu(\xi) < \infty,$$

one makes use of the formula

$$\frac{s'_\mu(r)}{s_\mu(r)} = -2 \int \frac{\xi}{(\xi - r)^2} d\mu(\xi)$$

and the estimate

$$\left| \frac{\xi}{(\xi - r)^2} \right| \leqslant \frac{4}{|1 - \xi|^2}$$

to prove that $s'_\mu(r)$ has a finite limit at $r \to 1^-$.

We will delay some specific examples of Theorem 2.17 and Theorem 2.21 until Chapter 7, where we frame this whole discussion in the context of boundary behavior of functions in model spaces.

2.6 Frostman's Theorem

For an inner function u and $w \in \mathbb{D}$, define a *Frostman shift* of u as

$$u_w(z) = \frac{w - u(z)}{1 - \overline{w}\, u(z)} = \tau_{1,w} \circ u.$$

Using the fact that $\tau_{1,w}(\zeta) \in \mathbb{T}$ for all $\zeta \in \mathbb{T}$ (see (2.2)), we see that the function u_w is also an inner function for any $w \in \mathbb{D}$. Frostman discovered that u_w is actually a Blaschke product for most values of $w \in \mathbb{D}$. More precisely, define the *exceptional set of u* to be

$$\mathcal{E}(u) = \{\, w \in \mathbb{D} : u_w \text{ is not a Blaschke product}\,\}.$$

We will show that $\mathcal{E}(u)$ is a very small set.

Theorem 2.22 (Frostman) *Let u be a non-constant inner function and fix $0 < \rho < 1$. Define*

$$\mathcal{E}_\rho(u) = \{\, \zeta \in \mathbb{T} : u_{\rho\zeta} \text{ is not a Blaschke product}\,\}.$$

Then $\mathcal{E}_\rho(u)$ has one-dimensional Lebesgue measure zero.

Proof First note that for each $\alpha \in \mathbb{D}$,

$$\int_{\mathbb{T}} \log\left| \frac{\rho\xi - \alpha}{1 - \rho\overline{\xi}\alpha} \right| dm(\xi) = \max(\log\rho, \log|\alpha|). \tag{2.23}$$

To see this, observe that the integrand in (2.23) is equal to

$$\log|\rho\xi - \alpha| - \log|1 - \overline{\rho\alpha}\xi|.$$

Since $|\rho\alpha| < 1$, the function $z \mapsto 1 - \overline{\alpha\rho}z$ is analytic and non-vanishing on $\mathbb{D}$. Thus

$$z \mapsto \log(1 - \overline{\rho\alpha}z)$$

is analytic and so

$$z \mapsto \log|1 - \overline{\rho\alpha}z|$$

is a harmonic function on $\mathbb{D}$. By the Mean Value Property for harmonic functions (Theorem 1.14),

$$\int_{\mathbb{T}} \log|1 - \overline{\rho\alpha}\xi|\, dm(\xi) = 0.$$

Depending on whether $|\alpha| > |\rho|$ or $|\alpha| < |\rho|$, one of the functions

$$z \mapsto \log(\rho z - \alpha) \quad \text{or} \quad z \mapsto \log(\overline{\rho} - \overline{\alpha}z)$$

will be analytic on $\mathbb{D}$. Now apply the Mean Value Property for harmonic functions to either of the functions

$$z \mapsto \log |\rho z - \alpha| \quad \text{or} \quad z \mapsto \log |\bar{\rho} - \bar{\alpha}z|$$

to obtain the identity in (2.23).

In (2.23), replace α by $u(r\zeta)$ and then integrate with respect to ζ to obtain

$$\int_{\mathbb{T}} \left(\int_{\mathbb{T}} \log |u_{\rho\xi}(r\zeta)|\, dm(\xi) \right) dm(\zeta) = \int_{\mathbb{T}} \max\left(\log \rho, \log |u(r\zeta)|\right) dm(\zeta).$$

The family of functions

$$f_r(\zeta) = \max(\log \rho, \log |u(r\zeta)|), \qquad r \in [0, 1),$$

satisfies $\log \rho \leqslant f_r(\zeta) \leqslant 0$ and, moreover,

$$\begin{aligned}
\lim_{r \to 1^-} f_r(\zeta) &= \max(\log \rho, \lim_{r \to 1^-} \log |u(r\zeta)|) \\
&= \max(\log \rho, 0) \\
&= 0
\end{aligned}$$

for almost every $\zeta \in \mathbb{T}$. Therefore, by the Dominated Convergence Theorem,

$$\lim_{r \to 1^-} \int_{\mathbb{T}} f_r\, dm = 0.$$

In other words,

$$\lim_{r \to 1^-} \int_{\mathbb{T}} \left(\int_{\mathbb{T}} \log |u_{\rho\xi}(r\zeta)|\, dm(\xi) \right) dm(\zeta) = 0.$$

Since the integrand $\log |u_{\rho\xi}|$ is negative, we can use Fubini's Theorem to deduce that

$$\lim_{r \to 1^-} \int_{\mathbb{T}} \left(\int_{\mathbb{T}} \log |u_{\rho\xi}(r\zeta)|\, dm(\zeta) \right) dm(\xi) = 0. \tag{2.24}$$

By Fatou's Lemma we see that

$$\int_{\mathbb{T}} \left(\varliminf_{r \to 1^-} \int_{\mathbb{T}} -\log |u_{\rho\xi}(r\zeta)|\, dm(\zeta) \right) dm(\xi)$$

is bounded above by

$$\varliminf_{r \to 1^-} \int_{\mathbb{T}} \left(\int_{\mathbb{T}} -\log |u_{\rho\xi}(r\zeta)|\, dm(\zeta) \right) dm(\xi).$$

Hence, by (2.24) and the positivity of the integrand, we must have

$$\varliminf_{r \to 1^-} \int_{\mathbb{T}} \log |u_{\rho\xi}(r\zeta)|\, dm(\zeta) = 0 \tag{2.25}$$

for almost all $\xi \in \mathbb{T}$. Now with ρ still fixed, we find a value of ξ for which the $\underline{\lim}$ in (2.25) is equal to zero. By the factorization theorem for inner functions (Theorem 2.14), we have

$$u = u_{\rho\xi} = Bs_\mu,$$

where B is a Blaschke product and

$$s_\mu(z) = \exp\left(-\int_{\mathbb{T}} \frac{w+z}{w-z} d\mu(w)\right)$$

is the singular inner factor of u. From here we observe, using the identity

$$\mathrm{Re}\left(\frac{w+z}{w-z}\right) = P_z(w)$$

(see (1.12)), that

$$\log|u(r\zeta)| = \log|B(r\zeta)| - \int_{\mathbb{T}} P_{r\zeta}(w)d\mu(w).$$

Now use fact that

$$\int_{\mathbb{T}} P_{r\zeta}(w)\,dm(\zeta) = \int_{\mathbb{T}} \frac{1-r^2}{|r\zeta - w|^2}\,dm(\zeta)$$

$$= \int_{\mathbb{T}} \frac{1-r^2}{|rw - \zeta|^2}\,dm(\zeta) \qquad (w, \zeta \in \mathbb{T})$$

$$= \int_{\mathbb{T}} P_{rw}(\zeta)\,dm(\zeta)$$

$$= 1 \qquad\qquad\qquad (\text{by } (1.15))$$

along with Fubini's Theorem to get

$$\int_{\mathbb{T}} \log|u(r\zeta)|\,dm(\zeta) = \int_{\mathbb{T}} \log|B(r\zeta)|\,dm(\zeta) - \mu(\mathbb{T}).$$

Using the fact that $\log|B(r\zeta)| < 0$, we see that

$$\int_{\mathbb{T}} \log|u(r\zeta)|\,dm(\zeta) \leqslant -\mu(\mathbb{T}).$$

Finally, from (2.25), we get

$$0 = \lim_{r \to 1^-} \int_{\mathbb{T}} \log|u(r\zeta)|dm(\zeta) \leqslant -\mu(\mathbb{T}),$$

which means that $\mu(\mathbb{T}) = 0$ and so u has no singular inner factor. $\qquad\square$

Remark 2.23 We proved the "measure theory" version of Frostman's Theorem in that the exceptional set $\mathcal{E}(u)$ has Lebesgue area measure zero. There is a potential theory version of Frostman's Theorem which says that $\mathcal{E}(u)$

has logarithmic capacity zero. The reader can refer to [50] for the details. In fact, one can also show that u_w is a Blaschke product having *simple* zeros except possibly for w belonging to a set of logarithmic capacity zero [142, Thm. 3.10.2].

While Theorem 2.22 is an interesting result all by itself, it has several useful consequences. A particularly useful one is the following.

Corollary 2.24 (Frostman)   *Let u be an inner function and $\varepsilon > 0$. Then there is a Blaschke product B such that*

$$\|u - B\|_\infty < \varepsilon.$$

Proof   Fix $\rho \in (0, 1)$. According to Theorem 2.22, there are numerous choices of ξ such that $u_{\rho\xi}$ is a Blaschke product. Pick any one you wish and set $B = -u_{\rho\xi}$. Then

$$|u(z) - B(z)| = \left| \frac{\rho\xi - \rho\overline{\xi}u(z)^2}{1 - \rho\overline{\xi}u(z)} \right| \leqslant \frac{2\rho}{1 - \rho}, \qquad z \in \mathbb{D}.$$

This means that $\|u - B\|_\infty < \varepsilon$ whenever ρ satisfies $2\rho/(1 - \rho) < \varepsilon$. □

2.7 Notes

2.7.1 Privalov's Uniqueness Theorem

A theorem of Privalov says that if f is analytic on $\mathbb{D}$ and has non-tangential limits equal to zero on a subset of $\mathbb{T}$ of positive Lebesgue measure, then $f \equiv 0$. A proof can be found in his original book [149] or in the well-known texts [50, 118]. Privalov's Uniqueness Theorem no longer holds if "non-tangential limits" are replaced by "radial limits" [21].

2.7.2 Fatou's Theorem

Non-tangential limits are the best one can do in Theorem 2.3 in two ways. First, a result of Lusin [50, p. 22] says that given any measurable set $E \subset \mathbb{T}$ with $m(E) = 0$, there exists a function in H^∞ that does not have a radial limit at any point of E. Second, a Stolz approach region is the best one can do in that if γ_ζ is the internally tangent circle passing through the origin and the point $\zeta \in \mathbb{T}$, then there is an H^∞ function that does not have a limit along γ_ζ for any $\zeta \in \mathbb{T}$ [50, p. 43].

For a simple example of the difficulties that arise with *tangential* limits, consider the atomic inner function

$$u(z) = \exp\left(\frac{z+1}{z-1}\right).$$

One can show that u has a radial, and hence non-tangential, limit of zero at $z = 1$. However, one can also show that u does not have a limit as $z \to 1$ along the internally tangent circle $r = \cos\theta$.

2.7.3 General Factorization Theorem

In this chapter we focused on inner functions and their factorization. It turns out that a bounded analytic function (and more generally a function in the Hardy space H^2) can be factored uniquely as a product of an inner function and a so-called outer function. We will discuss this further in Chapter 3.

2.7.4 Higher order derivatives

A paper of Ahern and Clark [5] discusses when higher order derivatives of inner function have non-tangential limits. A more detailed treatment of derivatives of inner functions can be found in the monograph [137].

2.7.5 Indestructible Blaschke products

Theorem 2.22 says that for a Blaschke product B, the inner function

$$\frac{a-B}{1-\overline{a}B}$$

is a Blaschke product for all $a \in \mathbb{D} \setminus \mathcal{E}(B)$ in which $\mathcal{E}(B)$ is a "small" set. When $\mathcal{E}(B) = \varnothing$, we say that B is an *indestructible Blaschke product*. This type of Blaschke product has several fascinating properties and make interesting connections to other theorems of Frostman. For example, if the zeros of B are $\{a_n\}_{n\geqslant 1}$ and

$$\sum_{n\geqslant 1} \frac{1-|a_n|}{|\zeta - a_n|} < \infty$$

for all $\zeta \in \mathbb{T}$, then B is indestructible. Attempts to classify the indestructible Blaschke products by means of their zeros have failed. For example, one can create a Blaschke product B for which $\mathcal{E}(B) \neq \varnothing$ (that is to say that B is destructible) but such that $B/\tau_{1,a}$ is indestructible for any zero a of B. References for further study can be found in the survey paper [155].

2.7.6 Uniform Frostman Blaschke products

Blaschke products for which

$$\sup_{|\zeta|=1} \sum_{n \geqslant 1} \frac{1 - |a_n|}{|\zeta - a_n|} < \infty$$

are called *uniform Frostman Blaschke products* and they play a role in several problems in complex analysis [42, 113].

3

Hardy spaces

Since model spaces are subspaces of the Hardy space, we require a review of the basics of this space. We will only cover what is needed for the purposes of this book. The interested reader can consult some of the well-known sources [63, 97, 112, 118] for more advanced topics and finer details.

3.1 Three approaches to the Hardy space

The Hardy space has several equivalent definitions. We will treat three of them here. Each of them emphasizes a different aspect of the space and all three of these aspects play a role in our discussion of model spaces.

3.1.1 First approach – square summable power series

Perhaps the simplest definition of the Hardy space is the following.

Definition 3.1 An analytic function f on $\mathbb{D}$ with Taylor series expansion

$$f(z) = \sum_{n \geqslant 0} a_n z^n$$

belongs to the *Hardy space*, denoted by H^2, if

$$\sum_{n \geqslant 0} |a_n|^2 < \infty.$$

We define an inner product on H^2 by

$$\langle f, g \rangle := \sum_{n \geqslant 0} a_n \overline{b_n},$$

where $f = \sum_{n \geq 0} a_n z^n$ and $g = \sum_{n \geq 0} b_n z^n$, which yields the corresponding norm

$$\|f\| := \Big(\sum_{n \geq 0} |a_n|^2 \Big)^{\frac{1}{2}}.$$

The reader might be concerned that we are using the same notation $\|f\|$ for the norm of a function in H^2 and for the norm of a function in L^2 from (1.10). There is no cause for alarm. We will see why momentarily.

Before moving on, it is important to show the following:

Proposition 3.2 H^2 *is a Hilbert space.*

Proof For any fixed $z \in \mathbb{D}$, the Cauchy–Schwarz Inequality (Theorem 1.23) shows that

$$\sum_{n \geq 0} |a_n| |z|^n \leq \Big(\sum_{n \geq 0} |a_n|^2 \Big)^{1/2} \Big(\sum_{n \geq 0} |z|^{2n} \Big)^{1/2}$$

$$= \Big(\sum_{n \geq 0} |a_n|^2 \Big)^{1/2} \frac{1}{\sqrt{1 - |z|^2}}.$$

This shows that if $\{a_n\}_{n \geq 0} \in \ell^2(\mathbb{N}_0)$, then the function $f = \sum_{n \geq 0} a_n z^n$ is analytic on $\mathbb{D}$.

Now define

$$\Phi : \ell^2(\mathbb{N}_0) \to H^2, \quad \Phi(\{a_n\}_{n \geq 0}) = \sum_{n \geq 0} a_n z^n$$

and note that Φ is linear bijection that preserves norms. Since $\ell^2(\mathbb{N}_0)$ is a Hilbert space, so is H^2. $\qquad \square$

Several things come out of this proof. First note that for each $\lambda \in \mathbb{D}$ we have

$$|f(\lambda)| \leq \frac{\|f\|}{\sqrt{1 - |\lambda|^2}}, \quad f \in H^2. \tag{3.1}$$

Thus we obtain an upper bound on how rapidly an H^2 function can grow as $|\lambda| \to 1$. We will see a refinement of this in Proposition 3.25 below. Second, we see that the linear functional

$$f \mapsto f(\lambda)$$

is bounded on H^2. By the Riesz Representation Theorem for bounded linear functionals on a Hilbert space (Theorem 1.26), this functional takes the form

$$f(\lambda) = \langle f, c_\lambda \rangle \tag{3.2}$$

for some unique $c_\lambda \in H^2$. We can provide an explicit formula for c_λ.

Proposition 3.3 *For each $\lambda \in \mathbb{D}$,*

$$c_\lambda(z) = \frac{1}{1 - \bar{\lambda}z}, \qquad z \in \mathbb{D},$$

and

$$\|c_\lambda\| = \frac{1}{\sqrt{1 - |\lambda|^2}}. \tag{3.3}$$

Proof   For each $\lambda \in \mathbb{D}$, the function $z \mapsto (1 - \bar{\lambda}z)^{-1}$ belongs to H^2 since

$$\frac{1}{1 - \bar{\lambda}z} = \sum_{n \geqslant 0} \bar{\lambda}^n z^n$$

and so

$$\left\| \frac{1}{1 - \bar{\lambda}z} \right\|^2 = \left\| \sum_{n \geqslant 0} \bar{\lambda}^n z^n \right\|^2 = \frac{1}{1 - |\lambda|^2} < \infty.$$

Furthermore, if $f = \sum_{n \geqslant 0} a_n z^n$ belongs to H^2, then

$$\left\langle f, \frac{1}{1 - \bar{\lambda}z} \right\rangle = \sum_{n \geqslant 0} a_n \lambda^n = f(\lambda)$$

and thus, by the uniqueness part of the Riesz Representation Theorem, we conclude that

$$c_\lambda(z) = \frac{1}{1 - \bar{\lambda}z}.$$

Moreover,

$$\|c_\lambda\|^2 = \langle c_\lambda, c_\lambda \rangle = c_\lambda(\lambda) = \frac{1}{1 - |\lambda|^2}. \qquad \square$$

The function c_λ is called the *Cauchy kernel* (sometimes called the *Szegő kernel*). The space H^2 is therefore a *reproducing kernel Hilbert space*. Along similar lines, for fixed $n \geqslant 0$ we have

$$f^{(n)}(\lambda) = \langle f, c_\lambda^{(n)} \rangle, \tag{3.4}$$

where

$$c_\lambda^{(n)}(z) = \frac{n! z^n}{(1 - \bar{\lambda}z)^{n+1}} \tag{3.5}$$

is the nth derivative of c_λ with respect to the variable $\bar{\lambda}$. The alert reader will recognize the identities (3.2) and (3.4) as the Cauchy Integral Formula

$$f^{(n)}(\lambda) = \frac{n!}{2\pi i} \int_{\mathbb{T}} \frac{f(\zeta)\,d\zeta}{(\zeta - \lambda)^{n+1}}.$$

Although initially we need to assume sufficient smoothness of the boundary function $\zeta \mapsto f(\zeta)$ on $\mathbb{T}$ in order for the integral above to make sense, we will show in a moment (Corollary 3.14) that the Cauchy Integral Formula is valid for all $f \in H^2$.

Proposition 3.4 *If $f_n \to f$ in the norm of H^2, then $f_n \to f$ uniformly on compact subsets of $\mathbb{D}$.*

Proof Using the reproducing property (3.2), the Cauchy–Schwarz Inequality, and (3.3), we find that for $|\lambda| \leq r < 1$

$$
\begin{aligned}
|f(\lambda) - f_n(\lambda)| &= |\langle f - f_n, c_\lambda \rangle| \\
&\leq \|f - f_n\| \|c_\lambda\| \\
&= \frac{\|f - f_n\|}{\sqrt{1 - |\lambda|^2}} \\
&\leq \frac{\|f - f_n\|}{\sqrt{1 - r^2}}.
\end{aligned}
$$

Since we are assuming $\|f - f_n\| \to 0$, we see that $f_n \to f$ uniformly on $\{\lambda : |\lambda| \leq r\}$. $\qquad\square$

Proposition 3.5 *The set $\{c_\lambda : \lambda \in \mathbb{D}\}$ is linearly independent.*

Proof If $\lambda_1, \lambda_2, \ldots, \lambda_n$ are distinct points in $\mathbb{D}$ and

$$
\sum_{1 \leq i \leq n} \alpha_i c_{\lambda_i} \equiv 0,
$$

then

$$
0 = \left\langle f, \sum_{1 \leq i \leq n} \alpha_i c_{\lambda_i} \right\rangle = \sum_{1 \leq i \leq n} \overline{\alpha_i} f(\lambda_i), \qquad \forall f \in H^2.
$$

The Lagrange Interpolation Theorem provides us with a polynomial f satisfying $f(\lambda_i) = \alpha_i$ for all $1 \leq i \leq n$ and so $\sum_{1 \leq i \leq n} |\alpha_i|^2 = 0$. Thus $\alpha_i = 0$ for all $1 \leq i \leq n$. $\qquad\square$

Proposition 3.6 $\bigvee \{c_\lambda : \lambda \in \mathbb{D}\} = H^2$.

Proof If $f \in H^2$ is orthogonal to c_λ for every $\lambda \in \mathbb{D}$, then f is identically zero by (3.2). Now use Theorem 1.27. $\qquad\square$

Corollary 3.7 *If $\Lambda \subset \mathbb{D}$ has an accumulation point in $\mathbb{D}$, then*

$$\bigvee \{c_\lambda : \lambda \subset \Lambda\} = H^2.$$

Proof If $f \in H^2$ is orthogonal to c_λ for every $\lambda \in \Lambda$, then the zeros of f have an accumulation point in $\mathbb{D}$. The Identity Theorem for analytic functions now says that $f \equiv 0$. To finish, use Theorem 1.27. □

The Identity Theorem for Hardy spaces (see Corollary 3.17 below) leads to a variation of Corollary 3.7 where the accumulation points of Λ are allowed to be on $\mathbb{T}$ (see Corollary 3.18 below).

When working with Hilbert spaces of analytic functions, it is always important to have dense subsets of concrete functions readily available. We have already seen that finite linear combinations of Cauchy kernels form a dense set. Here is another.

Proposition 3.8 *The polynomials are dense in H^2.*

Proof If $f = \sum_{n \geq 0} a_n z^n$ belongs to H^2, then

$$\left\| f - \sum_{0 \leq n \leq N} a_n z^n \right\|^2 = \sum_{n \geq N+1} |a_n|^2. \tag{3.6}$$

Being the tail of a convergent series, this expression goes to zero as $N \to \infty$. Thus the partial sums of the Taylor series of f approximate f in the H^2 norm. □

3.1.2 Second approach – bounded integral means

A second equivalent definition of the Hardy space involves the integral means

$$\int_{\mathbb{T}} |f(r\zeta)|^2 dm(\zeta), \qquad 0 < r < 1.$$

Recall from (2.3) that

$$\int_{\mathbb{T}} |f(r\zeta)|^2 dm(\zeta) = \sum_{n \geq 0} |a_n|^2 r^{2n}, \tag{3.7}$$

from which it follows that

$$r \mapsto \int_{\mathbb{T}} |f(r\zeta)|^2 dm(\zeta)$$

is an increasing function of r.

Proposition 3.9 *An analytic function f on $\mathbb{D}$ belongs to H^2 if and only if*

$$\sup_{0<r<1} \int_{\mathbb{T}} |f(r\zeta)|^2 dm(\zeta) < \infty.$$

Moreover,

$$\|f\|^2 = \sup_{0<r<1} \int_{\mathbb{T}} |f(r\zeta)|^2 dm(\zeta) = \lim_{r\to 1^-} \int_{\mathbb{T}} |f(r\zeta)|^2 dm(\zeta).$$

Proof   Suppose that $f = \sum_{n\geq 0} a_n z^n \in H^2$. Then

$$
\begin{aligned}
\sup_{0<r<1} \int_{\mathbb{T}} |f(r\zeta)|^2 dm(\zeta) &= \sup_{0<r<1} \sum_{n\geq 0} |a_n|^2 r^{2n} \\
&= \lim_{r\to 1^-} \sum_{n\geq 0} |a_n|^2 r^{2n} \\
&= \sum_{n\geq 0} |a_n|^2 \qquad \text{(Dominated Convergence)} \\
&= \|f\|^2 < \infty.
\end{aligned}
$$

Conversely, if the integral means are bounded in r, then

$$
\begin{aligned}
\|f\|^2 &= \sum_{n\geq 0} |a_n|^2 \\
&= \sum_{n\geq 0} \lim_{r\to 1^-} |a_n|^2 r^{2n} \\
&\leq \lim_{r\to 1^-} \sum_{n\geq 0} |a_n|^2 r^{2n} \qquad \text{(Fatou's Lemma)} \\
&= \lim_{r\to 1^-} \int_{\mathbb{T}} |f(r\zeta)|^2 dm(\zeta) < \infty.
\end{aligned}
$$

This completes the proof. $\square$

Corollary 3.10   *If $f \in H^2$ and, for $r \in (0, 1)$, $f_r(z) = f(rz)$, then $f_r \in H^2$ and $\|f_r - f\| \to 0$ as $r \to 1^-$.*

Proof   If $f \in H^2$ then

$$f = \sum_{n\geq 0} a_n z^n, \quad \|f\|^2 = \sum_{n\geq 0} |a_n|^2.$$

Thus

$$\|f_r\|^2 = \left\| \sum_{n\geq 0} a_n r^n z^n \right\| = \sum_{n\geq 0} |a_n|^2 r^{2n} \leq \sum_{n\geq 0} |a_n|^2 = \|f\|^2.$$

Moreover,

$$\|f_r - f\|^2 = \sum_{n \geqslant 0} |a_n|^2 (1 - r^{2n}).$$

By the Dominated Convergence Theorem, this last term goes to zero as $r \to 1^-$.   □

Corollary 3.11 *Every bounded analytic function on $\mathbb{D}$ belongs to H^2. In other words, $H^\infty \subset H^2$.*

Proof   For $f \in H^\infty$,

$$\|f\|^2 = \sup_{0 < r < 1} \int_{\mathbb{T}} |f(r\zeta)|^2 dm(\zeta) \leqslant \|f\|_\infty^2 < \infty.$$ □

The containment $H^\infty \subset H^2$ is strict. Indeed,

$$f(z) = \log \frac{1}{1 - z}$$

is an unbounded analytic function on $\mathbb{D}$ that belongs to H^2 since its Taylor series

$$\log \frac{1}{1 - z} = \sum_{n \geqslant 1} \frac{z^n}{n}$$

has square summable coefficients.

3.1.3 Third approach – vanishing negative Fourier coefficients

Let

$$f = \sum_{n \geqslant 0} a_n z^n$$

belong to H^2 (square summable power series definition). Parseval's Theorem (Theorem 1.8) says that the function h_f on $\mathbb{T}$ whose Fourier series is

$$\sum_{n \geqslant 0} a_n \zeta^n$$

belongs to L^2. Furthermore, by (1.17),

$$\mathscr{P}(h_f)(r\zeta) = \sum_{n \geqslant 0} a_n r^n \zeta^n = f(r\zeta).$$

Now apply Fatou's Theorem (Theorem 1.10) and Parseval's Theorem to obtain the following result.

Theorem 3.12 *For $f = \sum_{n \geqslant 0} a_n z^n$ in H^2 we have the following:*

(i) $\lim\limits_{r \to 1^-} f(r\zeta) = f(\zeta)$ *exists for m-almost every $\zeta \in \mathbb{T}$;*

(ii) The boundary function $\zeta \mapsto f(\zeta)$ belongs to L^2;

(iii) $\widehat{f}(n) = a_n$ *for $n \geqslant 0$ and $\widehat{f}(n) = 0$ for $n \leqslant -1$;*

(iv) $\sup\limits_{0 < r < 1} \int_{\mathbb{T}} |f(r\zeta)|^2 dm(\zeta) = \int_{\mathbb{T}} |f(\zeta)|^2 dm(\zeta) = \sum\limits_{n \geqslant 0} |a_n|^2 = \|f\|^2$;*

(v) In terms of boundary functions,

$$H^2 = \{f \in L^2 : \widehat{f}(n) = 0 \ \forall n \leqslant -1\}.$$

Since the norm of an H^2 function f equals the L^2 norm of its boundary function, there is no confusion in using the notation $\|f\|$ for both. Notice also how the inner product on the Hardy space H^2 is given by

$$\langle f, g \rangle = \int_{\mathbb{T}} f \overline{g} \, dm = \sum_{n \geqslant 0} a_n \overline{b_n},$$

where $f = \sum_{n \geqslant 0} a_n z^n$ and $g = \sum_{n \geqslant 0} b_n z^n$ denote generic elements of H^2. As mentioned earlier in the proof of Proposition 3.2, we have a natural identification of H^2 with the sequence space $\ell^2(\mathbb{N}_0)$, where each f in H^2 is identified with its sequence of Taylor coefficients $\{a_n\}_{n \geqslant 0}$.

We also re-emphasize the manner in which H^2 sits inside of L^2. If $f = \sum_{n \geqslant 0} a_n z^n$ belongs to H^2, then the almost everywhere defined boundary function f has an associated Fourier series

$$\sum_{n \geqslant 0} a_n \zeta^n$$

that belongs to the first component in the direct sum

$$L^2 = H^2 \oplus \overline{\zeta H^2}, \tag{3.8}$$

where

$$\overline{\zeta H^2} = \{\overline{\zeta h} : h \in H^2\}.$$

In terms of Fourier coefficients,

$$H^2 = \bigvee \{\zeta^n : n \geqslant 0\}, \qquad \overline{\zeta H^2} = \bigvee \{\zeta^n : n \leqslant -1\}.$$

In particular, note that

$$a_n = \begin{cases} \widehat{f}(n) & \text{if } n \geqslant 0, \\ 0 & \text{if } n < 0, \end{cases}$$

and that the polynomials in z are dense in H^2. These results are summarized in the following diagram:

$$
\begin{array}{ccc}
H^2 & \xleftrightarrow{\text{identified}} & \ell^2(\mathbb{N}_0) \\
\cap & & \cap \\
L^2 & \xleftrightarrow{\text{identified}} & \ell^2(\mathbb{Z})
\end{array}
\tag{3.9}
$$

3.2 The Riesz projection

In what follows, we let P denote the *Riesz projection*, the orthogonal projection from L^2 onto H^2. In terms of Fourier series, P is given by the formula

$$
P\Big(\sum_{n \in \mathbb{Z}} \widehat{f}(n) \zeta^n \Big) = \sum_{n \geqslant 0} \widehat{f}(n) \zeta^n. \tag{3.10}
$$

The Riesz projection returns the "analytic part" of a Fourier series in L^2. For instance,

$$
P(1 + 2 \cos \theta) = P(e^{-i\theta} + 1 + e^{i\theta}) = 1 + e^{i\theta} = 1 + \zeta.
$$

We also remark that, as an orthogonal projection, the operator P is self-adjoint and hence satisfies $\langle Pf, g \rangle = \langle f, Pg \rangle$ for all f, g in L^2.

Proposition 3.13 *The Riesz projection is given by the formula*

$$
[Pf](\lambda) = \langle f, c_\lambda \rangle = \int_{\mathbb{T}} \frac{f(\zeta)}{1 - \overline{\zeta}\lambda} \, dm(\zeta), \qquad f \in L^2, \lambda \in \mathbb{D}.
$$

Proof Since $c_\lambda = (1 - \overline{\lambda}z)^{-1}$ belongs to H^2 and satisfies the reproducing property (3.2), it follows that

$$
[Pf](\lambda) = \langle Pf, c_\lambda \rangle = \langle f, Pc_\lambda \rangle = \langle f, c_\lambda \rangle
$$

for all $f \in L^2$. Rewriting the expression $\langle f, c_\lambda \rangle$ as an integral yields the second formula. $\qquad\square$

Corollary 3.14 (Cauchy Integral Formula for H^2) *For $f \in H^2$,*

$$
f(\lambda) = \int_{\mathbb{T}} \frac{f(\zeta)}{1 - \overline{\zeta}\lambda} \, dm(\zeta), \qquad \lambda \in \mathbb{D}.
$$

3.3 Factorization

To obtain an important factorization theorem for functions H^2, we begin with a classical result of Jensen [158]. Let us set the following notation:

$$\log^+ x = \begin{cases} \log x & \text{if } x \geqslant 1, \\ 0 & \text{if } 0 < x < 1, \end{cases} \tag{3.11}$$

and

$$\log^- x = \begin{cases} -\log x & \text{if } 0 < x \leqslant 1, \\ 0 & \text{if } x > 1. \end{cases} \tag{3.12}$$

Note that

$$\log x = \log^+ - \log^- x \quad \text{and} \quad \log^+ x \leqslant x, \qquad x > 0. \tag{3.13}$$

Moreover,

$$|\log^+ x - \log^+ y| \leqslant |x - y|, \qquad x, y > 0. \tag{3.14}$$

Theorem 3.15 (Jensen's Formula) *Suppose that f is analytic on $\mathbb{D}$ with zeros $0 < |a_1| \leqslant |a_2| \leqslant |a_3| \leqslant \cdots < 1$. Then, for each $r \in (0, 1)$,*

$$\int_{\mathbb{T}} \log |f(r\zeta)| \, dm(\zeta) = \log |f(0)| + \sum_{|a_n| < r} \log \frac{r}{|a_n|}$$

Proof   We will follow [97, Ch. 2] and prove a slightly weaker version of Jensen's Formula (which is all we need for our purposes) and assume that $r \in (0, 1)$ is chosen so that $r \neq |a_n|$ for all n. A full proof can be found in [158, p. 300].

Since $r \neq |a_n|$ for any n, we let $a_1, \ldots, a_N$ be the zeros of f inside the open disk $\{|z| < r\}$. Let $f_r(z) = f(rz)$ and note that f_r is analytic in a neighborhood of $\mathbb{D}^-$ and has zeros $a_1/r, \ldots, a_N/r$. Let

$$B(z) = \prod_{|a_n| < r} \frac{\frac{a_n}{r} - z}{1 - \frac{\overline{a_n}}{r} z}$$

be the finite Blaschke product whose zeros are precisely those of f_r and let $g = f_r/B$. Then g is analytic and never equal to zero on $\mathbb{D}^-$.

This means that $\log |g|$ is harmonic in a neighborhood of $\mathbb{D}^-$. By the Mean Value Principle for harmonic functions (Theorem 1.14),

$$\log |g(0)| = \int_{\mathbb{T}} \log |g| \, dm.$$

Notice that $|g| = |f_r|$ on $\mathbb{T}$, and so the formula in the preceding line yields

$$\int_{\mathbb{T}} |f_r|\, dm - \log|g(0)| = \log\left|\frac{f_r(0)}{B(0)}\right|$$

$$= \log|f(0)| - \log|B(0)|$$

$$= \log|f(0)| - \log\prod_{|a_n|<r}\frac{a_n}{r}$$

$$= \log|f(0)| - \sum_{|a_n|<r}\log\frac{|a_n|}{r},$$

which is Jensen's Formula. $\qquad\square$

Corollary 3.16 *The zeros $\{a_n\}_{n\geqslant 1}$, repeated according to multiplicity, of an $f \in H^2 \setminus \{0\}$ satisfy the Blaschke condition*

$$\sum_{n\geqslant 1}(1 - |a_n|) < \infty.$$

Proof   Without loss of generality, assume that $f(0) \neq 0$. Exponentiating Jensen's formula we get

$$|f(0)|\prod_{|a_n|<r}\frac{r}{|a_n|} = \exp\left(\int_{\mathbb{T}}\log|f(r\zeta)|\,dm(\zeta)\right)$$

$$= \exp\left(\frac{1}{2}\int_{\mathbb{T}}\log|f(r\zeta)|^2 dm(\zeta)\right)$$

$$\leqslant \exp\left(\frac{1}{2}\int_{\mathbb{T}}\log^+|f(r\zeta)|^2 dm(\zeta)\right)$$

$$\leqslant \exp\left(\frac{1}{2}\int_{\mathbb{T}}|f(r\zeta)|^2 dm(\zeta)\right) \qquad\text{(by (3.13))}$$

$$\leqslant \exp\left(\frac{1}{2}\|f\|^2\right). \qquad\text{(by (3.12))}$$

Letting $r \to 1^-$ and noting that $f(0) \neq 0$, it follows that $\prod_{n\geqslant 1}|a_n| > 0$ from which we conclude that $\sum_{n\geqslant 1}(1 - |a_n|) < \infty$. $\qquad\square$

Corollary 3.17 (Identity Theorem for Hardy spaces) *Suppose $\{a_n\}_{n\geqslant 1} \subset \mathbb{D}$ and $\sum_{n\geqslant 1}(1 - |a_n|) = \infty$. If $f \in H^2$ with $f(a_n) = 0$ for all n, then $f \equiv 0$.*

Using annihilators as in the proof of Corollary 3.7, one can prove the following:

Corollary 3.18 *If $\{a_n\}_{n\geqslant 1} \subset \mathbb{D}$ and $\sum_{n\geqslant 1}(1 - |a_n|) = \infty$, then*

$$\bigvee\{c_{a_n} : n \geqslant 1\} = H^2.$$

We have discussed the zeros of H^2 functions. We now discuss what is left when we divide out these zeros.

Definition 3.19 An *outer function* is an analytic function F on $\mathbb{D}$ of the form

$$F(z) = \xi \exp\left(\int_{\mathbb{T}} \frac{\zeta + z}{\zeta - z} \varphi(\zeta)\, dm(\zeta)\right), \tag{3.15}$$

where $\xi \in \mathbb{T}$ and $\varphi \in L^1$ is real-valued.

The significance of the formula (3.15) lies in the fact that

$$\log|F(z)| = \int_{\mathbb{T}} P_z(\zeta)\varphi(\zeta)\, dm(\zeta) = \mathscr{P}(\varphi)(z),$$

and so, by Fatou's Theorem (Theorem 1.10), $\varphi = \log|F|$ almost everywhere on $\mathbb{T}$.

Theorem 3.20 *Each function f in $H^2 \backslash \{0\}$ has a factorization of the form*

$$f = BsF \tag{3.16}$$

where B is a Blaschke product, s is a singular inner function, and F is an outer function in H^2. Conversely, any product of the form (3.16) belongs to H^2. Furthermore, up to unimodular constant factors, this factorization is unique.

Proof   Let us first show that we can write any $f \in H^2$ as a product

$$f = Bg, \tag{3.17}$$

where B is a Blaschke product and g is a function in H^2 having no zeros in $\mathbb{D}$. To do this, we let B be the Blaschke product corresponding to the zeros of f (Corollary 3.16) and let B_N denote the Nth partial product. Then f/B_N is analytic on $\mathbb{D}$. Since B_N is continuous on $\mathbb{D}^-$, there is a $\delta_N > 0$ and an $r_N \in (0, 1)$ such that $|B_N(z)| \geqslant \delta_N > 0$ for all $|z| > r > r_N$. From here we see that for any $r > r_N$,

$$\int_{\mathbb{T}} \left|\frac{f}{B_N}(r\zeta)\right|^2 dm(\zeta) \leqslant \frac{1}{\delta_N} \int_{\mathbb{T}} |f(r\zeta)|^2 dm(\zeta) \leqslant \frac{1}{\delta_N}\|f\|^2.$$

Thus $f/B_N \in H^2$. Since $|B_N(\zeta)| = 1$ for all $\zeta \in \mathbb{T}$, we see that $\|f/B_N\| = \|f\|$. However, B_N converges uniformly on compact subsets of $\mathbb{D}$ to B, from which we conclude that f/B is analytic on $\mathbb{D}$. We just need to check that $f/B \in H^2$. For fixed $r \in (0, 1)$ we use the fact that $B_N \to B$ uniformly on $r\mathbb{T}$ to get

$$\int_{\mathbb{T}} \left|\frac{f}{B}(r\zeta)\right|^2 dm(\zeta) = \lim_{N\to\infty} \int_{\mathbb{T}} \left|\frac{f}{B_N}(r\zeta)\right|^2 dm(\zeta)$$

$$\leqslant \varlimsup_{N\to\infty} \|f/B_N\| = \|f\|.$$

This means that the integral means of f/B are uniformly bounded and so $g = f/B$ belongs to H^2. Furthermore, g has no zeros in $\mathbb{D}$, proving (3.17).

Next we show that the (zero free) function g from (3.17) satisfies

$$\int_{\mathbb{T}} \log |g| \, dm > -\infty. \tag{3.18}$$

To this end, we may assume that $g(0) = 1$. Since g has no zeros in $\mathbb{D}$, we see that $\log |g(z)|$ is harmonic on $\mathbb{D}$. The Mean Value Property for harmonic functions (Theorem 1.14) tells us that for any $r \in (0, 1)$, we have

$$0 = \log |g(0)| = \int_{\mathbb{T}} \log |g(r\zeta)| \, dm(\zeta)$$
$$= \int_{\mathbb{T}} \log^+ |g(r\zeta)| \, dm(\zeta) - \int_{\mathbb{T}} \log^- |g(r\zeta)| \, dm(\zeta).$$

Thus

$$\int_{\mathbb{T}} \log^- |g(r\zeta)| \, dm(\zeta) = \int_{\mathbb{T}} \log^+ |g(r\zeta)| \, dm(\zeta)$$
$$\leqslant \int_{\mathbb{T}} |g(r\zeta)| \, dm(\zeta) \qquad \text{(by (3.13))}$$
$$\leqslant \|g\|. \qquad \text{(Cauchy–Schwarz)}$$

Since $g \in H^2$ then g, along with the functions $\log^+ |g|$ and $\log^- |g|$, have radial limits almost everywhere on $\mathbb{T}$. By Fatou's Lemma

$$\int_{\mathbb{T}} \log^- |g| \, dm \leqslant \varliminf_{r \to 1^-} \int_{\mathbb{T}} \log^- |g(r\zeta)| dm(\zeta) \leqslant \|g\|$$

and so $\log^- |g|$ is integrable on $\mathbb{T}$. In a similar way, $\log^+ |g|$ and hence $\log |g|$ is integrable on $\mathbb{T}$. This confirms (3.18).

We now define

$$F(z) = \exp\left(\int_{\mathbb{T}} \frac{\zeta + z}{\zeta - z} \log |f(\zeta)| \, dm(\zeta) \right)$$

and note that this expression makes sense since $|f| = |g|$ and $\log |g|$ is integrable on $\mathbb{T}$. We now want to show that

$$|g(z)| \leqslant |F(z)|, \quad z \in \mathbb{D}. \tag{3.19}$$

To verify (3.19), it suffices to show, using the definition of F, that

$$\log |g(z)| \leqslant \int_{\mathbb{T}} P_z(\zeta) \log |g(\zeta)| \, dm(\zeta). \tag{3.20}$$

To prove (3.20), we let $g_r(z) = g(rz)$ and note that since $\log |g_r|$ is harmonic,

$$\log |g_r(z)| = \mathscr{P}(\log^+ |g_r|)(z) - \mathscr{P}(\log^- |g_r|)(z), \quad z \in \mathbb{D}. \tag{3.21}$$

Apply the estimate in (3.14), along with the fact (from Corollary 3.10) that

$$\lim_{r \to 1^-} \|g_r - g\|_{L^2} = 0$$

(and hence in the norm of L^1), to the equation in (3.21) to see that for each $z \in \mathbb{D}$,

$$\lim_{r \to 1^-} \mathscr{P}(\log^+ |g_r|)(z) = \mathscr{P}(\log^+ |g|)(z).$$

Finally, by Fatou's Lemma we have, for each $z \in \mathbb{D}$,

$$\mathscr{P}(\log^- |g|)(z) \leqslant \varliminf_{r \to 1^-} \mathscr{P}(\log^- |g_r|)(z)$$
$$= \mathscr{P}(\log^+ |g|)(z) - \log |g(z)|$$

which, after rearranging the terms, is the desired inequality in (3.20).

This will show that g/F is a bounded analytic function on $\mathbb{D}$. Moreover, we also get

$$\lim_{r \to 1^-} |F(r\xi)| = \lim_{r \to 1^-} \exp\left(\int_{\mathbb{T}} \mathrm{Re}\left(\frac{\zeta + r\xi}{\zeta - r\xi}\right) \log |f(\zeta)| \, dm(\zeta)\right)$$
$$= \lim_{r \to 1^-} \exp\left(\int_{\mathbb{T}} P_{r\xi}(\zeta) \log |g(\zeta)| \, dm(\zeta)\right)$$
$$= |g(\xi)|$$

for m-almost every $\xi \in \mathbb{T}$. This means that $s = g/F$ is an inner function without zeros and thus must be a singular inner function (Theorem 2.14). □

As a consequence of (3.18) we have the following:

Corollary 3.21 *If $f \in H^2 \setminus \{0\}$, then*

$$\int_{\mathbb{T}} \log |f| \, dm > -\infty.$$

This shows that the boundary function of an $f \in H^2 \setminus \{0\}$ cannot vanish on a set of positive measure.

Proposition 3.22 *Suppose $f \in H^2$. Then the following are equivalent:*

(i) f is outer, that is,

$$f(z) = e^{i\gamma} \exp\left(\int_{\mathbb{T}} \frac{\zeta + z}{\zeta - z} \log |f(\zeta)| \, dm(\zeta)\right), \qquad z \in \mathbb{D}, \qquad (3.22)$$

for some real γ;

(ii)

$$\log|f(0)| = \int_{\mathbb{T}} \log|f(\zeta)|\, dm(\zeta).$$

Proof   Assuming (i), we use (1.12) to obtain the identity

$$|f(z)| = \exp\left(\int_{\mathbb{I}} P_z(\zeta) \log|f(\zeta)|\, dm(\zeta)\right).$$

Take logarithms of both sides of the equation above and evaluate the result at $z = 0$ to get the integral formula in (ii).

On the other hand, suppose that (ii) holds. Then the analytic function F on $\mathbb{D}$ defined by

$$F(z) = \exp\left(\int_{\mathbb{T}} \frac{\zeta + z}{\zeta - z} \log|f(\zeta)|\, dm(\zeta)\right)$$

is an outer function (by definition). By (3.20), f/F is a bounded analytic function on $\mathbb{D}$ with $|f/F| \leq 1$. However, $f(0) = e^{i\theta} F(0)$ and so, by the Maximum Modulus Theorem (Theorem 2.4), $F = e^{i\theta} f$ for some $\theta \in [0, 2\pi)$. Thus f is outer.   □

Certainly if f is analytic and zero free on a neighborhood of $\mathbb{D}^-$, then $\log|f|$ is harmonic. The Mean Value Principle for harmonic functions (Theorem 1.14) says that

$$\log|f(0)| = \int_{\mathbb{T}} \log|f|\, dm$$

and so f is outer. The following two results give other examples of outer functions.

Corollary 3.23   *If $f \in H^2$ and $\operatorname{Re} f > 0$ on $\mathbb{D}$, then f is outer.*

Proof   We follow [97]. Let us first prove this result when $\operatorname{Re} f \geq \delta > 0$ on $\mathbb{D}$. By Proposition 3.22 we see that

$$\log|f(0)| \leq \int_{\mathbb{T}} \log|f(\zeta)|\, dm(\zeta).$$

We also have (Theorem 1.14)

$$\log\frac{|f(0)|}{\delta} = \int_{\mathbb{T}} \log\frac{|f(r\zeta)|}{\delta}\, dm(\zeta), \qquad 0 < r < 1.$$

By Fatou's Lemma, upon letting $r \to 1^-$, we obtain

$$\int_{\mathbb{T}} \log\frac{|f(\zeta)|}{\delta}\, dm(\zeta) \leq \log\frac{|f(0)|}{\delta}.$$

Note the use of the fact that Re $f \geqslant \delta$ in order to guarantee the integrands above are positive for each δ. Thus,

$$\log|f(0)| = \int_{\mathbb{T}} \log|f(\zeta)| \, dm(\zeta)$$

and so f is outer.

If Re $f > 0$ on $\mathbb{D}$, then, for each $n \geqslant 1$, the previous discussion shows that the function $f + 1/n$ is outer and thus by Proposition 3.22,

$$\log\left|f(0) + \tfrac{1}{n}\right| = \int_{\mathbb{T}} \log|f + \tfrac{1}{n}| \, dm.$$

A combination of the Monotone Convergence Theorem on $\{|f| \leqslant 1/2\}$ (note the use of the fact that Re $f > 0$) and the Dominated Convergence Theorem on $\{|f| > 1/2\}$ shows that

$$
\begin{aligned}
\log|f(0)| &= \lim_{n\to\infty} \log\left|f(0) + \tfrac{1}{n}\right| \\
&= \lim_{n\to\infty} \int_{\mathbb{T}} \log|f + \tfrac{1}{n}| \, dm \\
&= \lim_{n\to\infty} \int_{|f|\leqslant 1/2} \log|f + \tfrac{1}{n}| \, dm + \lim_{n\to\infty} \int_{|f|>1/2} \log|f + \tfrac{1}{n}| \, dm \\
&= \int_{|f|\leqslant 1/2} \log|f| \, dm + \int_{|f|>1/2} \log|f| \, dm \\
&= \int_{\mathbb{T}} \log|f| \, dm.
\end{aligned}
$$

Apply Proposition 3.22 one more time to see that f is outer. $\qquad\square$

A particularly useful application of this result is the following.

Corollary 3.24 *If φ is analytic on $\mathbb{D}$ with $\varphi(\mathbb{D}) \subset \mathbb{D}$, then $1 + \varphi$ is outer.*

3.4 A growth estimate

Recall that (3.1) tells us that any $f \in H^2$ satisfies

$$|f(r\zeta)| \leqslant \frac{\|f\|}{\sqrt{1 - r^2}}, \qquad \zeta \in \mathbb{T}, \ 0 < r < 1.$$

In other words,

$$|f(r\zeta)| = O\left(\frac{1}{\sqrt{1 - r}}\right);$$

or equivalently,

$$\sup_{0<r<1} \sqrt{1-r}\,|f(r\zeta)| < \infty.$$

It turns out that we can say a bit more.

Proposition 3.25 *For $f \in H^2$ and $\zeta \in \mathbb{T}$,*

$$|f(r\zeta)| = o\left(\frac{1}{\sqrt{1-r}}\right) \text{ as } r \to 1^-,$$

or equivalently,

$$\lim_{r\to1^-} \sqrt{1-r}\,|f(r\zeta)| = 0.$$

Proof   Without loss of generality we can assume that $\zeta = 1$. If $g \in H^\infty$, apply the Cauchy Integral Formula (Corollary 3.14) to get

$$g(r) = \int_{\mathbb{T}} \frac{g(\xi)}{1 - r\bar{\xi}} dm(\xi), \qquad 0 < r < 1.$$

From here we see that

$$|g(r)| \leqslant \|g\|_\infty \int_{\mathbb{T}} \frac{1}{|\xi - r|} \, dm(\xi) = O\left(\log\frac{1}{1-r}\right)$$

(see [42, p. 40]). Now, given $\varepsilon > 0$ and $f \in H^2$, choose a $g \in H^\infty$ with $\|f - g\| \leqslant \varepsilon$ (Proposition 3.8). Then

$$\sqrt{1-r}\,|f(r)| = \sqrt{1-r}\,|f(r) - g(r)| + \sqrt{1-r}\,|g(r)|$$
$$\leqslant \sqrt{1-r}\,\frac{1}{\sqrt{1-r}}\,\|f-g\| + O\left(\sqrt{1-r}\,\log\frac{1}{1-r}\right)$$
$$\leqslant \varepsilon + o(1).$$

Since the inequality above holds for every $\varepsilon > 0$, this proves the result.   □

3.5 Associated classes of functions

Some closely related classes of functions will come into play when discussing model spaces.

Definition 3.26   A meromorphic function on $\mathbb{D}$ is of *bounded type* if it can be written as a quotient of two bounded analytic functions on $\mathbb{D}$. The set of meromorphic functions of bounded type is denoted by $\mathfrak{N}$. The *Smirnov class* N^+ is the set of all *analytic* functions on $\mathbb{D}$ that are expressible as the quotient of two bounded analytic functions, the denominator of which is an outer function. In other words,

$$\mathfrak{N} = \left\{ \frac{h_1}{h_2} : h_1 \in H^\infty, h_2 \in H^\infty \setminus \{0\} \right\}$$

and

$$N^+ = \left\{ \frac{h_1}{h_2} : h_1, h_2 \in H^\infty, h_2 \text{ outer} \right\}.$$

Let us also point out that by Fatou's Theorem (Theorem 1.10), every $f \in H^\infty$ has non-tangential limits almost everywhere on $\mathbb{T}$ and, by Corollary 3.21, the boundary function satisfies

$$\int_\mathbb{T} \log |f| \, dm > -\infty.$$

This means that the non-tangential limits of f are non-zero almost everywhere. From here we see the following:

Proposition 3.27 *Every function in $\mathfrak{N}$ has a non-tangential limit almost everywhere on $\mathbb{T}$.*

Proposition 3.28 $H^2 \subset N^+$.

Proof By the canonical factorization theorem (Theorem 3.20), each $f \in H^2 \setminus \{0\}$ can be written as

$$f = BsF, \tag{3.23}$$

where B is a Blaschke product, s is a singular inner function, and $F \in H^2$ is the outer function

$$F(z) = e^{i\gamma} \exp\left(\int_\mathbb{T} \frac{\zeta + z}{\zeta - z} \log |F(\zeta)| \, dm(\zeta) \right).$$

Recall the definitions of $\log^+$ and $\log^-$ from (3.11) and (3.12) and note they are both non-negative functions. Define

$$F_1(z) = e^{i\gamma} \exp\left(\int_\mathbb{T} \frac{\zeta + z}{\zeta - z} (-\log^- |F(\zeta)|) \, dm(\zeta) \right),$$

$$F_2(z) = \exp\left(\int_\mathbb{T} \frac{\zeta + z}{\zeta - z} (-\log^+ |F(\zeta)|) \, dm(\zeta) \right),$$

and note that since $-\log^- |F| \leqslant 0$,

$$|F_1(z)| = \exp\left(\int_\mathbb{T} P_z(\zeta)(-\log^- |F(\zeta)|) \, dm(\zeta) \right) \leqslant 1,$$

with a similar estimate for F_2. Thus the outer factor F from (3.23) is equal to F_1/F_2. Hence

$$f = BsF = \frac{BsF_1}{F_2}$$

can be written as the quotient of two bounded analytic functions in which the denominator is outer.   □

Theorem 3.29 (Smirnov)   *If $f \in N^+$ and*

$$\int_{\mathbb{T}} |f(\zeta)|^2 dm(\zeta) < \infty,$$

then $f \in H^2$. If the boundary function for f belongs to L^∞, then $f \in H^\infty$.

Proof   The proof depends on the Arithmetic–Geometric Mean Inequality [158, p. 63],

$$\exp\left(\int_{\mathbb{T}} \log h\, d\sigma\right) \leq \int_{\mathbb{T}} h\, d\sigma, \tag{3.24}$$

where h is a non-negative function on $\mathbb{T}$ that is integrable with respect to the probability measure σ on $\mathbb{T}$.

If $f \in N^+$, then $f = g_1/g_2$ where $g_1, g_2 \in H^\infty$ and g_2 is outer. Since the presence of an inner factor in g_1 will not affect whether or not $f \in H^2$, we can assume that g_1 is also outer. Using the definition of an outer function (see (3.22)) applied to the functions g_1 and g_2, we see that

$$\frac{g_1}{g_2}(z) = \exp\left(\int_{\mathbb{T}} \frac{\zeta+z}{\zeta-z} \log \frac{|g_1(\zeta)|}{|g_2(\zeta)|}\, dm(\zeta)\right).$$

Furthermore, for each $r \in (0,1)$ and $w \in \mathbb{T}$,

$$\left|\frac{g_1}{g_2}(rw)\right|^2 = \exp\left(\int_{\mathbb{T}} P_{rw}(\zeta) \log \frac{|g_1(\zeta)|^2}{|g_2(\zeta)|^2}\, dm(\zeta)\right).$$

Now apply the Arithmetic–Geometric Mean Inequality (3.24) to the non-negative function $|g_1/g_2|$ and the probability measure $\sigma = P_{rw}\, dm$ (see (1.15)) to obtain the inequality

$$\left|\frac{g_1}{g_2}(rw)\right|^2 \leq \int_{\mathbb{T}} \left|\frac{g_1}{g_2}(\zeta)\right|^2 P_{rw}(\zeta)\, dm(\zeta). \tag{3.25}$$

Integrate both sides of the preceding to obtain

$$\int_{\mathbb{T}} |f(rw)|^2 dm(w) = \int_{\mathbb{T}} \left| \frac{g_1}{g_2}(rw) \right|^2 dm(w)$$

$$\leqslant \int_{\mathbb{T}} \left(\int_{\mathbb{T}} \left| \frac{g_1}{g_2}(\zeta) \right|^2 P_{rw}(\zeta) \, dm(\zeta) \right) dm(w)$$

$$= \int_{\mathbb{T}} \left(\int_{\mathbb{T}} |f(\zeta)|^2 P_{rw}(\zeta) \, dm(\zeta) \right) dm(w)$$

$$= \int_{\mathbb{T}} |f(\zeta)|^2 \left(\int_{\mathbb{T}} P_{rw}(\zeta) \, dm(w) \right) dm(\zeta)$$

$$= \int_{\mathbb{T}} |f|^2 dm.$$

Thus

$$\sup_{0<r<1} \int_{\mathbb{T}} |f(rw)|^2 dm(w) \leqslant \int_{\mathbb{T}} |f|^2 dm,$$

which means that $f \in H^2$.

To prove the second statement of the theorem, observe that if $f \in N^+$ and $f|_{\mathbb{T}} \in L^\infty$ then, as before, we can assume that $f = g_1/g_2$ and g_1, g_2 are bounded outer functions. By (3.25) we see that

$$|f(rw)|^2 = \left| \frac{g_1}{g_2}(rw) \right|^2 \leqslant \int_{\mathbb{T}} \left| \frac{g_1}{g_2}(\zeta) \right|^2 P_{rw}(\zeta) \, dm(\zeta)$$

$$= \int_{\mathbb{T}} |f(\zeta)|^2 P_{rw}(\zeta) \, dm(\zeta)$$

$$\leqslant \|f|_{\mathbb{T}}\|_\infty^2 \int_{\mathbb{T}} P_{rw}(\zeta) \, dm(\zeta)$$

$$= \|f|_{\mathbb{T}}\|_\infty^2,$$

which implies that $f \in H^\infty$. $\qquad\square$

Smirnov's Theorem is no longer true for $f \in \mathfrak{N}$, even when f is analytic on $\mathbb{D}$. For instance, the function

$$f(z) = \exp\left(\frac{1+z}{1-z} \right)$$

(which is the reciprocal of the atomic inner function in (2.8)) belongs to $\mathfrak{N}$, is analytic on $\mathbb{D}$, and has boundary values of unit modulus a.e. on $\mathbb{T}$. However, it does not belong to H^2 since, as shown in Remark 2.8,

$$|f(r)| = \exp\left(\frac{1+r}{1-r} \right), \qquad r \in (0, 1),$$

which does not satisfy the necessary growth condition in (3.1) for an H^2 function.

3.6 Notes

3.6.1 Hardy spaces H^p

For $0 < p < \infty$, the Hardy space H^p is defined to be the space of analytic functions f on $\mathbb{D}$ for which

$$\|f\|_p := \lim_{r \to 1^-} \left(\int_{\mathbb{T}} |f(r\zeta)|^p dm(\zeta) \right)^{1/p} < \infty. \qquad (3.26)$$

If $p \geqslant 1$, then H^p is a Banach space, while for $0 < p < 1$, H^p can be made into a topological vector space. Almost all of the basic function-theoretic results that hold for H^2 functions carry over to H^p *mutatis mutandis* (for example, each H^p function has an almost everywhere defined non-tangential boundary function that belongs to L^p).

For $p \neq 2$, the spaces H^p are not Hilbert spaces, although they do have identifiable duals. When $1 < p < \infty$, the dual space $(H^p)^*$ can be identified with H^q, where q is the Hölder conjugate index to p (that is, $\frac{1}{p} + \frac{1}{q} = 1$). When $p = 1$, the dual space $(H^1)^*$ can be identified with the space of all ana-lytic functions of bounded mean oscillation. When $0 < p < 1$, $(H^p)^*$ can be identified with certain classes of smooth functions on $\mathbb{D}^-$ (Lipschitz or Zyg-mund classes). In each of these cases, the dual pairing is given by the *Cauchy pairing*

$$\lim_{r \to 1^-} \int_{\mathbb{T}} f(r\zeta)\overline{g(r\zeta)}\, dm(\zeta), \quad f \in H^p, g \in (H^p)^*. \qquad (3.27)$$

Standard references for this material are [63, 97].

3.6.2 Outer functions

There is another proof of Corollary 3.23 that is worthy of note, although it requires knowledge of the more general H^p spaces described above. If f and g belong to some H^p space and fg is outer, then both f and g are outer. This is a consequence of the uniqueness part of the canonical factor-ization theorem for H^p. If f is such that $\operatorname{Re} f > 0$, then $f \in H^p$ for all $0 < p < 1$ [97, p. 109]. If $\operatorname{Re} f > 0$, then also $\operatorname{Re}(1/f) > 0$ so that $1/f \in H^p$ for $0 < p < 1$. Now let $g = 1/f$ and observe that $fg = 1$ is an outer function.

3.6.3 The Hardy space of the upper half plane

Let $\mathbb{C}_+ := \{z \in \mathbb{C} : \operatorname{Im} z > 0\}$ denote the upper half plane and let $\mathscr{H}^2$ denote the *Hardy space of the upper half plane*. This is the space of analytic functions f on $\mathbb{C}_+$ for which

$$\sup_{y>0} \int_{-\infty}^{\infty} |f(x + iy)|^2 \, dx < \infty.$$

This is the analogue of the "bounded integral mean" condition from the Hardy space H^2 on the unit disk. As it turns out, the majority of the theory from H^2 (on the disk) transfers over *mutatis mutandis* to $\mathscr{H}^2$. For example, every $f \in \mathscr{H}^2$ has an almost everywhere well-defined "radial" boundary function

$$f(x) := \lim_{y \to 0^+} f(x + iy)$$

and this function belongs to $L^2(dx) := L^2(\mathbb{R}, dx)$. Moreover,

$$\int_{-\infty}^{\infty} |f(x)|^2 \, dx = \sup_{y>0} \int_{-\infty}^{\infty} |f(x + iy)|^2 dx = \lim_{y \to 0^+} \int_{-\infty}^{\infty} |f(x + iy)|^2 dx.$$

This allows us to endow $\mathscr{H}^2$ with an inner product

$$\langle f, g \rangle := \int_{-\infty}^{\infty} f(x) \, \overline{g(x)} dx,$$

where, in a manner that is analogous to the H^2 case, $f(x)$ and $g(x)$ are the almost everywhere defined boundary functions of f, g in $\mathscr{H}^2$. We also have the Fourier transform characterization

$$\mathscr{H}^2 := \{\widehat{f} : f \in L^2(dx), f|_{(-\infty,0)} = 0\},$$

where

$$\widehat{f}(t) = \int_{-\infty}^{\infty} f(x) \, e^{-2\pi i x t} \, dx$$

is the *Fourier–Plancherel transform* of f.

The reason that most of the Hardy space theory on the disk can be imported into $\mathscr{H}^2$ is because the operator

$$\mathcal{U} : L^2(m) \to L^2(dx), \quad (\mathcal{U}f)(x) := \frac{1}{\sqrt{\pi}(x + i)} f(w(x)),$$

where

$$w(z) := \frac{z - i}{z + i}$$

is a conformal map from $\mathbb{C}_+$ onto $\mathbb{D}$, is a unitary map from $L^2(\mathbb{T}, m)$ onto $L^2(\mathbb{R}, dx)$. In particular, $\mathcal{U}$ maps H^2 unitarily onto $\mathscr{H}^2$ and maps $L^\infty(\mathbb{T}, m)$ contractively into $L^\infty(\mathbb{R}, dx)$.

We say that an analytic function Θ on $\mathbb{C}_+$ is *inner* if $|\Theta(z)| \leqslant 1$ for all $z \in \mathbb{C}_+$ and if the almost everywhere defined boundary function $\Theta(x)$ is unimodular almost everywhere on $\mathbb{R}$. The two most basic types of inner functions on $\mathbb{C}_+$ are

$$S_c(z) := e^{icz}, \qquad c > 0$$

and

$$b_\lambda(z) := \frac{z - \lambda}{z - \bar{\lambda}}, \qquad \lambda \in \mathbb{C}_+.$$

The first class of examples are the most basic type of *singular inner functions* on $\mathbb{C}_+$ while the second are the most basic type of *Blaschke products*.

Further examples of singular inner functions are given by

$$S_{\mu,c}(z) := e^{icz} \exp\left(-\frac{1}{\pi i} \int_{-\infty}^{\infty} \left(\frac{1}{x - z} - \frac{x}{1 + x^2}\right) d\mu(x)\right),$$

where $c > 0$ and μ is a positive measure on $\mathbb{R}$ which is singular with respect to Lebesgue measure and is also *Poisson finite* in that

$$\int_{-\infty}^{\infty} \frac{1}{1 + x^2} d\mu(x) < \infty.$$

The general Blaschke product is formed by specifying its zeros

$$\Lambda = \{\lambda_n\}_{n \geqslant 1} \subset \mathbb{C}_+ \setminus \{i\}$$

and forming the product

$$B_\Lambda(z) = b_i(z)^m \prod_{n \geqslant 1} \epsilon_n b_{\lambda_n}(z),$$

where the unimodular constant constants

$$\epsilon_n := \frac{|\lambda_n^2 + 1|}{\lambda_n^2 + 1}$$

are chosen so that $\epsilon_n b_{\lambda_n}(i) > 0$. It is well known [97] that the product above converges uniformly on compact subsets of $\mathbb{C}_+$ to an inner function if and only if the zeros $\{\lambda_n\}_{n \geqslant 1}$ satisfy the *Blaschke condition*

$$\sum_{n \geqslant 1} \frac{\operatorname{Im} \lambda_n}{1 + |\lambda_n|^2} < \infty.$$

Furthermore, the resulting product forms an inner function. Every inner function Θ for the upper half plane can be factored (uniquely up to unimodular constant factors) as

$$\Theta = B_\Lambda S_{\mu,c}.$$

There is also an inner–outer factorization theorem for $\mathscr{H}^2$ analogous to Theorem 3.20. Several good sources for the Hardy space of the upper half plane are [97, 118, 136].

3.6.4 Reproducing kernel Hilbert spaces

A Hilbert space $\mathcal{H}$ of analytic functions on a domain $\Omega \subset \mathbb{C}$ is called a *reproducing kernel Hilbert space* if for each $\lambda \in \Omega$ there is a function $k_\lambda \in \mathcal{H}$ for which $f(\lambda) = \langle f, k_\lambda \rangle_{\mathcal{H}}$ for all $f \in \mathcal{H}$. All of the classical Hilbert spaces of analytic functions on the disk (for example, Hardy, Bergman, Dirichlet, etc.) are reproducing kernel Hilbert spaces. Understanding the properties of their kernels often yields valuable information about the functions in $\mathcal{H}$. A few good sources for this are [2, 18, 144].

3.7 For further exploration

3.7.1 The Bergman space

The *Bergman space* L_a^2 is the Hilbert space of analytic functions

$$f(z) = \sum_{n \geq 0} a_n z^n$$

on $\mathbb{D}$ such that

$$\|f\|_{L_a^2} := \left(\sum_{n \geq 0} \frac{|a_n|^2}{n+1} \right)^{\frac{1}{2}} < \infty. \tag{3.28}$$

A calculation with power series and polar coordinates shows that

$$\|f\|_{L_a^2} = \left(\int_{\mathbb{D}} |f|^2 \frac{dA}{\pi} \right)^{\frac{1}{2}},$$

where dA is Lebesgue area measure. The inner product on L_a^2 is

$$\langle f, g \rangle_{L_a^2} = \int_{\mathbb{D}} f\overline{g}\, \frac{dA}{\pi}.$$

Note that $H^2 \subset L_a^2$. Functions in H^2 have non-tangential boundary values almost everywhere on $\mathbb{T}$. In contrast to this, functions in L_a^2 can be wild near $\mathbb{T}$. The classification of the zero sets for functions in the Hardy space is well understood; they are the Blaschke sequences (Corollary 3.16). For the Bergman spaces, the zero sets are not completely understood and they are worthy of further exploration. Although Bergman spaces have been studied intensively in the last several decades, there are plenty of open problems. Two excellent sources to learn more are [64, 109].

3.7.2 The Dirichlet space

The *Dirichlet space* $\mathcal{D}$ is the Hilbert space of analytic functions

$$f(z) = \sum_{n \geqslant 0} a_n z^n$$

on $\mathbb{D}$ such that

$$\|f\|_{\mathcal{D}} := \left(\sum_{n \geqslant 0} (n+1)|a_n|^2 \right)^{\frac{1}{2}} < \infty. \tag{3.29}$$

A calculation with power series and polar coordinates shows that

$$\|f\|_{\mathcal{D}}^2 = \|f\|_{H^2}^2 + \int_{\mathbb{D}} |f'|^2 \frac{dA}{\pi}.$$

Note that $\mathcal{D} \subset H^2$. Functions in $\mathcal{D}$ have even stronger regularity near $\mathbb{T}$ than functions in H^2 but their zero sets, though satisfying the Blaschke condition, have more restrictions than those for H^2. A good source for more on the Dirichlet space, including open problems, is [70].

4

Operators on the Hardy space

4.1 The shift operator

Of supreme importance in the world of operator-related function theory are the shift operators. Chief among these is the *unilateral shift* $S : H^2 \to H^2$ defined by

$$[Sf](z) = zf(z);$$

or, in terms of the Taylor coefficients $\{a_n\}_{n \geqslant 0}$ of $f = \sum_{n \geqslant 0} a_n z^n \in H^2$, by

$$S(a_0, a_1, \ldots) = (0, a_0, a_1, \ldots).$$

Observe that

$$\ker S = \{0\} \quad \text{and} \quad \operatorname{ran} S = H_0^2 := \{f \in H^2 : f(0) = 0\}. \tag{4.1}$$

Identifying the orthonormal basis element z^n, for $n \geqslant 0$, in H^2 with the orthonormal basis element $\mathbf{e}_n = (0, 0, \ldots, 1, 0, \ldots)$ (where the 1 is in the nth slot) in $\ell^2(\mathbb{N}_0)$ as in (3.9), we see that the matrix representation of S with respect to the basis $\{\mathbf{e}_n\}_{n \geqslant 0}$ is

$$\begin{bmatrix} 0 & 0 & 0 & 0 & 0 & \cdots \\ 1 & 0 & 0 & 0 & 0 & \cdots \\ 0 & 1 & 0 & 0 & 0 & \cdots \\ 0 & 0 & 1 & 0 & 0 & \cdots \\ 0 & 0 & 0 & 1 & 0 & \cdots \\ \vdots & \vdots & \vdots & \vdots & \vdots & \ddots \end{bmatrix}.$$

Because of its ubiquity in the realm of operator theory, one often refers to S as *the* shift operator. Before we discuss Beurling's characterization of the invariant subspaces of S, one of the crowning achievements in this area, let us first collect some elementary facts about the shift.

Proposition 4.1 *Let S denote the unilateral shift on H^2. Then:*

(i) S is an isometry,
(ii) The spectrum $\sigma(S)$ of S is equal to $\mathbb{D}^-$;
(iii) The point spectrum $\sigma_p(S)$ of S is the empty set.

Proof To prove (i), notice that

$$\|S f\|^2 = \int_{\mathbb{T}} |\zeta f(\zeta)|^2 dm(\zeta) = \int_{\mathbb{T}} |f|^2 dm = \|f\|^2.$$

Thus S is an isometry.

To prove (ii), first observe that since $\|S\| = 1$ from (i), we know that $\sigma(S) \subset \mathbb{D}^-$ (Proposition 1.30). To prove the reverse inclusion, notice that for each $\lambda \in \mathbb{D}$,

$$(S - \lambda I)H^2 \subset \{f \in H^2 : f(\lambda) = 0\} \neq H^2$$

and so $S - \lambda I$ is not onto. Thus $\mathbb{D} \subset \sigma(S)$. But since $\sigma(S)$ must be closed, we get $\mathbb{D}^- \subset \sigma(S)$ and equality follows.

For (iii), observe that if $(S - \lambda I)f \equiv 0$, then $(z - \lambda)f(z) = 0$ for all $z \in \mathbb{D}$. This means that f is vanishes on $\mathbb{D} \setminus \{\lambda\}$. By continuity, $f \equiv 0$. Thus S has no eigenvalues, that is to say, $\sigma_p(S) = \varnothing$. $\qquad\square$

Remark 4.2 We will show in Corollary 9.27 that the essential spectrum $\sigma_e(S)$ is equal to $\mathbb{T}$.

If u is an inner function, we can use the fact that $|u| = 1$ a.e. on $\mathbb{T}$ to see that for every $f \in H^2$,

$$\|u f\|^2 = \int_{\mathbb{T}} |u f|^2 dm = \int_{\mathbb{T}} |f|^2 dm = \|f\|^2.$$

This means that the operator $f \mapsto uf$ on H^2 is isometric and thus the linear manifold

$$uH^2 = \{uh : h \in H^2\}$$

is closed in H^2. In other words, uH^2 is a subspace of H^2. Moreover,

$$S(uH^2) = zuH^2 = u(zH^2) \subset uH^2;$$

and so uH^2 is a non-zero S-invariant subspace of H^2. Beurling's Theorem [30] says these are all of the non-zero S-invariant subspaces of H^2.

Theorem 4.3 (Beurling)   *If M is a non-zero subspace of H^2 that is invariant under S, then there is an inner function u such that*

$$M = uH^2.$$

Proof   First notice that $SM \neq M$. If this were not the case, then $f/z \in M$ whenever $f \in M$. Applying this k times, we conclude that $f/z^k \in H^2$ for all $k \in \mathbb{N}$. In particular, this means that f/z^k is an analytic function on $\mathbb{D}$ for all $k \geq 0$. By a power series argument, this can only happen when $f \equiv 0$, which contradicts the assumption that $M \neq \{0\}$.

Second, since $SM \neq M$, one observes that

$$M \ominus SM \neq \{0\} \tag{4.2}$$

and so $M \cap (SM)^\perp$ contains a function u that is not identically zero. We now argue that there exists a $c > 0$ such that

$$|u(\zeta)| = c \qquad \text{a.e. } \zeta \in \mathbb{T}.$$

Indeed,

$$\int_{\mathbb{T}} |u(\zeta)|^2 \bar{\zeta}^n \, dm(\zeta) = \langle u, S^n u \rangle = 0, \qquad \forall n \geq 1.$$

Taking complex conjugates of both sides of the preceding equation, we also see that

$$\int_{\mathbb{T}} |u(\zeta)|^2 \zeta^n \, dm(\zeta) = 0, \qquad \forall n \geq 1.$$

This means that the Fourier coefficients $\widehat{|u|^2}(n)$ of $|u|^2$ satisfy

$$\widehat{|u|^2}(n) = 0, \qquad \forall n \in \mathbb{Z} \setminus \{0\};$$

and so $|u|^2 = c$ a.e. on $\mathbb{T}$ (Proposition 1.16) for some $c > 0$. Without loss of generality, we can assume that $|u| = 1$ a.e. on $\mathbb{T}$. Thus u is an H^2 function with unimodular boundary values on $\mathbb{T}$ and so, by Smirnov's Theorem (Theorem 3.29), we conclude that $u \in H^\infty$. Hence u is an inner function.

Third, let

$$[u] = \bigvee \{u, Su, S^2u, \ldots\} = \bigvee \{pu : p \text{ is a polynomial}\}. \tag{4.3}$$

We now prove that

$$[u] = uH^2.$$

To see this, observe that $[u] \subset uH^2$ (we have used the fact that uH^2 is closed since u is inner). For the other containment, let $g = uG \in uH^2$ and let G_N be the Nth partial sum of the Taylor series of G. Notice that $uG_N \in [u]$ since G_N is a polynomial. From (3.6), $G_N \to G$ in H^2 and so, since $|u| = 1$ a.e. on $\mathbb{T}$,

$$\|uG_N - uG\|^2 = \int_{\mathbb{T}} |uG_N - uG|^2 dm$$

$$= \int_{\mathbb{T}} |G_N - G|^2 dm$$

$$= \|G_N - G\|^2 \to 0.$$

This means that $uG \in [u]$.

Finally, observe that

$$[u] = \mathcal{M}.$$

Indeed, since $u \in \mathcal{M}$ and $\mathcal{M}$ is S-invariant, $[u] \subset \mathcal{M}$. To show the reverse inclusion, suppose that $f \in \mathcal{M}$ and $f \perp [u]$. In order to show that $f \equiv 0$, we proceed as follows. Since $f \perp [u]$,

$$\int_{\mathbb{T}} f(\zeta)\overline{u(\zeta)}\zeta^n \, dm(\zeta) = \langle f, S^n u \rangle = 0, \qquad \forall n \geqslant 0.$$

However, since $u \perp S\mathcal{M}$ by (4.2), we also know that

$$\int_{\mathbb{T}} f(\zeta)\overline{u(\zeta)}\zeta^n \, dm(\zeta) = \langle S^n f, u \rangle = 0, \qquad \forall n \geqslant 1.$$

The previous two equations imply that all of the Fourier coefficients of $f\overline{u}$ vanish. Thus $f\overline{u} = 0$ a.e. on $\mathbb{T}$ (Proposition 1.16). But we have already shown that $|u| = 1$ a.e. on $\mathbb{T}$ and so $f = 0$ a.e. on $\mathbb{T}$. This proves that $f \equiv 0$ and thus, by Theorem 1.27, $[u] = \mathcal{M}$. □

Remark 4.4 The inner function u in Beurling's Theorem is unique up to a unimodular constant multiple. Indeed, if $u_1 H^2 = u_2 H^2$, then $u_1 | u_2$ and $u_2 | u_1$ (recall the definition of divisibility from Definition 2.16) so that u_1 and u_2 are unimodular constant multiples of each other by Theorem 2.14.

As a consequence of Theorem 4.3, Beurling identified the cyclic vectors for S. Recall the definition of $[f]$ from (4.3).

Corollary 4.5 *A vector $f \in H^2$ is cyclic for S if and only if f is an outer function.*

Proof Suppose that $f \in H^2$ is an outer function. By Beurling's Theorem, we know that $[f]$ is a non-zero S-invariant subspace of H^2 and is therefore equal to uH^2 for some inner function u. Since $f \in uH^2$ and f has no non-trivial inner factor, it must be the case that u is a unimodular constant. Thus $[f] = H^2$.

Conversely, suppose that $[f] = H^2$. Let $f = uF$ where u is inner and $F \in H^2$ is outer. Notice that

$$[f] \subset uH^2.$$

Indeed, if $g \in [f]$, there is a sequence $\{p_n\}_{n \geqslant 1}$ of analytic polynomials such that $p_n f = uF p_n \to g$ in H^2. Since $|u| = 1$ almost everywhere on $\mathbb{T}$, it follows that $\{F p_n\}_{n \geqslant 1}$ is a Cauchy sequence in H^2 and so $F p_n \to h$ in H^2. Again use the fact that $|u| = 1$ almost everywhere on $\mathbb{T}$ to see that $f p_n = uF p_n \to uh$ in H^2. Thus $g = uh \in uH^2$.

The argument above shows that

$$H^2 = [f] \subset uH^2$$

and thus $H^2 = uH^2$. This means that the constant function 1 belongs to uH^2. By the uniqueness of the inner–outer factorization (Theorem 3.20), u must be a unimodular constant and hence $f = F$ is an outer function. □

We leave it to the reader to make adjustments to the proof above to obtain the following.

Corollary 4.6 *Suppose $f \in H^2 \setminus \{0\}$ with $f = uF$, where u is inner and F is outer. Then $[f] = uH^2$.*

This next theorem shows that the non-zero invariant subspaces of S are in an order preserving correspondence with the inner functions, ordered by divisibility.

Theorem 4.7 *Let u_1, u_2 be two inner functions. Then the following are equivalent:*

(i) $u_1 | u_2$;
(ii) $u_2 H^2 \subset u_1 H^2$.

Proof (i) $\implies$ (ii): By assumption, there is an inner function u such that $u_2 = u_1 u$. Hence,

$$u_2 H^2 = u_1 u H^2 \subset u_1 H^2.$$

(ii) $\implies$ (i): Since $u_2 \in u_2 H^2$ and we are assuming that $u_2 H^2 \subset u_1 H^2$, we also have $u_2 \in u_1 H^2$. Therefore, there is a function $u \in H^2$ such that $u_2 = u_1 u$. This identity implies that $|u| = 1$ a.e. on $\mathbb{T}$, that is to say, u is an inner function. Thus $u_1 | u_2$. □

Corollary 4.8 *Let u_1, u_2 be two inner functions. Then there is an inner function u, unique up to a unimodular constant factor, such that*

$$u_1 H^2 \cap u_2 H^2 = u H^2. \tag{4.4}$$

Moreover, the function u satisfies the following properties:

(i) $u_1 | u$ and $u_2 | u$;
(ii) if $u_1 | v$ and $u_2 | v$, where v is an inner function, then $u | v$.

Proof The set $\mathcal{M} = u_1 H^2 \cap u_2 H^2$ is a closed subspace of H^2 which is invariant under S. Moreover, it contains $u_1 u_2$ but does not contain 1 and so $\{0\} \subsetneq \mathcal{M} \subsetneq H^2$. By Beurling's Theorem, there is an inner function u, unique up to a unimodular constant factor (see Remark 4.4), such that (4.4) holds.

Since (4.4) implies $u_1 H^2 \supset u H^2$ and $u_2 H^2 \supset u H^2$, property (i) follows from Theorem 4.7. If both u_1 and u_2 divide the inner function v, then we have

$$v H^2 \subset u_1 H^2 \cap u_2 H^2 = u H^2,$$

which, by the same theorem, implies that u divides v. This proves (ii).   □

The inner function u from Corollary 4.8 is called the *least common multiple* of u_1 and u_2 and it is denoted by

$$u = \mathrm{lcm}(u_1, u_2).$$

We can also define and prove the existence of the *greatest common divisor* of two inner functions. This is denoted by

$$u = \gcd(u_1, u_2) \tag{4.5}$$

and explored in the following result.

Corollary 4.9 *Let u_1, u_2 be two inner functions. Then there is an inner function u, unique up to a unimodular constant factor, such that*

$$u_1 H^2 \vee u_2 H^2 = u H^2. \tag{4.6}$$

Moreover, the function u satisfies the following properties:

(i) $u | u_1$ and $u | u_2$;
(ii) if $v | u_1$ and $v | u_2$, then $v | u$.

Proof The subspace $\mathcal{M} = u_1 H^2 \vee u_2 H^2$ is a non-zero S-invariant subspace of H^2. By Beurling's Theorem, there is an inner function u, uniquely determined up to a unimodular constant factor, such that (4.6) holds.

As in the preceding corollary, property (i) follows immediately from Theorem 4.7. Moreover, if v divides both u_1 and u_2, then

$$uH^2 = u_1 H^2 \vee u_2 H^2 \subset vH^2.$$

By appealing to the same theorem, we deduce that v divides u. This proves (ii).

□

We now describe $\mathrm{lcm}(u_1, u_2)$ and $\gcd(u_1, u_2)$ in terms of the canonical factorization

$$u_1 = z^{n_1} B_1 s_{\mu_1}, \quad u_2 = z^{n_2} B_2 s_{\mu_2};$$

where $n_j \geqslant 0$, $B_j = B_{\Lambda_j}$ are Blaschke products with zeros Λ_j (repeated according to multiplicity), and s_{μ_j} is the singular inner function with positive singular measure μ_j. We leave it to the reader to fill in the details here. Let

$$n_1 \vee n_2 = \max(n_1, n_2), \quad n_1 \wedge n_2 = \min(n_1, n_2);$$

$$B_1 \vee B_2 = B_{\Lambda_1 \cup \Lambda_2}, \quad B_1 \wedge B_2 = B_{\Lambda_1 \cap \Lambda_2}.$$

With $\Lambda_1 \cup \Lambda_2$ and $\Lambda_1 \cup \Lambda_2$ we need to take multiplicities into account. So for example if

$$\Lambda_1 = \{0, 0, a_3, a_4, \ldots\}, \quad \Lambda_2 = \{0, b_2, b_3, \ldots\},$$

where a_j and b_j are zeros of multiplicity one, then

$$\Lambda_1 \cup \Lambda_2 = \{0, 0, 0\} \cup \{a_j\}_{j \geqslant 3} \cup \{b_j\}_{j \geqslant 2},$$

$$\Lambda_1 \cap \Lambda_2 = \{0\} \cup \left(\{a_j\}_{j \geqslant 3} \cap \{b_j\}_{j \geqslant 3} \right).$$

We also define

$$\mu_1 \vee \mu_2 = \max \left(\frac{d\mu_1}{d(\mu_1 + \mu_2)}, \frac{d\mu_2}{d(\mu_1 + \mu_2)} \right)(\mu_1 + \mu_2);$$

$$\mu_1 \wedge \mu_2 = \min \left(\frac{d\mu_1}{d(\mu_1 + \mu_2)}, \frac{d\mu_2}{d(\mu_1 + \mu_2)} \right)(\mu_1 + \mu_2).$$

Proposition 4.10 *For inner functions $u_1 = z^{n_1} B_1 s_{\mu_1}$ and $u_2 = z^{n_2} B_2 s_{\mu_2}$,*

$$\mathrm{lcm}(u_1, u_2) = z^{n_1 \vee n_2}(B_1 \vee B_2) s_{\mu_1 \vee \mu_2},$$

$$\gcd(u_1, u_2) = z^{n_1 \wedge n_2}(B_1 \wedge B_2) s_{\mu_1 \wedge \mu_2}.$$

4.2 Toeplitz operators

In the later chapters of this book, we will examine truncations of Toeplitz operators that act on model spaces. Thus we need to familiarize ourselves with the basics of Toeplitz operators on H^2. This is a vast subject; and we refer the reader to the texts [33, 62, 134] for a more detailed treatment.

Before proceeding, we recall from Theorem 3.12 that we can regard H^2, via non-tangential boundary values, as a closed subspace of L^2 in the sense that

$$H^2 = \{f \in L^2 : \widehat{f}(n) = 0 \ \forall n \leqslant -1\}.$$

Definition 4.11 For φ in L^∞, the *Toeplitz operator* $T_\varphi : H^2 \to H^2$ with *symbol* φ is defined by

$$T_\varphi(f) = P(\varphi f), \qquad f \in H^2.$$

In the preceding definition, P denotes the Riesz projection of L^2 onto H^2 (see Section 3.2). One can write an integral formula for T_φ using the integral representation of the Riesz projection P from Proposition 3.13. Indeed,

$$(T_\varphi f)(\lambda) = \int_{\mathbb{T}} \frac{f(\zeta)\varphi(\zeta)}{1 - \overline{\zeta}\lambda} dm(\zeta), \quad \lambda \in \mathbb{D}, f \in H^2.$$

When $\varphi \in H^\infty$, observe that

$$T_\varphi f = P(\varphi f) = \varphi f, \qquad f \in H^2; \tag{4.7}$$

since $\varphi f \in H^2$ whenever $f \in H^2$. In other words, when $\varphi \in H^\infty$, the Toeplitz operator $T_\varphi f = \varphi f$ is just the multiplication (Laurent) operator with symbol φ. These types of Toeplitz operators are often called *analytic Toeplitz operators* due to the fact that the symbol φ is the boundary function of a bounded analytic function. Toeplitz operators of the form $T_{\overline{\varphi}}$, where $\varphi \in H^\infty$, are often called *conjugate analytic Toeplitz operators*.

Since

$$aT_\varphi + T_\psi = T_{a\varphi + \psi}, \quad a \in \mathbb{C}, \varphi, \psi \in L^\infty,$$

we see that the set of all (bounded) Toeplitz operators on H^2 is a linear manifold in $\mathcal{B}(H^2)$, although it is not an algebra (see Theorem 4.22 below).

If one computes the matrix representation of T_φ with respect to the standard orthonormal basis $\{1, z, z^2, z^3, \ldots\}$ for H^2, one sees that the (m, n) entry of this matrix is

$$\langle T_\varphi \zeta^n, \zeta^m \rangle = \langle P(\varphi \zeta^n), \zeta^m \rangle = \langle \varphi \zeta^n, P\zeta^m \rangle = \langle \varphi \zeta^n, \zeta^m \rangle = \widehat{\varphi}(m - n). \tag{4.8}$$

With $\alpha_k = \widehat{\varphi}(k)$, this yields an infinite *Toeplitz matrix*

$$\begin{bmatrix} \alpha_0 & \alpha_{-1} & \alpha_{-2} & \alpha_{-3} & \alpha_{-4} & \cdots \\ \alpha_1 & \alpha_0 & \alpha_{-1} & \alpha_{-2} & \alpha_{-3} & \cdots \\ \alpha_2 & \alpha_1 & \alpha_0 & \alpha_{-1} & \alpha_{-2} & \cdots \\ \alpha_3 & \alpha_2 & \alpha_1 & \alpha_0 & \alpha_{-1} & \cdots \\ \alpha_4 & \alpha_3 & \alpha_2 & \alpha_1 & \alpha_0 & \cdots \\ \vdots & \vdots & \vdots & \vdots & \vdots & \ddots \end{bmatrix}.$$

The operator norm of T_φ takes a very pleasant form [34].

Proposition 4.12 (Brown–Halmos) $\|T_\varphi\| = \|\varphi\|_\infty$.

Proof   For any $f \in H^2$, we use the fact that P is an orthogonal projection (and hence a contraction) to see that

$$\|T_\varphi f\| = \|P(\varphi f)\| \leqslant \|\varphi f\|.$$

Furthermore,

$$\|\varphi f\|^2 = \int_{\mathbb{T}} |\varphi f|^2 dm \leqslant \|\varphi\|_\infty^2 \int_{\mathbb{T}} |f|^2 dm \leqslant \|\varphi\|_\infty^2 \|f\|^2.$$

Thus

$$\|T_\varphi\| = \sup_{\|f\|=1} \|T_\varphi f\| \leqslant \|\varphi\|_\infty.$$

For the lower bound, note that for any $\lambda \in \mathbb{D}$, Proposition 3.3 tells us that the reproducing kernel c_λ for H^2 satisfies

$$\|c_\lambda\| = \frac{1}{\sqrt{1 - |\lambda|^2}}.$$

Define the normalized reproducing kernel $\widetilde{c}_\lambda = c_\lambda/\|c_\lambda\|$ and note, by the Cauchy–Schwarz Inequality, that

$$\left| \langle T_\varphi \widetilde{c}_\lambda, \widetilde{c}_\lambda \rangle \right| \leqslant \left\| T_\varphi \widetilde{c}_\lambda \right\| \|\widetilde{c}_\lambda\| \leqslant \|T_\varphi\| \|\widetilde{c}_\lambda\| \|\widetilde{c}_\lambda\| = \|T_\varphi\|.$$

Using this inequality, we get

$$\|T_\varphi\| \geqslant |\langle T_\varphi \widetilde{c}_\lambda, \widetilde{c}_\lambda \rangle| = |\langle P(\varphi \widetilde{c}_\lambda), \widetilde{c}_\lambda \rangle| = |\langle \varphi \widetilde{c}_\lambda, P\widetilde{c}_\lambda \rangle|$$

$$= |\langle \varphi \widetilde{c}_\lambda, \widetilde{c}_\lambda \rangle| = \left| \int_{\mathbb{T}} \frac{1 - |\lambda|^2}{|\zeta - \lambda|^2} \varphi(\zeta) \, dm(\zeta) \right|$$

$$= |\mathscr{P}(\varphi)(\lambda)|.$$

Finally, we let $\lambda = r\zeta$, where $\zeta \in \mathbb{T}$ and $r \in (0, 1)$, and use Fatou's Theorem (Theorem 1.10) to see that $|\varphi(\zeta)| \leqslant \|T_\varphi\|$ for almost every $\zeta \in \mathbb{T}$. This yields the desired lower bound $\|\varphi\|_\infty \leqslant \|T_\varphi\|$. ⊔

The symbol of a Toeplitz operator is unique. Indeed, we have the following corollary.

Corollary 4.13 $T_\varphi = T_\psi$ *if and only if* $\varphi = \psi$.

Proof   If $T_\varphi = T_\psi$, then $T_{\varphi-\psi} = T_\varphi - T_\psi = 0$. Thus $\|\varphi - \psi\|_\infty = 0$ by Proposition 4.12. □

Proposition 4.14   *For* $\varphi \in L^\infty$, $T_\varphi^* = T_{\overline{\varphi}}$.

Proof   For any $f, g \in H^2$ we have

$$\langle T_\varphi f, g \rangle = \langle P(\varphi f), g \rangle = \langle \varphi f, Pg \rangle = \langle \varphi f, g \rangle$$
$$= \langle f, \overline{\varphi} g \rangle = \langle Pf, \overline{\varphi} g \rangle = \langle f, P(\overline{\varphi} g) \rangle$$
$$= \langle f, T_{\overline{\varphi}} g \rangle.$$

This proves that $T_\varphi^* = T_{\overline{\varphi}}$. □

Recall the definition of a compact operator from Section 1.3. This next result from [34] characterizes the compact Toeplitz operators.

Theorem 4.15 (Brown–Halmos) T_φ *is compact if and only if* $\varphi = 0$.

Proof   The zero operator is certainly compact. To prove the converse, first observe that $z^n \to 0$ weakly in H^2 as $n \to \infty$. To see this, let $f = \sum_{n \geqslant 0} a_n z^n \in H^2$, that is $\sum_{n \geqslant 0} |a_n|^2 < \infty$. Then $\langle f, z^n \rangle = a_n \to 0$ and so $z^n \to 0$ weakly in H^2. Assume T_φ is compact, then

$$\|T_\varphi z^n\|^2 = \left\| P\left(\sum_{k \in \mathbb{Z}} \widehat{\varphi}(k) \zeta^{k+n} \right) \right\|^2 = \left\| \sum_{k \geqslant -n} \widehat{\varphi}(k) \zeta^{k+n} \right\|^2 = \sum_{k \geqslant -n} |\widehat{\varphi}(k)|^2.$$

As $n \to \infty$, the right-hand side approaches $\|\varphi\|^2$ (via Parseval's Theorem) while the left-hand side approaches zero (Proposition 1.37). Thus $\|\varphi\| = 0$. □

4.3 A characterization of Toeplitz operators

The following is a characterization of Toeplitz operators from [34] which has some useful corollaries and an analogue for truncated Toeplitz operators (see Theorem 13.8).

Theorem 4.16 (Brown–Halmos) *A bounded operator A on H^2 is a Toeplitz operator if and only if $S^*AS = A$.*

A portion of the proof needs the following real analysis result which characterizes the commutant of the *bilateral shift*

$$Z : L^2 \to L^2, \quad (Zf)(\zeta) = \zeta f(\zeta).$$

Lemma 4.17 *The bilateral shift Z is unitary. Furthermore, if $V \in \mathcal{B}(L^2)$ and $VZ = ZV$, then there is a $\varphi \in L^\infty$ such that $Vf = \varphi f$ for all $f \in L^2$ and $\|V\| = \|\varphi\|_\infty$.*

Proof A calculation with adjoints, which we leave as a short exercise, shows that $(Z^*f)(\zeta) = \bar{\zeta}f(\zeta)$. Thus

$$I = ZZ^* = Z^*Z \tag{4.9}$$

and so Z is unitary. In addition to the identity $VZ = ZV$, we also have

$$VZ^* = Z^*V; \tag{4.10}$$

since $Z^{-1} = Z^*$. For any trigonometric polynomial p, define M_p on L^2 by $M_p f = pf$ and note from the identities (4.9) and (4.10) that $M_p V = VM_p$. If $\varphi = V1$, then $p\varphi = Vp$. If $f \in L^2$ and $\{p_n\}_{n \geqslant 1}$ is a sequence of trigonometric polynomials with $\|p_n - f\| \to 0$ then, using the fact that $\|Vp_n - Vf\| \to 0$, we can pass to a subsequence if necessary and also assume that $Vp_n \to Vf$ almost everywhere on $\mathbb{T}$. This means that $Vf = \varphi f$ for all $f \in L^2$.

To show that $\varphi \in L^\infty$, note that

$$\|\varphi^n\| = \|V^n 1\| \leqslant \|V\|^n \|1\| = \|V\|^n$$

whence

$$\left(\int_{\mathbb{T}} |\varphi|^{2n} dm \right)^{\frac{1}{2n}} \leqslant \|V\|.$$

The left-hand side of the previous equation approaches $\|\varphi\|_\infty$ as $n \to \infty$. Thus $\|\varphi\|_\infty \leqslant \|V\|$ and so $\varphi \in L^\infty$. Since V is the operator of multiplication by φ on L^2, we see that $\|V\| \leqslant \|\varphi\|_\infty$. $\square$

Proof of Theorem 4.16   We follow a mixture of the original proof in [34] and the one in [134]. Suppose that A is a Toeplitz operator. Use (4.8) to see that for any $m, n \geq 0$,

$$\langle S^* T_\varphi S z^n, z^m \rangle = \langle T_\varphi z^{n+1}, z^{m+1} \rangle = \widehat{\varphi}(m - n).$$

By (4.8) again, the last expression is equal to $\langle T_\varphi z^n, z^m \rangle$ and so $S^* T_\varphi S = T_\varphi$.

Conversely suppose that $A \in \mathcal{B}(H^2)$ satisfies $S^* A S = A$. Then, by induction, we have $S^{*n} A S^n = A$ for every $n \geq 1$ and so

$$\langle A z^M, z^N \rangle = \langle S^{*m} A S^m z^M, z^N \rangle = \langle A z^{M+m}, z^{N+m} \rangle. \qquad (4.11)$$

This says that the matrix representation of A is a infinite Toeplitz matrix. Now we need to show that $A = T_\varphi$ for some $\varphi \in L^\infty$.

Let $Z : L^2 \to L^2$ be the operator of multiplication by the independent variable $(Zf)(\zeta) = \zeta f(\zeta)$ from Lemma 4.17. For each $n \geq 0$, define

$$A_n : L^2 \to L^2, \quad A_n := Z^{*n} \iota A P Z^n,$$

where $\iota : H^2 \to L^2$ is the inclusion map. Observe that for $k, \ell \geq 0$ we have

$$\begin{aligned}
\langle A_n z^k, z^\ell \rangle &= \langle Z^{*n} \iota A P Z^n z^k, z^\ell \rangle \\
&= \langle \iota A P z^{k+n}, z^{\ell+n} \rangle \\
&= \langle A z^{k+n}, z^{\ell+n} \rangle \\
&= \langle A z^k, z^\ell \rangle. \qquad \text{(by (4.11))}
\end{aligned}$$

From here we see that $\langle A_n z^k, z^\ell \rangle$ is independent of n for all $n \geq 0$. From the previous set of equations, we have

$$\langle A_n z^k, z^\ell \rangle = \langle A z^{k+n}, z^{\ell+n} \rangle.$$

Thus, even for possibly negative values of k and ℓ, we see that for large enough values of n (say $k + n \geq 0$ and $\ell + n \geq 0$), the quantity $\langle A_n z^k, z^\ell \rangle$ is eventually independent of n. Thus for any pair p, q of trigonometric polynomials, the sequence $\langle A_n p, q \rangle$ is constant for large enough values of n.

Furthermore,

$$\begin{aligned}
|\langle A_n p, q \rangle| &\leq \|A_n\| \, \|p\| \, \|q\| \\
&\leq \|Z^{*n} \iota A P Z^n\| \, \|p\| \, \|q\| \\
&\leq \|Z^{*n}\| \, \|A\| \, \|P\| \, \|Z^n\| \, \|p\| \, \|q\| \\
&\leq \|A\| \, \|p\| \, \|q\|.
\end{aligned}$$

This says that the sesquilinear form

$$\Lambda(p, q) = \lim_{n \to \infty} \langle A_n p, q \rangle,$$

defined initially for trigonometric polynomials p and q, can be extended to a continuous sesquilinear form on L^2. By the Riesz Representation Theorem for sesquilinear forms [152, Sec. 84], we have $\Lambda(f, g) = \langle Vf, g \rangle$ for some bounded operator V on L^2.

From the definition of A_n, and the fact that $Z^*Z = ZZ^* = I$ (Lemma 4.17), we obtain $A_{n+1} = Z^* A_n Z$. Thus for any $f, g \in L^2$ we have

$$\langle A_{n+1} f, g \rangle = \langle Z^* A_n Z f, g \rangle = \langle A_n Z f, Z g \rangle.$$

Take limits as $n \to \infty$ to deduce the formula

$$\langle Vf, g \rangle = \langle VZf, Zg \rangle = \langle Z^* VZf, g \rangle.$$

This means that the operator V commutes with Z and so $V = M_\varphi$ for some $\varphi \in L^\infty$ (Lemma 4.17). Finally, note that for $k, \ell \geqslant 0$ we can use the fact proved earlier that for large enough n, $\langle A_n z^k, z^\ell \rangle = \langle A z^k, z^\ell \rangle$ to conclude that

$$\langle P M_\varphi z^k, z^\ell \rangle = \langle P V z^k, z^\ell \rangle = \langle V z^k, z^\ell \rangle = \lim_{n \to \infty} \langle A_n z^k, z^\ell \rangle = \langle A z^k, z^\ell \rangle.$$

From here we get the identity $A = P M_\varphi | H^2$, that is, $A = T_\varphi$.   □

Corollary 4.18   *The set of Toeplitz operators $\{T_\varphi : \varphi \in L^\infty\}$ is closed in the weak operator topology.*

Proof   Suppose $\{A_\alpha\}_{\alpha \in I}$ is a net of Toeplitz operators that converges in the weak operator topology to $A \in \mathcal{B}(H^2)$. This means that

$$\lim_\alpha \langle A_\alpha f, g \rangle = \langle Af, g \rangle \quad \forall f, g \in H^2.$$

Thus for all $f, g \in H^2$,

$$
\begin{aligned}
\langle (S^*AS - A)f, g \rangle &= \langle S^*AS f, g \rangle - \langle Af, g \rangle \\
&= \langle AS f, S g \rangle - \langle Af, g \rangle \\
&= \lim_\alpha \langle A_\alpha S f, S g \rangle - \lim_\alpha \langle A_\alpha f, g \rangle \\
&= \lim_\alpha \langle S^* A_\alpha S f, g \rangle - \lim_\alpha \langle A_\alpha f, g \rangle \\
&= \lim_\alpha (\langle A_\alpha f, g \rangle - \langle A_\alpha f, g \rangle) \qquad \text{(Theorem 4.16)} \\
&= 0.
\end{aligned}
$$

This yields the identity $S^*AS = A$ and so, by Theorem 4.16, A is indeed a Toeplitz operator.   □

The analytic Toeplitz operators $\{T_\varphi : \varphi \in H^\infty\}$ form a special class of operators. For instance, by (4.8), the matrix representation of T_φ with respect to the basis $\{1, z, z^2, \ldots\}$ is lower triangular, that is,

$$\begin{bmatrix} \alpha_0 & 0 & 0 & 0 & 0 & \cdots \\ \alpha_1 & \alpha_0 & 0 & 0 & 0 & \cdots \\ \alpha_2 & \alpha_1 & \alpha_0 & 0 & 0 & \cdots \\ \alpha_3 & \alpha_2 & \alpha_1 & \alpha_0 & 0 & \cdots \\ \alpha_4 & \alpha_3 & \alpha_2 & \alpha_1 & \alpha_0 & \cdots \\ \vdots & \vdots & \vdots & \vdots & \vdots & \ddots \end{bmatrix},$$

where $\alpha_k = \widehat{\varphi}(k)$. Note that the set of analytic Toeplitz operators form a commutative algebra, as does the set of conjugate analytic Toeplitz operators $\{T_{\overline{\varphi}} : \varphi \in H^\infty\}$.

One also has the following useful formula for the eigenvalues of T_φ^* for $\varphi \in H^\infty$.

Proposition 4.19   *For $\varphi \in H^\infty$, $T_\varphi^* c_\lambda = \overline{\varphi(\lambda)}\, c_\lambda$ for each $\lambda \in \mathbb{D}$.*

Proof   Note that

$$(T_\varphi^* c_\lambda)(z) = \langle T_\varphi^* c_\lambda, c_z \rangle = \langle c_\lambda, T_\varphi c_z \rangle$$
$$= \langle c_\lambda, \varphi c_z \rangle = \overline{\langle \varphi c_z, c_\lambda \rangle}$$
$$= \overline{\varphi(\lambda) c_z(\lambda)} = \overline{\varphi(\lambda)} c_\lambda(z).$$

In summary, we see that $T_\varphi^* c_\lambda = \overline{\varphi(\lambda)} c_\lambda$.   □

Corollary 4.20   *For $\varphi \in H^\infty \setminus \{0\}$, T_φ^* has dense range.*

Proof   By Proposition 4.19,

$$T_\varphi^* c_\lambda = \overline{\varphi(\lambda)} c_\lambda, \qquad \lambda \in \mathbb{D}.$$

In addition, since φ analytic on $\mathbb{D}$ and $\varphi \not\equiv 0$, the set $\{\lambda \in \mathbb{D} : \varphi(\lambda) \neq 0\}$ has an accumulation point in $\mathbb{D}$. By Corollary 3.7 we see that

$$(\operatorname{ran} T_\varphi^*)^- \supset \bigvee \{c_\lambda : \varphi(\lambda) \neq 0\} = H^2.$$ □

4.4 The commutant of the shift

The analytic Toeplitz operators also form the *commutant*

$$\{S\}' := \{A \in \mathcal{B}(H^2) : AS = SA\}$$

of the unilateral shift [34].

Theorem 4.21 (Brown–Halmos) *A bounded operator A on H^2 satisfies $SA = AS$ if and only if $A = T_\varphi$ for some $\varphi \in H^\infty$.*

Proof For any $\varphi \in H^\infty$, observe that

$$S T_\varphi = T_{z\varphi} = T_\varphi S.$$

For the other direction, let $\varphi = A1$ and note that the identity $AS = SA$ implies that $AS^n = S^nA$ for all $n \geqslant 0$. Thus $AS^n1 = S^nA1$ which implies $Az^n = z^n\varphi$. Using linearity, we get $Ap = \varphi p$ for any analytic polynomial p. Given $f \in H^2$, there is a sequence of analytic polynomials p_n with $\|p_n - f\| \to 0$ (Proposition 3.8). Since A is a bounded linear operator, it follows that A is continuous and so $Ap_n \to Af$ in the norm of H^2. By (3.1), $p_n(z) \to f(z)$ and $(Ap_n)(z) \to (Af)(z)$ for each $z \in \mathbb{D}$. From here we see that $(Af)(z) = \varphi(z)f(z)$. To finish we just need to show that $\varphi \in H^\infty$ so that $A = T_\varphi$.

For each $n \geqslant 1$, note that $A^n1 = \varphi^n$, which yields

$$|\varphi(\lambda)^n| = |\langle A^n1, c_\lambda\rangle| \leqslant \|A^n1\|\,\|c_\lambda\| \leqslant \|A^n\|\,\|1\|\,\|c_\lambda\| \leqslant \|A\|^n\,\|c_\lambda\|.$$

Now take nth roots followed by a limit as $n \to \infty$ to see that

$$|\varphi(\lambda)| \leqslant \lim_{n\to\infty} \|A\|\,\|c_\lambda\|^{1/n} = \|A\|.$$

Since $|\varphi(\lambda)| \leqslant \|A\|$ for all $\lambda \in \mathbb{D}$, we see that $\varphi \in H^\infty$. $\qquad\square$

Theorem 4.22 *For $\psi, \varphi \in L^\infty$ the operator $T_\psi T_\varphi$ is a Toeplitz operator if and only if either T_ψ is conjugate analytic or T_φ is analytic. In both cases, $T_\psi T_\varphi = T_{\psi\varphi}$.*

We follow the proof of [134] for which we need the following two lemmas.

Lemma 4.23 *For $\psi \in L^\infty$, $P(\overline{\zeta}\psi) = S^*T_\psi 1$.*

Proof For $g \in H^2$,

$$\begin{aligned}
\langle P(\overline{\zeta}\psi), g\rangle &= \langle \overline{\zeta}\psi, Pg\rangle = \langle \overline{\zeta}\psi, g\rangle = \langle \psi, \zeta g\rangle \\
&= \langle \psi, Sg\rangle = \langle \psi, PSg\rangle = \langle P\psi, Sg\rangle \\
&= \langle T_\psi 1, Sg\rangle \\
&= \langle S^*T_\psi 1, g\rangle.
\end{aligned}$$

Since $g \in H^2$ was arbitrary, this proves the desired identity. $\qquad\square$

We refer the reader to Proposition 1.32 to recall some facts about rank-one operators

$$(g \otimes h)(f) = \langle f, h \rangle g.$$

Lemma 4.24 *For $\psi, \varphi \in L^\infty$,*

$$S^* T_\psi T_\varphi S - T_{\psi\varphi} = P(\overline{\zeta}\psi) \otimes P(\overline{\zeta}\overline{\varphi}).$$

Proof   Note that $I = S S^* + 1 \otimes 1$ and so

$$\begin{aligned}
S^* T_\psi T_\varphi S &= S^* T_\psi (S S^* + 1 \otimes 1) T_\varphi S \\
&= S^* T_\psi S S^* T_\varphi S + S^* T_\psi (1 \otimes 1) T_\varphi S \\
&= T_\psi T_\varphi + S^* T_\psi (1 \otimes 1) T_\varphi S & \text{(Theorem 4.16)} \\
&= T_\psi T_\varphi + (S^* T_\psi 1) \otimes (S^* T_{\overline{\varphi}} 1) & \text{(Theorem 1.32)} \\
&= T_\psi T_\varphi + P(\overline{\zeta}\psi) \otimes P(\overline{\zeta}\overline{\varphi}). & \text{(Lemma 4.23)} \qquad \square
\end{aligned}$$

Proof of Theorem 4.22   If $\varphi \in H^\infty$, then for all $f \in H^2$,

$$T_\psi T_\varphi f = T_\psi(\varphi f) = P(\psi \varphi f) = T_{\psi\varphi} f.$$

If $\psi \in \overline{H^\infty}$, then

$$(T_\psi T_\varphi)^* f = T_{\overline{\varphi}} T_{\overline{\psi}} f = T_{\overline{\varphi}}(\overline{\psi} f) = P(\overline{\varphi}\,\overline{\psi} f) = T_{\overline{\varphi}\overline{\psi}} f.$$

Thus $T_\psi T_\varphi = T_{\psi\varphi}$ by Proposition 4.14.

To prove the converse, suppose that $T_\psi T_\varphi$ is a Toeplitz operator. Then by Lemma 4.24 and Theorem 4.16,

$$T_\psi T_\varphi = S^* T_\psi T_\varphi S = T_\psi T_\varphi + P(\overline{\zeta}\psi) \otimes P(\overline{\zeta}\overline{\varphi}).$$

Thus we have

$$P(\overline{\zeta}\psi) \otimes P(\overline{\zeta}\overline{\varphi}) = 0,$$

which means that at least one of $P(\overline{\zeta}\psi)$ or $P(\overline{\zeta}\overline{\varphi})$ must be zero. If $P(\overline{\zeta}\psi) = 0$, then $\overline{\zeta}\psi \in \overline{\zeta}H^2$ and so $\psi \in \overline{H^2}$, that is, T_ψ is conjugate analytic. If $P(\overline{\zeta}\overline{\varphi}) = 0$, then $\overline{\zeta}\overline{\varphi} \in \overline{\zeta}H^2$ and so $\varphi \in H^2$, that is, T_φ is analytic. $\square$

Corollary 4.25 $T_\psi T_\varphi = 0$ *if and only if at least one of ψ or φ is zero.*

Proof   One direction is obvious. For the other, suppose $T_\varphi T_\psi = 0 = T_0$. Then, by Theorem 4.22, either $\psi \in \overline{H^\infty}$ or $\varphi \in H^\infty$ and $0 = T_\psi T_\varphi = T_{\psi\varphi}$. Theorem 4.13 implies that $\psi\varphi = 0$. Since, for any $f \in H^2 \setminus \{0\}$, we have $m(\{f = 0\}) = 0$ (Corollary 3.21), we conclude that at least one of ψ or φ must be the zero function. $\square$

4.5 The backward shift

The adjoint S^* of the unilateral shift S, which, by definition, satisfies

$$\langle Sf, g\rangle = \langle f, S^*g\rangle, \quad f, g \in H^2;$$

is called the *backward shift operator* (or simply the *backward shift*). The reason for this terminology stems from the following formula.

Proposition 4.26 *The operator $S^* : H^2 \to H^2$ satisfies*

$$[S^*f](z) = \frac{f(z) - f(0)}{z}, \tag{4.12}$$

or, in terms of Taylor coefficients $\{a_n\}_{n \geqslant 0}$ of $f \in H^2$,

$$S^*(a_0, a_1, \ldots) = (a_1, a_2, \ldots). \tag{4.13}$$

Proof First note that if f belongs to H^2, then $Sf = zf$ is orthogonal to any constant function. Therefore, for any $g \in H^2$,

$$\begin{aligned}
\langle f, S^*g\rangle = \langle Sf, g\rangle &= \langle zf, g\rangle \\
&= \langle zf, g - g(0)\rangle \\
&= \langle f, z(g - g(0))\rangle \\
&= \left\langle f, \frac{g - g(0)}{z}\right\rangle.
\end{aligned}$$

Since the preceding holds for all f, g in H^2, it follows that S^* is given by (4.12). The identity (4.13) follows immediately from (4.12). □

As we discussed with the shift, equating $z^n \in H^2$ with $\mathbf{e}_n \in \ell^2(\mathbb{N}_0)$, we see that the matrix representation of S^* with respect to the basis $\{\mathbf{e}_n\}_{n \geqslant 0}$ is

$$\begin{bmatrix}
0 & 1 & 0 & 0 & 0 & \cdots \\
0 & 0 & 1 & 0 & 0 & \cdots \\
0 & 0 & 0 & 1 & 0 & \cdots \\
0 & 0 & 0 & 0 & 1 & \cdots \\
0 & 0 & 0 & 0 & 0 & \cdots \\
\vdots & \vdots & \vdots & \vdots & \vdots & \ddots
\end{bmatrix}.$$

With this representation (or from (4.1)) we obtain

$$\ker S^* = \mathbb{C} \quad \text{and} \quad \operatorname{ran} S^* = H^2.$$

Corollary 4.27 $S = T_z$ and $S^* = T_{\bar{z}}$.

Proof   Note that $Sf = zf = P(zf) = T_z f$ for all $f \in H^2$. By Proposition 4.14, $S^* = T_z^* = T_{\bar{z}}$.   □

Like the unilateral shift, the backward shift is a constant source of counterexamples to naïve conjectures made by budding young operator theorists. First of all, S^* is right invertible but not left invertible since $S^*S = I$ while SS^* is the orthogonal projection onto the proper subspace $\bigvee\{z, z^2, \ldots\}$ of H^2. Moreover, S^* is a surjection that fails to be injective since S^* annihilates the constant functions. Proposition 4.19 combined with Corollary 4.27 imply that

$$S^* c_\lambda = \bar{\lambda} c_\lambda \tag{4.14}$$

for every λ in $\mathbb{D}$. This implies that S^* possesses uncountably many linearly independent eigenvectors, something that beginning graduate students often assume cannot occur in a Hilbert space of countable dimension (since no two Cauchy kernels are orthogonal, (4.14) does not provide us with an uncountable collection of mutually orthogonal vectors).

4.6 Difference quotient operator

For $f \in H^2$ and $\lambda \in \mathbb{D}$ the function

$$z \mapsto \frac{f(z) - f(\lambda)}{z - \lambda}$$

is certainly analytic on $\mathbb{D}$. Does it belong to H^2? It turns out that the answer is yes. Although the difference quotient above might seem elementary, it will play an important role when we discuss the spectrum of the compressed shift in Chapter 9.

For each $z, \lambda \in \mathbb{D}$, we have

$$(I - \bar{\lambda} S) c_\lambda(z) = \frac{1}{1 - \bar{\lambda} z} - \bar{\lambda} z \frac{1}{1 - \bar{\lambda} z} = 1 = c_0(z).$$

Hence, $(I - \bar{\lambda} S) c_\lambda = c_0$. Since $\|S\| = 1$ (Proposition 4.1), we know from Proposition 1.30 that $\sigma(S) \subset \mathbb{D}^-$ and so the operator $I - \bar{\lambda} S$ is invertible. Thus

$$c_\lambda = (I - \bar{\lambda} S)^{-1} c_0. \tag{4.15}$$

By (4.15), for each $f \in H^2$, we have

$$f(\lambda) = \langle f, c_\lambda \rangle = \langle f, (I - \bar{\lambda} S)^{-1} c_0 \rangle = \langle (I - \lambda S^*)^{-1} f, c_0 \rangle.$$

Define

$$Q_\lambda = (I - \lambda S^*)^{-1} S^*, \qquad \lambda \in \mathbb{D}. \tag{4.16}$$

Note that Q_λ is a bounded operator on H^2 and, in particular, $Q_0 = S^*$. It is also easy to see that the Neumann series

$$Q_\lambda = \sum_{n \geqslant 1} \lambda^{n-1} S^{*n} \qquad (4.17)$$

is convergent in the operator norm.

The following result shows why Q_λ is referred to as the *difference quotient operator*.

Lemma 4.28 *For each $f \in H^2$ and $\lambda \in \mathbb{D}$,*

$$(Q_\lambda f)(z) = \frac{f(z) - f(\lambda)}{z - \lambda}, \qquad z \in \mathbb{D} \setminus \{\lambda\}.$$

Proof We start with the following resolvent computation:

$$
\begin{aligned}
(I - zS^*)^{-1} - (I - \lambda S^*)^{-1} &= (I - zS^*)^{-1} I - I(I - \lambda S^*)^{-1} \\
&= (I - zS^*)^{-1}(I - \lambda S^*)(I - \lambda S^*)^{-1} \\
&\quad - (I - zS^*)^{-1}(I - zS^*)(I - \lambda S^*)^{-1} \\
&= (I - zS^*)^{-1}(-\lambda S^* + zS^*)(I - \lambda S^*)^{-1} \\
&= (z - \lambda)(I - zS^*)^{-1}(I - \lambda S^*)^{-1} S^*,
\end{aligned}
$$

which we write as

$$(I - zS^*)^{-1}(I - \lambda S^*)^{-1} S^* = \frac{(I - zS^*)^{-1} - (I - \lambda S^*)^{-1}}{z - \lambda}. \qquad (4.18)$$

We now have

$$
\begin{aligned}
(Q_\lambda f)(z) &= \langle Q_\lambda f, c_\lambda \rangle \\
&= \langle f, (I - \bar{\lambda} S)^{-1} c_0 \rangle \quad \text{(by (4.15))} \\
&= \langle (I - zS^*)^{-1}(Q_\lambda f), c_0 \rangle \\
&= \langle (I - zS^*)^{-1}(I - \lambda S^*)^{-1} S^* f, c_0 \rangle \\
&= \frac{1}{z - \lambda}\left(\langle (I - zS^*)^{-1} f, c_0 \rangle - \langle (I - \lambda S^*)^{-1} f, c_0 \rangle\right) \\
&= \frac{f(z) - f(\lambda)}{z - \lambda}. \quad \text{(by (4.18))} \qquad \qquad \square
\end{aligned}
$$

This next corollary is reminiscent of the "product rule" from calculus.

Corollary 4.29 *Let $\lambda \in \mathbb{D}$ and let $f, g \in H^2$ be such that $fg \in H^2$. Then*

$$Q_\lambda(fg) = f Q_\lambda g + g(\lambda) Q_\lambda f.$$

Proof   For each $z \in \mathbb{D}$, we have

$$Q_\lambda(fg)(z) = \frac{f(z)g(z) - f(\lambda)g(\lambda)}{z - \lambda}$$

$$= f(z)\frac{g(z) - g(\lambda)}{z - \lambda} + g(\lambda)\frac{f(z) - f(\lambda)}{z - \lambda}$$

$$= f(z)(Q_\lambda g)(z) + g(\lambda)(Q_\lambda f)(z) \quad \text{(by Lemma 4.28).} \qquad \square$$

Taking $\lambda = 0$ in Corollary 4.29 gives us the useful formula

$$S^*(fg) = fS^*g + g(0)S^*f. \tag{4.19}$$

4.7 Notes

4.7.1 Generalizations of Beurling's Theorem

Beurling's Theorem (Theorem 4.3) characterizes the non-zero S-invariant sub-spaces M of H^2 as $M = uH^2$ for some inner function u. The keys to proving this are (i) the factorization theorem (Theorem 3.20); (ii) the fact that $M \ominus S M$ is one dimensional and equal to $\mathbb{C}u$ for some inner u; (iii) the fact that $[M \ominus S M]$, the smallest S-invariant subspace containing $M \ominus S M$, is equal to M.

This last phenomenon appears to be ubiquitous. Recall the definitions of the Bergman space L_a^2 and the Dirichlet space $\mathcal{D}$ from the end notes of the previous chapter. Also recall that the zero sets of these spaces are not so well understood (unlike H^2, where the zero sets are the Blaschke sequences). The S-invariant subspaces are not well understood either. For the Dirichlet space, it is true that for any S-invariant subspace M of $\mathcal{D}$, the space $M \ominus S M$ is still one dimensional. However, it is not generated by an inner function but rather by a multiplier for $\mathcal{D}$. It is the case that $M = [M \ominus S M]$.

For the Bergman space L_a^2, the S-invariant subspaces can show exceptional behavior. Unlike H^2 and $\mathcal{D}$, where $M \ominus S M$ is one dimensional, for the Bergman space, given any $n \in \mathbb{N} \cup \{\infty\}$, there are S-invariant subspaces M of L_a^2 such that $\dim(M \ominus S M) = n$. It is still the case that $M = [M \ominus S M]$. Good references for the shift on the Bergman and Dirichlet spaces, along with more of a discussion of the ubiquity of $M = [M \ominus S M]$, are [64, 109, 151, 175].

4.8 For further exploration

4.8.1 Invariant subspaces of Toeplitz operators

Since $T_z = S$, the invariant subspaces of the Toeplitz operator T_z are understood via Beurling's theorem. For other Toeplitz operators T_φ, the description

of the invariant subspaces is wide open. In fact, it is unclear whether or not
every Toeplitz operator has a proper, nontrivial invariant subspace. Some of
them do [145, 150]. An even more difficult, but exciting, problem is to describe
the invariant subspaces for a particular Toeplitz operator. Even for an analytic
univalent symbol, this problem is challenging. Indeed, if $\varphi \in H^\infty$ is univalent,
then by means of conformal mapping, one can show that T_φ is unitarily equivalent to M_z (multiplication by z) on the Hardy space of $\varphi(\mathbb{D})$. Characterizing
the M_z-invariant subspaces of the Hardy space of a general domain is a difficult and open problem which has only been worked out in a handful of cases
[13, 16, 111].

5

Model spaces

In this chapter, we introduce our primary object of study, the model subspaces of H^2. We first identify these subspaces as the proper invariant subspaces of H^2 for the backward shift operator and then proceed to investigate their reproducing kernels and other basic properties.

5.1 Model spaces as invariant subspaces

The following definition of a model space is, at first, somewhat underwhelming; for it does not indicate the presence of any interesting phenomena from the perspective of function theory, nor does it suggest the existence of any interesting operators defined on these spaces. The rest of this book is dedicated to fleshing out these two lines of inquiry.

Definition 5.1 If u is an inner function, the corresponding *model space* $\mathcal{K}_u$ is defined to be

$$\mathcal{K}_u := (uH^2)^\perp = \{f \in H^2 : \langle f, uh \rangle = 0 \ \forall h \in H^2\}. \tag{5.1}$$

Just as the subspaces uH^2 constitute the non-trivial invariant subspaces for the unilateral shift $Sf = zf$ on H^2 (Theorem 4.3), the subspaces $\mathcal{K}_u$ play an analogous role for the backward shift

$$S^*f = \frac{f - f(0)}{z}.$$

Proposition 5.2 *The model spaces $\mathcal{K}_u$ are precisely the proper S^*-invariant subspaces of H^2.*

Proof If M is a proper (that is, $M \neq H^2$) S^*-invariant subspace of H^2, then $\langle f, Sg \rangle = \langle S^*f, g \rangle = 0$ whenever $f \in M$ and $g \in M^\perp$. Since Sg is orthogonal to every vector in $\mathcal{M}$, it follows that $Sg \in \mathcal{M}^{\perp}$. In other words, $\mathcal{M}^{\perp}$ is a non-trivial S-invariant subspace and hence, in light of Beurling's Theorem (Theorem 4.3), it must be of the form uH^2 for some inner function u. Taking orthogonal complements shows that $\mathcal{M} = (uH^2)^{\perp}$. □

Remark 5.3 Note from Remark 4.4 that $\mathcal{K}_{u_1} = \mathcal{K}_{u_2}$ if and only if $u_1 = e^{i\theta}u_2$. Also note that $\mathcal{K}_u = \{0\}$ if and only if u is a constant function.

The term *model space* originates in the theory of model operators, developed by Sz.-Nagy and Foiaș, where it is shown that certain types of Hilbert space contractions are unitarily equivalent to the compressions of the unilateral shift to a model space. This underscores the importance of model spaces in developing concrete, function-theoretic realizations of abstract Hilbert space operators. We will explore this important subject in Chapter 9.

Unlike the Beurling subspaces uH^2, for which membership is easily determined using inner–outer factorization (Theorem 3.20), it is often difficult to tell whether or not a given function belongs to a particular (or any) model space. We will discuss this in great detail in Chapter 7. For the moment, we content ourselves with the following important description of $\mathcal{K}_u$ in terms of boundary values.

Proposition 5.4 *For an inner function u, the model space $\mathcal{K}_u$ is the set of all $f \in H^2$ such that*

$$f = \overline{gz}u$$

a.e. on $\mathbb{T}$ for some $g \in H^2$. In other words,

$$\mathcal{K}_u = H^2 \cap u\overline{zH^2},$$

where the right-hand side above is regarded, via non-tangential boundary values, as a set of functions on $\mathbb{T}$.

Proof For each $f \in H^2$, we see that

$$\begin{aligned} f \in (uH^2)^{\perp} &\iff \langle f, uh \rangle = 0 \; \forall h \in H^2 \\ &\iff \langle \overline{u}f, h \rangle = 0 \; \forall h \in H^2 \\ &\iff \overline{u}f \in \overline{zH^2}. \qquad \text{(by (3.8))} \end{aligned}$$

Since $u\overline{u} = 1$ a.e. on $\mathbb{T}$, it follows that $f \in H^2$ belongs to $(uH^2)^{\perp}$ if and only if $f \in u\overline{zH^2}$ (that is, $f = \overline{gz}u$ for some g in H^2). □

In fact, it turns out that the function g from the preceding proposition belongs to $\mathcal{K}_u$ as well. Indeed, the condition $f = \overline{gz}u$ is equivalent to $g = \overline{fz}u$ since $\overline{z}u$ is of unit modulus a.e. on $\mathbb{T}$. This observation allows us to define a conjugation (a certain type of conjugate-linear operator) on $\mathcal{K}_u$ that interacts in surprisingly fruitful ways with a number of linear operators defined on model spaces. We will examine the ramifications of this observation in Chapter 8.

5.2 Stability under conjugate analytic Toeplitz operators

Being $T_{\overline{z}}$-invariant already (see Corollary 4.27), it should come as no surprise that the model spaces $\mathcal{K}_u$ are also invariant under conjugate analytic Toeplitz operators $\{T_{\overline{\varphi}} : \varphi \in H^\infty\}$.

Proposition 5.5 *If $\varphi \in H^\infty$ and u is inner, then $T_{\overline{\varphi}}\mathcal{K}_u \subset \mathcal{K}_u$.*

Proof   For each $f \in \mathcal{K}_u$ and $h \in H^2$, we use the identity $T_\varphi^* = T_{\overline{\varphi}}$ (Proposition 4.14) to obtain

$$\langle T_{\overline{\varphi}}f, uh \rangle = \langle T_\varphi^* f, uh \rangle = \langle f, T_\varphi(uh) \rangle = \langle f, P(\varphi uh) \rangle = \langle f, u(\varphi h) \rangle = 0.$$

Thus $T_{\overline{\varphi}}f \in (uH^2)^\perp = \mathcal{K}_u$. $\square$

Using Proposition 5.5, one can show that the spaces $\mathcal{K}_u$ enjoy the so-called F-property. In general, a subset $\mathscr{C}$ of functions contained in H^2 has the F-*property* if whenever an inner function θ divides an $f \in \mathscr{C}$ (here we mean that θ divides, in the sense of Definition 2.16, the inner factor of f), the quotient f/θ also belongs to $\mathscr{C}$. Good sources for this are [176, 177]. Observe that H^2 enjoys the F-property (Theorem 3.20). It is precisely the stability of $\mathcal{K}_u$ under the action of conjugate analytic Toeplitz operators that allows us to remove the inner factor of any function in $\mathcal{K}_u$ without leaving $\mathcal{K}_u$.

Proposition 5.6 *If $f \in \mathcal{K}_u$ and θ is an inner function that divides f, then $f/\theta \in \mathcal{K}_u$. In particular, the outer factor of any function in $\mathcal{K}_u$ also belongs to $\mathcal{K}_u$.*

Proof   For any $f \in \mathcal{K}_u$, observe that

$$T_{\overline{\theta}}f = P(\overline{\theta}f) = P\left(\frac{f}{\theta}\right) = \frac{f}{\theta}$$

since $f/\theta \in H^2$. By Proposition 5.5, $f/\theta = T_{\overline{\theta}}f \in \mathcal{K}_u$. $\square$

In Proposition 7.15 we will give a description of the cyclic vectors for S^*, that is, those $f \in H^2$ for which

$$\bigvee \{S^{*n} f : n \geqslant 0\} = H^2.$$

For now though, we mention this next result which links the concepts of cyclicity and "stability under conjugate analytic Toeplitz operators."

Corollary 5.7 *Let $f \in H^2$ and $\varphi \in H^\infty \setminus \{0\}$. Then the following are equivalent:*

(i) f is cyclic for S^*;
(ii) $T_{\bar{\varphi}} f$ is cyclic for S^*.

Proof   We will prove that f is not cyclic for S^* if and only if $T_{\bar{\varphi}} f$ is not cyclic for S^*. Since $S^{*n} = T_{\bar{z}^n}$ for each $n \geqslant 0$, we have (Theorem 4.22)

$$S^{*n} T_{\bar{\varphi}} = T_{\bar{\varphi}} S^{*n}. \tag{5.2}$$

This is used below.

First assume that $T_{\bar{\varphi}} f$ is not cyclic for S^*, that is,

$$\bigvee \{S^{*n} T_{\bar{\varphi}} f : n \geqslant 0\} \neq H^2.$$

By (5.2) we see that if

$$g = \sum_{0 \leqslant j \leqslant N} c_j S^{*j} f,$$

then

$$T_{\bar{\varphi}} g = T_{\bar{\varphi}} \left(\sum_{0 \leqslant j \leqslant N} c_j S^{*j} f \right) = \sum_{0 \leqslant j \leqslant N} c_j T_{\bar{\varphi}} S^{*j} f = \sum_{0 \leqslant j \leqslant N} c_j S^{*j} T_{\bar{\varphi}} f.$$

This implies that

$$T_{\bar{\varphi}} \left(\bigvee \{S^{*n} f : n \geqslant 0\} \right) \subset \bigvee \{S^{*n} T_{\bar{\varphi}} f : n \geqslant 0\} \neq H^2. \tag{5.3}$$

At this point, (5.3), along with the fact that $T_{\bar{\varphi}}$ has dense range (Corollary 4.20), combine to give us

$$\bigvee \{S^{*n} f : n \geqslant 0\} \neq H^2,$$

that is, f is not a cyclic vector for S^*.

Conversely, assume that f is not cyclic for S^*; in other words,

$$\mathcal{M} = \bigvee \{S^{*n} f : n \geqslant 0\} \neq H^2.$$

Since $\mathcal{M}$ is a proper S^*-invariant subspace, there is an inner function u such that $\mathcal{M} = \mathcal{K}_u$ (Proposition 5.2). By Theorem 5.5, it follows that $T_{\overline{\varphi}}\mathcal{M} \subset \mathcal{M}$. In particular, $T_{\overline{\varphi}}f \in \mathcal{M}$. Therefore, $T_{\overline{\varphi}}f$ is not cyclic for S^*.   □

Here is another interesting relationship between model spaces and conjugate analytic Toeplitz operators.

Proposition 5.8 *Let $\varphi \in H^\infty \setminus \{0\}$, and let u be the inner factor of φ. Then*

$$\ker T_{\overline{\varphi}} = \mathcal{K}_u.$$

In particular, if φ is an outer function, then $T_{\overline{\varphi}}$ is injective.

Proof   First assume that φ is outer and let $f \in \ker T_{\overline{\varphi}}$. This means that $\overline{\varphi}f \in \overline{zH^2}$, which is equivalent to $\varphi\overline{f} \in zH^2$. Since φ is outer and $\overline{f} \in L^2$, we deduce that $\overline{f} \in zH^2$. To be more specific, since $\varphi\overline{f} = zh$ a.e. on $\mathbb{T}$ for some $h \in H^2$, we see that $zh/\varphi \in N^+$ and has L^2 boundary values. By Smirnov's Theorem (Theorem 3.29), $zh/\varphi \in zH^2$, whence $\overline{f} \in zH^2$. Putting this all together, we see that $f \in H^2 \cap zH^2 = \{0\}$. That is to say, $T_{\overline{\varphi}}$ is injective whenever φ is an outer function. We will use this fact below.

For the general case, let $\varphi = uh$, where u is inner and h is outer (Theorem 3.20). Then

$$
\begin{aligned}
\ker T_{\overline{\varphi}} &= \{f \in H^2 : T_{\overline{\varphi}}f = 0\} \\
&= \{f \in H^2 : T_{\overline{h}}T_{\overline{u}}f = 0\} \\
&= \{f \in H^2 : T_{\overline{u}}f = 0\} \qquad\qquad (T_{\overline{h}} \text{ is injective}) \\
&= \{f \in H^2 : \overline{u}f \in \overline{zH^2}\} \\
&= H^2 \cap u\overline{zH^2} \\
&= \mathcal{K}_u,
\end{aligned}
$$

which completes the proof.   □

5.3 Containment and lattice operations

As a consequence of Proposition 5.2, we now show that the lattice of non-trivial S^*-invariant subspaces mirrors the lattice of inner functions, partially ordered by divisibility. Recall Definition 2.16 and the discussion of sums and intersections of S-invariant subspaces from Chapter 4. In particular, recall that $A \vee B$ denotes the closed linear span of the subspaces A and B.

Corollary 5.9 *If u and v are inner functions, then*

(i) $\mathcal{K}_u \subset \mathcal{K}_v \iff u|v$;
(ii) $\mathcal{K}_u \cap \mathcal{K}_v = \mathcal{K}_{\gcd(u,v)}$;
(iii) $\mathcal{K}_u \vee \mathcal{K}_v = \mathcal{K}_{\mathrm{lcm}(u,v)}$.

Proof   For the proof of (i), note that $u|v$ if and only if $vH^2 \subset uH^2$ (Theorem 4.7), which occurs if and only if

$$\mathcal{K}_u = (uH^2)^\perp \subset (vH^2)^\perp = \mathcal{K}_v.$$

To prove (ii), observe that

$$(\mathcal{K}_u \cap \mathcal{K}_v)^\perp = \mathcal{K}_u^\perp \vee \mathcal{K}_v^\perp = uH^2 \vee vH^2 = wH^2,$$

where $w = \gcd(u,v)$ (Corollary 4.9). Taking orthogonal complements completes the proof of (ii).

To prove (iii), observe that

$$(\mathcal{K}_u \vee \mathcal{K}_v)^\perp = \mathcal{K}_u^\perp \cap \mathcal{K}_v^\perp = uH^2 \cap vH^2$$

and that $uH^2 \cap vH^2 \neq \{0\}$ since it contains uv. By Corollary 4.8, this space is equal to wH^2 where $w = \mathrm{lcm}(u,v)$. To finish, note that

$$\mathcal{K}_u \vee \mathcal{K}_v = (uH^2 \cap vH^2)^\perp = (wH^2)^\perp = \mathcal{K}_w. \qquad \square$$

5.4 A decomposition for $\mathcal{K}_u$

How is the factorization of an inner function realized in the corresponding model space? We start with a simple result.

Lemma 5.10 *Let u_1 and u_2 be inner. Then*

$$\mathcal{K}_{u_1 u_2} = \mathcal{K}_{u_1} \oplus u_1 \mathcal{K}_{u_2},$$

where $\oplus$ denotes the orthogonal sum.

Proof   Since u_1 divides $u_1 u_2$, we have $\mathcal{K}_{u_1} \subset \mathcal{K}_{u_1 u_2}$ (Corollary 5.9). Moreover, since $u_1 \mathcal{K}_{u_2} \subset u_1 H^2$ and $\mathcal{K}_{u_1} \perp u_1 H^2$, we conclude that $\mathcal{K}_{u_1} \perp u_1 \mathcal{K}_{u_2}$. Let us now verify that $u_1 \mathcal{K}_{u_2} \subset \mathcal{K}_{u_1 u_2}$. This will imply

$$\mathcal{K}_{u_1} \oplus u_1 \mathcal{K}_{u_2} \subset \mathcal{K}_{u_1 u_2}.$$

Let $f \in u_1 \mathcal{K}_{u_2}$ and observe that $f = u_1 g$, where $g \in \mathcal{K}_{u_2}$. Therefore, for each $h \in H^2$, we have

$$\langle f, u_1 u_2 h \rangle = \langle u_1 g, u_1 u_2 h \rangle = \langle g, u_2 h \rangle = 0.$$

The last equality is a consequence of the assumption $g \in \mathcal{K}_{u_2} = (u_2 H^2)^{\perp}$. Hence $f \perp u_1 u_2 H^2$, which means $f \in \mathcal{K}_{u_1 u_2}$.

To establish the reverse inclusion $\mathcal{K}_{u_1 u_2} \subset \mathcal{K}_{u_1} \oplus u_1 \mathcal{K}_{u_2}$, take any $f \in \mathcal{K}_{u_1 u_2}$. By definition, $\langle f, u_1 u_2 g \rangle = 0$ for all $g \in H^2$. Since $H^2 = u_1 H^2 \oplus \mathcal{K}_{u_1}$, there are functions $h \in H^2$ and $k \in \mathcal{K}_{u_1}$ such that $f = u_1 h + k$. Therefore, for each $g \in H^2$,

$$\begin{aligned} 0 = \langle f,\, u_1 u_2 g \rangle &= \langle u_1 h,\, u_1 u_2 g \rangle + \langle k,\, u_1 u_2 g \rangle \\ &= \langle h,\, u_2 g \rangle + \langle k,\, u_1(u_2 g) \rangle \\ &= \langle h,\, u_2 g \rangle. \end{aligned}$$

This means that $h \in \mathcal{K}_{u_2}$. In other words, $f \in \mathcal{K}_{u_1} \oplus u_1 \mathcal{K}_{u_2}$. $\square$

A more sophisticated extension of Lemma 5.10 is due to Ahern and Clark [3] and Kriete [120].

Theorem 5.11 *Let $\{u_n\}_{n \geqslant 1}$ be a sequence of inner functions such that the product*

$$u = \prod_{n \geqslant 1} u_n$$

converges uniformly on compact sets subsets of $\mathbb{D}$ and is an inner function. Then

$$\mathcal{K}_u = \bigoplus_{n \geqslant 1} (u_1 u_2 \cdots u_{n-1}) \mathcal{K}_{u_n}.$$

Proof   By induction, Lemma 5.10, and Corollary 5.9, we have

$$\mathcal{K}_{u_1} \oplus u_1 \mathcal{K}_{u_2} \oplus u_1 u_2 \mathcal{K}_{u_3} \oplus \cdots \oplus (u_1 u_2 \cdots u_{n-1}) \mathcal{K}_{u_n} = \mathcal{K}_{u_1 u_2 \cdots u_n} \subset \mathcal{K}_u.$$

Hence

$$\bigoplus_{n \geqslant 1} (u_1 u_2 \cdots u_{n-1}) \mathcal{K}_{u_n} \subset \mathcal{K}_u. \tag{5.4}$$

Write

$$u = u_1 \cdots u_n v_n$$

where

$$v_n = \frac{u}{u_1 u_2 \cdots u_n},$$

and observe that v_n is inner. Apply the argument above to see that

$$\mathcal{K}_u = \mathcal{K}_{u_1} \oplus u_1\mathcal{K}_{u_2} \oplus u_1 u_2 \mathcal{K}_{u_3} \oplus \cdots \oplus (u_1 u_2 \cdots u_{n-1})\mathcal{K}_{u_n} \oplus (u_1 u_2 \cdots u_n)\mathcal{K}_{v_n}.$$

This representation shows that if $f \subset \mathcal{K}_u$ satisfies

$$f \perp (u_1 u_2 \cdots u_{n-1})\mathcal{K}_{u_n}, \qquad \forall n \geqslant 1,$$

then $f \in (u_1 u_2 \cdots u_n)\mathcal{K}_{v_n}$ for all $n \geqslant 1$. This implies that $(u_1 u_2 \cdots u_n)|f$ for all $n \geqslant 1$. Passing to the limit, we see that $u|f$ and thus $f \in uH^2$. However, we are already assuming that $f \in \mathcal{K}_u$ and so $f \in \mathcal{K}_u^{\perp} \cap \mathcal{K}_u = \{0\}$. Hence equality holds in (5.4). $\qquad\square$

5.5 Reproducing kernels

As a subspace of the reproducing kernel Hilbert space H^2 (whose reproducing kernel is the Cauchy kernel c_λ), each model space $\mathcal{K}_u$ itself possesses a reproducing kernel, which we now compute.

If $f \in \mathcal{K}_u$ and $\lambda \in \mathbb{D}$, it follows that

$$f(\lambda) = \langle f, c_\lambda \rangle = \langle f, c_\lambda \rangle - u(\lambda)\langle f, uc_\lambda \rangle = \langle f, (1 - \overline{u(\lambda)}u)c_\lambda \rangle.$$

Furthermore, $(1 - \overline{u(\lambda)}u)c_\lambda \in \mathcal{K}_u$ since, for any $h \in H^2$,

$$\begin{aligned}
\langle uh, (1 - \overline{u(\lambda)}u)c_\lambda \rangle &= u(\lambda)h(\lambda) - u(\lambda)\langle uh, uc_\lambda \rangle \\
&= u(\lambda)h(\lambda) - u(\lambda)\langle h, c_\lambda \rangle \\
&= u(\lambda)h(\lambda) - u(\lambda)h(\lambda) \\
&= 0.
\end{aligned}$$

In other words, for each $\lambda \in \mathbb{D}$, $k_\lambda = (1 - \overline{u(\lambda)}u)c_\lambda \in \mathcal{K}_u$ and satisfies the reproducing formula

$$f(\lambda) = \langle f, k_\lambda \rangle, \qquad f \in \mathcal{K}_u.$$

More explicitly, the function

$$k_\lambda(z) = \frac{1 - \overline{u(\lambda)}u(z)}{1 - \overline{\lambda}z}, \qquad \lambda, z \in \mathbb{D}, \tag{5.5}$$

is called the *reproducing kernel* for $\mathcal{K}_u$. When dealing with more than one model space at a time, the notation k_λ^u is often used to denote the reproducing kernel for $\mathcal{K}_u$. Observe that $k_\lambda(z)$ is analytic in the variable z and conjugate analytic in the variable λ.

If $\{e_n\}_{n\geqslant 1}$ is any orthonormal basis for $\mathcal{K}_u$, then

$$k_\lambda = \sum_{n\geqslant 1} \langle k_\lambda, e_n \rangle e_n,$$

where the sum converges in the norm of $\mathcal{K}_u$. Since norm convergence implies pointwise convergence on $\mathbb{D}$ (Proposition 3.4), we conclude that

$$k_\lambda(z) = \sum_{n\geqslant 1} \overline{e_n(\lambda)} e_n(z), \qquad \lambda, z \in \mathbb{D}. \tag{5.6}$$

The preceding argument, which uses no specific information about $\mathcal{K}_u$, applies to any reproducing kernel Hilbert space [2, 18, 144].

Example 5.12　If $u = z^n$, then

$$\mathcal{K}_u = \bigvee \{1, z, z^2, \ldots, z^{n-1}\},$$

since, via Fourier series, the orthogonal complement of $z^n H^2$ in H^2 is the space of polynomials of degree at most $n-1$. In light of (5.5), we see that

$$k_\lambda(z) = \frac{1 - \overline{\lambda}^n z^n}{1 - \overline{\lambda} z}. \tag{5.7}$$

On the other hand, applying (5.6) to the orthonormal basis

$$\{1, z, z^2, \ldots, z^{n-1}\}$$

for $\mathcal{K}_u$, tells us that

$$k_\lambda(z) = 1 + \overline{\lambda}z + \overline{\lambda}^2 z^2 + \cdots + \overline{\lambda}^{n-1} z^{n-1} = \frac{1 - (\overline{\lambda}z)^n}{1 - \overline{\lambda}z},$$

which is precisely the expression in (5.7).

5.6 The projection P_u

Via non-tangential boundary values, we can regard $\mathcal{K}_u$ as a closed subspace of L^2. We let P_u denote the orthogonal projection of L^2 onto $\mathcal{K}_u$.

Proposition 5.13　*For each $f \in L^2$ and $\lambda \in \mathbb{D}$,*

$$(P_u f)(\lambda) = \langle f, k_\lambda \rangle. \tag{5.8}$$

Moreover, $k_\lambda = P_u c_\lambda$.

Proof   Since $k_\lambda \in \mathcal{K}_u$, we have $P_u k_\lambda = k_\lambda$ and so

$$\langle f, k_\lambda \rangle = \langle f, P_u k_\lambda \rangle = \langle P_u f, k_\lambda \rangle = (P_u f)(\lambda).$$

For the second statement, note that for any $f \in \mathcal{K}_u$,

$$\langle f, k_\lambda \rangle = f(\lambda) = \langle f, c_\lambda \rangle = \langle P_u f, c_\lambda \rangle = \langle f, P_u c_\lambda \rangle,$$

which proves that $k_\lambda = P_u c_\lambda$. $\square$

The following simple relationship between the Riesz projection P (the orthogonal projection of L^2 onto H^2) and P_u often comes in handy.

Proposition 5.14 $P_u f = f - u P(\overline{u} f)$ *for all* $f \in H^2$.

Proof   For $f \in H^2$ and $\lambda \in \mathbb{D}$, Proposition 5.13 tells us that

$$
\begin{aligned}
(P_u f)(\lambda) &= \langle f, k_\lambda \rangle \\
&= \langle f, (1 - \overline{u(\lambda)}u) c_\lambda \rangle \\
&= \langle f, c_\lambda \rangle - u(\lambda) \langle f, u c_\lambda \rangle \\
&= \langle f, c_\lambda \rangle - u(\lambda) \langle \overline{u} f, c_\lambda \rangle \\
&- \langle f, c_\lambda \rangle - u(\lambda) \langle \overline{u} f, P c_\lambda \rangle \\
&= \langle f, c_\lambda \rangle - u(\lambda) \langle P(\overline{u} f), c_\lambda \rangle \\
&= f(\lambda) - u(\lambda) P(\overline{u} f)(\lambda).
\end{aligned}
$$
     $\square$

Using the projection P_u, we can prove the following curious result.

Proposition 5.15   *If u is an inner function, then $S^{*n} u \in \mathcal{K}_u$ for all $n \geqslant 1$. Furthermore, for each $f \in \mathcal{K}_u$ we have*

$$f = \sum_{n \geqslant 1} \langle f, S^{*n} u \rangle S^{*n} u, \tag{5.9}$$

where the sum above converges in the L^2 norm, and

$$\|f\|^2 = \sum_{n \geqslant 1} |\langle f, S^{*n} u \rangle|^2. \tag{5.10}$$

In particular,

$$\mathcal{K}_u = \bigvee \{S^{*n} u : n \geqslant 1\}, \tag{5.11}$$

so that $\mathcal{K}_u$ is precisely the S^*-invariant subspace generated by $S^* u$.

Proof   First note that the set $\{\overline{\zeta}^n u\}_{n\in\mathbb{Z}}$ is an orthonormal basis for L^2. Indeed, using the fact that $u\overline{u} = 1$ a.e. on $\mathbb{T}$ yields

$$\langle \overline{\zeta}^n u, \overline{\zeta}^m u\rangle = \langle \overline{\zeta}^n, \overline{\zeta}^m u\overline{u}\rangle = \int_{\mathbb{T}} \zeta^{m-n} dm(\zeta) - \delta_{mn}. \tag{5.12}$$

Thus $\{\overline{\zeta}^n u\}_{n\in\mathbb{Z}}$ is an orthonormal set. To prove it is a basis, suppose that $f \in L^2$ and

$$\langle f, \overline{\zeta}^n u\rangle = 0, \qquad n \in \mathbb{Z}.$$

By the uniqueness of Fourier coefficients (Proposition 1.16), $f\overline{u} = 0$ a.e. on $\mathbb{T}$. However, since u is unimodular a.e. on $\mathbb{T}$ we have $f = 0$.

Second, we have

$$S^{*n}u \in \mathcal{K}_u, \qquad n \geqslant 1. \tag{5.13}$$

To see this, observe that for any $h \in H^2$,

$$\langle S^*u, uh\rangle = \langle u, \zeta uh\rangle = \langle 1, \zeta h\rangle = \overline{\widehat{h}(-1)} = 0$$

(Theorem 3.12). Thus $S^*u \in (uH^2)^\perp = \mathcal{K}_u$. Now use the S^*-invariance of $\mathcal{K}_u$ to conclude that $S^{*n}u \in \mathcal{K}_u$ for all $n \geqslant 1$.

Third, notice the identity

$$S^{*n}u = P_u(\overline{\zeta}^n u), \qquad n \geqslant 1. \tag{5.14}$$

Indeed, we have already shown $S^{*n}u \in \mathcal{K}_u$ for all $n \geqslant 1$ and so we can apply Proposition 5.13 to see that for any $\lambda \in \mathbb{D}$,

$$(S^{*n}u)(\lambda) = \langle S^{*n}u, k_\lambda\rangle = \langle u, S^n k_\lambda\rangle$$
$$= \langle u, \zeta^n k_\lambda\rangle = \langle \overline{\zeta}^n u, k_\lambda\rangle$$
$$= \langle \overline{\zeta}^n u, P_u k_\lambda\rangle$$
$$= P_u(\overline{\zeta}^n u)(\lambda).$$

To prove (5.9) note that for $f \in \mathcal{K}_u$ we have

$$f = P_u f$$
$$= P_u\Big(\sum_{n\in\mathbb{Z}}\langle f, \overline{\zeta}^n u\rangle \overline{\zeta}^n u\Big)$$
$$= \sum_{n\geqslant 1}\langle f, \overline{\zeta}^n u\rangle P_u(\overline{\zeta}^n u)$$
$$= \sum_{n\geqslant 1}\langle P_u f, \overline{\zeta}^n u\rangle S^{*n}u \qquad \text{(by (5.14))}$$
$$= \sum_{n\geqslant 1}\langle f, P_u(\overline{\zeta}^n u)\rangle S^{*n}u$$
$$= \sum_{n\geqslant 1}\langle f, S^{*n}u\rangle S^{*n}u, \qquad \text{(by (5.14))}$$

where the convergence is in the H^2 norm. In the first step above, notice how we used the fact that $\{\overline{\zeta}^n u\}_{n \in \mathbb{Z}}$ is an orthonormal basis of L^2. This establishes (5.9).

We now prove (5.10). By Proposition 5.4, we can write $f = \overline{gz}u$ for some g in H^2. Therefore, the coefficients in (5.9) are given by

$$\langle f, S^{*n} u \rangle = \langle \zeta^n f, u \rangle = \langle \zeta^{n-1} \overline{g} u, u \rangle = \langle \zeta^{n-1}, g \rangle, \qquad (5.15)$$

which is precisely the complex conjugate of the $(n-1)$st Taylor coefficient of g. Thus

$$\begin{aligned}
\|f\|^2 &= \|\overline{gz}u\|^2 \\
&= \|g\|^2 \\
&= \sum_{n \geqslant 1} |\langle g, \zeta^{n-1} \rangle|^2 \qquad \text{(Parseval's Theorem)} \\
&= \sum_{n \geqslant 1} |\langle f, S^{*n} u \rangle|^2, \qquad \text{(by (5.15))}
\end{aligned}$$

which proves (5.10). The identity in (5.11) now follows.   □

Although the expansion in (5.9) closely resembles the expansion of a vector with respect to an orthonormal basis, we are dealing with something that is altogether different. First of all, the vectors $\{S^{*n} u : n \geqslant 1\}$ do not form an orthornormal set. For instance, all but finitely many of them have norm strictly less than 1 (see Lemma 9.12). Even more concerning is the fact that the vectors $S^{*n} u$ are not, in general, even linearly independent (this occurs when $\mathcal{K}_u$ is finite dimensional, see Section 5.7). Nevertheless, we still have the equality (5.10) that one would expect when dealing with an orthonormal basis. In the terminology of frame theory, the vectors $\{S^{*n} : n \geqslant 1\}$ form a *tight frame* for $\mathcal{K}_u$ [40].

We also point out that if one just wanted to prove (5.11), one could argue that if $f \in \mathcal{K}_u$ with $\langle S^{*n} u, f \rangle = 0$ for all $n \geqslant 1$, then

$$0 = \langle S^{*n} u, f \rangle = \langle u, \zeta^n f \rangle = \langle u \overline{f}, \zeta^n \rangle, \quad n \geqslant 1.$$

This means that $f\overline{u} \in H^2$ and so $f \in \mathcal{K}_u \cap u H^2 = \{0\}$. The significance of Proposition 5.15 is that it actually gives an explicit formula for f in terms of the vectors $\{S^{*n} u : n \geqslant 1\}$.

5.7 Finite-dimensional model spaces

Perhaps the simplest examples of model spaces are those corresponding to finite Blaschke products. For a finite Blaschke product u, one can dispense with the unimodular constants attached to each Blaschke factor (since convergence is not an issue here) and, for our present purposes, assume that u takes the simpler form

$$u(z) = \prod_{1 \leqslant j \leqslant N} \frac{z - \lambda_j}{1 - \overline{\lambda}_j z}, \qquad \lambda_j \in \mathbb{D}.$$

These inner functions yield the only model spaces that are finite dimensional (see Proposition 5.19). Moreover, the vectors in $\mathcal{K}_u$ can be completely characterized in an explicit fashion.

Recall from (3.4) and (3.5) the formula

$$f^{(n)}(\lambda) = \langle f, c_\lambda^{(n)} \rangle, \quad f \in H^2, \ \lambda \in \mathbb{D},$$

where for each integer $n \geqslant 0$,

$$c_\lambda^{(n)}(z) = \frac{n! z^n}{(1 - \overline{\lambda} z)^{n+1}}.$$

Proposition 5.16 *If u is a finite Blaschke product with distinct zeros*

$$\lambda_1, \lambda_2, \ldots, \lambda_n$$

having corresponding multiplicities $m_1, m_2, \ldots, m_n$, then

$$\mathcal{K}_u = \bigvee \{c_{\lambda_i}^{(\ell_i - 1)} : 1 \leqslant i \leqslant n, \ 1 \leqslant \ell_i \leqslant m_i\}. \tag{5.16}$$

In particular, $\dim \mathcal{K}_u = m_1 + m_2 + \cdots + m_n$.

Proof   First observe that each of the functions

$$c_{\lambda_i}, c'_{\lambda_i}, c''_{\lambda_i}, \ldots, c_{\lambda_i}^{(m_i - 1)}$$

belongs to $\mathcal{K}_u$. Indeed, when $1 \leqslant \ell_i \leqslant m_i$,

$$\begin{aligned}
\langle uh, c_{\lambda_i}^{(\ell_i - 1)} \rangle &= (uh)^{(\ell_i - 1)}(\lambda_i) \\
&= \sum_{0 \leqslant j \leqslant \ell_i - 1} \binom{\ell_i - 1}{j} u^{(j)}(\lambda_i) h^{(\ell_i - 1 - j)}(\lambda_i) \\
&= 0,
\end{aligned}$$

since $u^{(j)}(\lambda_i) = 0$ for $0 \leqslant j \leqslant m_i - 1$. This establishes the containment $\supset$ in (5.16). On the other hand, if $f \in H^2$ is orthogonal to $c_{\lambda_i}^{(\ell_i - 1)}$ for all $1 \leqslant i \leqslant n$ and $1 \leqslant \ell_i \leqslant m_i$, then f has a zero of order m_i at λ_i (see (3.4)) and so $u | f$. In other words, $f \in uH^2 = \mathcal{K}_u^{\perp}$, yielding, via orthogonal complements, the containment $\subset$ in (5.16). $\square$

Example 5.17 If $u = z^n$, then u has a zero of multiplicity n at $z = 0$. Thus

$$\mathcal{K}_u = \bigvee \{c_0^{(0)}, c_0^{(1)}, \ldots, c_0^{(n-1)}\}.$$

However,

$$c_0^{(j)}(z) = j! z^j$$

and so $\mathcal{K}_u = \bigvee \{1, z, \ldots, z^{n-1}\}$. One can also see this directly by observing, via Fourier series, that $(z^n H^2)^\perp$ consists of the polynomials of degree at most $n - 1$. We already observed this fact in Example 5.12.

Another explicit description of $\mathcal{K}_u$ is now within easy reach.

Corollary 5.18 *If u is a finite Blaschke product with distinct zeros*

$$\lambda_1, \lambda_2, \ldots, \lambda_n$$

having corresponding multiplicities $m_1, m_2, \ldots, m_n$, then

$$\mathcal{K}_u = \left\{ \frac{a_0 + a_1 z + \cdots + a_{N-1} z^{N-1}}{(1 - \overline{\lambda_1} z)^{m_1} (1 - \overline{\lambda_2} z)^{m_2} \cdots (1 - \overline{\lambda_n} z)^{m_n}} : a_j \in \mathbb{C} \right\}. \tag{5.17}$$

where $N = m_1 + m_2 + \cdots + m_n$.

Proof Any linear combination of the functions

$$c_{\lambda_i}^{(\ell_i - 1)}, \quad 1 \leqslant i \leqslant n, 1 \leqslant \ell_i \leqslant m_i, \tag{5.18}$$

can be expressed as a rational function of the type prescribed in (5.17). Conversely, any expression of the type encountered in (5.17) can be decomposed, via partial fractions, into a linear combination of the functions in (5.18). □

We are now in a position to prove that finite Blaschke products are the only inner functions that give rise to finite-dimensional model spaces.

Proposition 5.19 $\dim \mathcal{K}_u < \infty$ *if and only if u is a finite Blaschke product.*

Proof We see that $\dim \mathcal{K}_u < \infty$ whenever u is a finite Blaschke product (Proposition 5.16), so it suffices to show that when u is not a finite Blaschke product, then $\dim \mathcal{K}_u = \infty$. If u contains an infinite Blaschke factor, then Corollary 5.9 and Proposition 3.5 ensure that $\mathcal{K}_u$ contains an infinite, linearly independent set, namely the Cauchy kernels corresponding to the distinct zeros of u. On the other hand, if u contains a singular inner factor v, then $v^{1/n}$ is an inner function that divides u (see the discussion following Definition 2.16) so that $\mathcal{K}_{v^{1/(n+1)}} \subsetneq \mathcal{K}_{v^{1/n}} \subsetneq \mathcal{K}_v$ for $n \geqslant 1$ (Corollary 5.9). We now find an infinite orthonormal sequence in $\mathcal{K}_u$ by selecting unit vectors f_n from $\mathcal{K}_{v^{1/n}} \ominus \mathcal{K}_{v^{1/(n+1)}}$, from which it follows that dim $\mathcal{K}_u = \infty$,   ⊔

5.8 Density results

As mentioned earlier, the model spaces $\mathcal{K}_u$ tend to contain few readily identifiable functions. Only when u is a finite Blaschke product can we completely characterize, in an explicit manner, the elements of $\mathcal{K}_u$ (see Section 5.7). Therefore, one often relies heavily on the kernel functions k_λ, since they are among the few concrete functions in $\mathcal{K}_u$.

As the following proposition shows, one can always find a collection of kernels with dense linear span.

Proposition 5.20 *If Λ is a subset of $\mathbb{D}$ such that either*

(i) Λ has an accumulation point in $\mathbb{D}$, or
(ii) $\sum_{\lambda \in \Lambda} (1 - |\lambda|)$ diverges,

then for any inner function u,

$$\bigvee \{k_\lambda : \lambda \in \Lambda\} = \mathcal{K}_u.$$

Proof The containment $\subset$ is automatic. To show the reverse containment, let $f \in \mathcal{K}_u$ with $f \perp k_\lambda$ for all $\lambda \in \Lambda$. Since k_λ is a reproducing kernel, we see that f vanishes on Λ. (i) If Λ has an accumulation point in $\mathbb{D}$, the Identity Theorem for analytic functions implies that $f \equiv 0$. (ii) If $\sum_{\lambda \in \Lambda}(1 - |\lambda|)$ diverges, the Identity Theorem for Hardy spaces (Corollary 3.17) implies that $f \equiv 0$. An application of Theorem 1.27 completes the proof.   □

Since, for fixed $\lambda \in \mathbb{D}$,

$$|k_\lambda(z)| \leqslant \frac{2}{1 - |\lambda|}, \quad z \in \mathbb{D}, \tag{5.19}$$

it follows that each k_λ belongs to $\mathcal{K}_u \cap H^\infty$. This observation yields the following useful result.

Proposition 5.21 *The linear manifold $\mathcal{K}_u \cap H^\infty$ is dense in $\mathcal{K}_u$.*

Proof Use Proposition 5.20 with $\Lambda = \mathbb{D}$ and (5.19).   □

A second proof of Proposition 5.21 can be fashioned from (5.11) by using the fact that $S^{*n}u \in \mathcal{K}_u \cap H^\infty$ for all $n \geqslant 1$.

For certain sophisticated applications, one may require a dense subset of $\mathcal{K}_u$ consisting of functions having a certain degree of smoothness on $\mathbb{D}^-$. Functions in H^∞ need not be continuous on $\mathbb{D}^-$ and so Proposition 5.21 is of no use in this regard. This next result is a restatement of Proposition 5.16 for infinite Blaschke products.

Proposition 5.22 *If u is a Blaschke product with distinct zeros $\lambda_1, \lambda_2, \ldots$ having corresponding multiplicities $m_1, m_2, \ldots$, then*

$$\mathcal{K}_u = \bigvee \{c_{\lambda_i}^{(\ell_i-1)} : i \geqslant 1, \ 1 \leqslant \ell_i \leqslant m_i\}. \tag{5.20}$$

Thus $\mathcal{K}_u$ contains a dense subset whose elements are continuous on $\mathbb{D}^-$ and whose boundary functions are infinitely differentiable on $\mathbb{T}$.

Proof   The $\supset$ containment in (5.22) is automatic since $c_{\lambda_i}^{(\ell_i-1)} \in \mathcal{K}_u$. To show the reverse containment, suppose that $f \in \mathcal{K}_u$ is orthogonal to each of the functions on the right side of (5.20). Then, for each i, f has a zero of order m_i at λ_i (Corollary 3.14). Thus $u|f$ so that $f \in uH^2 \cap \mathcal{K}_u = \{0\}$ and, via Theorem 1.27, (5.20) holds. The second statement follows from the formula (3.5) for $c_{\lambda_i}^{(\ell_i-1)}$. $\square$

When u is not a Blaschke product, finding a dense set of functions in $\mathcal{K}_u$, each of which is continuous on $\mathbb{D}^-$, is much more difficult. A deep result in this direction is due to A. B. Aleksandrov [9] (see [42] for a detailed exposition).

Definition 5.23   The *disk algebra* $\mathcal{A}$ is the algebra of all analytic functions on $\mathbb{D}$ that extend to be continuous on $\mathbb{D}^-$.

Theorem 5.24 (Aleksandrov)   *For any inner function u, the set $\mathcal{K}_u \cap \mathcal{A}$ is dense in $\mathcal{K}_u$.*

This theorem is remarkable due to the fact that $\mathcal{K}_u$ often does not seem to contain a single readily identifiable function that is continuous on $\mathbb{D}^-$. For example, if u is the singular inner function

$$u(z) = \exp\left(\frac{z+1}{z-1}\right),$$

then it is not at all obvious that $\mathcal{K}_u$ contains any functions that are continuous on $\mathbb{D}^-$, let alone a dense set of them. See [65] for some related results concerning when $\mathcal{K}_u$ contains functions belonging to other smoothness classes on $\mathbb{D}^-$.

5.9 Takenaka–Malmquist–Walsh bases

Unlike the Hardy space H^2, which has the natural orthonormal basis $\{z^n\}_{n \geq 0}$, a model space $\mathcal{K}_u$ does not come pre-equipped with a canonical orthonormal basis. Indeed, finding a concrete orthonormal basis for a general model space is a difficult problem. However, when u is a Blaschke product, an orthonormal basis can be obtained by orthogonalizing the kernel functions corresponding to the zeros of u.

The terminology here is not completely standard. It seems that these bases first appeared in Takenaka's 1925 paper involving finite Blaschke products [187]. The classic text [141, V.1] considers the case where u is an arbitrary Blaschke product, referring to this result as the Malmquist–Walsh Lemma. In light of Takenaka's early contribution to the subject, various authors (see [91] for example) refer to such a basis for $\mathcal{K}_u$ as a *Takenaka–Malmquist–Walsh (TMW) basis*.

For w in $\mathbb{D}$, recall that

$$b_w(z) = \frac{z - w}{1 - \overline{w}z}. \tag{5.21}$$

If u is a Blaschke product with zeros $\lambda_1, \lambda_2, \ldots$, observe that

$$\langle b_{\lambda_1} k_{\lambda_2}, k_{\lambda_1} \rangle = b_{\lambda_1}(\lambda_1) k_{\lambda_2}(\lambda_1) = 0.$$

Similarly, we have

$$\langle b_{\lambda_1} b_{\lambda_2} k_{\lambda_3}, k_{\lambda_1} \rangle = \langle b_{\lambda_1} b_{\lambda_2} k_{\lambda_3}, k_{\lambda_2} \rangle = 0.$$

This process suggests the following result.

Proposition 5.25 *If u is a Blaschke product with zeros $\lambda_1, \lambda_2, \ldots$, then the vectors*

$$v_1(z) = \frac{\sqrt{1 - |\lambda_1|^2}}{1 - \overline{\lambda_1}z},$$

$$v_\ell(z) = \Big(\prod_{1 \leq i \leq \ell - 1} b_{\lambda_i} \Big) \frac{\sqrt{1 - |\lambda_\ell|^2}}{1 - \overline{\lambda_\ell}z}, \qquad \ell \geq 2 \tag{5.22}$$

form an orthonormal basis for $\mathcal{K}_u$.

Proof   A short computation based on the reproducing property, and the facts that $b_w \overline{b_w} = 1$ on $\mathbb{T}$ and $b_{\lambda_j}(\lambda_j) = 0$, shows that for $j < k$ we have

$$\langle v_k, v_j \rangle = \Big\langle \Big(\prod_{j \leq i \leq k - 1} b_{\lambda_i} \Big) \frac{\sqrt{1 - |\lambda_k|^2}}{1 - \overline{\lambda_k}z}, \frac{\sqrt{1 - |\lambda_j|^2}}{1 - \overline{\lambda_j}z} \Big\rangle = 0.$$

For $j = k$ we have

$$\langle v_j, v_j \rangle = \int_{\mathbb{T}} \frac{1 - |\lambda_j|^2}{|1 - \overline{\lambda_j}\zeta|^2} \, dm(\zeta) = \int_{\mathbb{T}} P_{\lambda_j}(\zeta) \, dm(\zeta) = 1$$

by (1.15).

For $n \geqslant 1$, let $u_n = b_{\lambda_1} b_{\lambda_2} \cdots b_{\lambda_n}$ and observe that

$$\bigvee \{v_1, v_2, \ldots, v_n\} \subset \mathcal{K}_{u_n} \subset \mathcal{K}_u$$

by Corollary 5.18 and Corollary 5.9. However, since $\{v_1, v_2, \ldots, v_n\}$ is an orthonormal set and dim $\mathcal{K}_{u_n} = n$, it follows that

$$\bigvee \{v_1, v_2, \ldots, v_n\} = \mathcal{K}_{u_n}$$

by Proposition 5.16. Putting this all together, we see that

$$\bigvee \{v_1, v_2, \ldots\} = \mathcal{K}_u$$

by Proposition 5.22.   □

Remark 5.26 An alternate proof of Proposition 5.25 can be fashioned from Theorem 5.11.

5.10 Notes

5.10.1 Containment

Another proof of Lemma 5.10 can be obtained using reproducing kernels. We sketch the argument here. First divide the identity

$$1 - \overline{u(\lambda)v(\lambda)}u(z)v(z) = (1 - \overline{u(\lambda)}u(z)) + \overline{u(\lambda)}u(z)(1 - \overline{v(\lambda)}v(z))$$

by $1 - \overline{\lambda}z$ to get the formula

$$k_\lambda^{uv} = k_\lambda^u + \overline{u(\lambda)}u k_\lambda^v. \tag{5.23}$$

This is simply a restatement, in terms of reproducing kernels, of the fact that $\mathcal{K}_{uv} = \mathcal{K}_u \oplus u\mathcal{K}_v$. In fact, taking the inner product of an L^2 function with the function (5.23) and using (5.8) reveals that

$$P_{uv}f = P_u f + u P_v(\overline{u}f).$$

We leave the details of these computations to the interested reader.

5.10.2 Kernels and boundary behavior

The kernel functions k_λ are among the few readily identifiable functions in $\mathcal{K}_u$ and they often provide insight into the structure and properties of model spaces. In fact, kernel functions often wind up being "test functions" for various statements about model spaces. For example, suppose that the quantity $\|k_\lambda\|$ remains bounded as λ approaches a point $\zeta \in \mathbb{T}$ along some continuous path $\Gamma \subset \mathbb{D}$ ending at ζ. Since

$$|f(\lambda)| = |\langle f, k_\lambda \rangle| \leqslant \|f\| \, \|k_\lambda\|,$$

it follows that every f in $\mathcal{K}_u$ is bounded along Γ. This basic observation barely hints at the many important results concerning the boundary behavior of functions in model spaces. We will explore this important topic further in Chapter 7.

5.10.3 Model spaces inside H^p

The shift $S : H^p \to H^p$, defined by $Sf = zf$, is continuous on H^p for $1 \leqslant p < \infty$ (see the end notes for Chapter 3 for the definition of H^p) and there is a version of Beurling's Theorem which says that the non-trivial S-invariant subspaces are of the form uH^p for some inner u. The backward shift operator

$$Bf = \frac{f - f(0)}{z}$$

is also continuous on H^p and its B-invariant subspaces take the form

$$\mathcal{K}_u^p := H^p \cap u\overline{z H^p}.$$

In the above notation, we need to think of $\mathcal{K}_u^p$ as the subspace of H^p consisting of functions f such that $f = u\overline{z}\overline{g}$ almost everywhere on $\mathbb{T}$ for some $g \in H^p$. The proof of this relies on the duality between H^p and H^q via the Cauchy pairing from (3.27) to compute the annihilator of uH^q. For $p = 1$, the B-invariant subspaces take the expected form $\mathcal{K}_u^1 = H^1 \cap u\overline{z H^1}$ but the proof needs to deal with some complications arising from the more difficult dual space of H^1. When $0 < p < 1$, the B-invariant subspaces of H^p are too involved to describe in this survey. A full account of all this is found in [41]. Many of the function-theoretic results that hold for the spaces $\mathcal{K}_u$ can be appropriately restated for $\mathcal{K}_u^p$, often with nearly the same proof.

5.10.4 Model spaces on the upper half plane

For an inner function Θ on the upper half plane $\mathbb{C}_+$, that is to say,

$$\Theta \in H^\infty(\mathbb{C}_+) \quad \text{and} \quad \lim_{y \to 0^+} \Theta(x + iy) \in \mathbb{T} \quad \text{a.e. } x \in \mathbb{R},$$

define a *model space* $\mathscr{K}_\Theta$ on $\mathbb{C}_+$ as

$$\mathscr{K}_\Theta := \mathscr{H}^2 \ominus \Theta \mathscr{H}^2, \tag{5.24}$$

where, as discussed in the end notes of Chapter 3, $\mathscr{H}^2$ is the Hardy space of $\mathbb{C}_+$. Much of the theory for the model spaces $\mathcal{K}_u$ on the disk carries over to the upper half plane. For instance, via Proposition 5.4, we can regard $\mathscr{K}_\Theta$ as a space of boundary functions on $\mathbb{R}$ by noting that

$$\mathscr{K}_\Theta = \mathscr{H}^2 \cap \Theta \overline{\mathscr{H}^2}. \tag{5.25}$$

Of particular importance here are the reproducing kernels

$$K_\lambda(z) = \frac{i}{2\pi} \frac{1 - \overline{\Theta(\lambda)}\Theta(z)}{z - \overline{\lambda}}, \qquad \lambda, z \in \mathbb{C}_+.$$

These belong to $\mathscr{K}_\Theta$ and satisfy

$$f(\lambda) = \langle f, K_\lambda \rangle, \qquad \lambda \in \mathbb{C}_+, \ f \in \mathscr{K}_\Theta.$$

5.10.5 de Branges–Rovnyak spaces

There is an important generalization of model spaces, the *de Branges–Rovnyak spaces*, that play an increasingly prominent role in analysis. Unlike the model spaces $\mathcal{K}_u$, which are parameterized by inner functions u, de Branges–Rovnyak spaces are parameterized by $u \in H_1^\infty$, the closed unit ball of H^∞, where

$$H_1^\infty = \{u \in H^\infty : \|u\|_\infty \leqslant 1\}.$$

These spaces have properties similar to those of model spaces but there are many differences. References for this are [73, 74, 165].

These spaces are formed via reproducing kernels. Indeed, for $u \in H_1^\infty$, define the kernel function

$$k_\lambda(z) := \frac{1 - \overline{u(\lambda)}u(z)}{1 - \overline{\lambda}z}, \qquad \lambda, z \in \mathbb{D}.$$

When u is an inner function, the reader will recognize this kernel as the reproducing kernel for the model space $\mathcal{K}_u$. We construct a reproducing kernel Hilbert space $\mathscr{H}(u)$ of analytic functions on $\mathbb{D}$ in the following way. First notice that $k_\lambda(z)$ is positive definite in the sense that

$$\sum_{1 \leqslant \ell, j \leqslant n} c_j \overline{c_\ell} k_{\lambda_j}(\lambda_\ell) \geqslant 0$$

for every choice of $c_1, c_2, \ldots, c_n \in \mathbb{C}$ and $\lambda_1, \lambda_2, \ldots, \lambda_n \in \mathbb{D}$. Then initially populate $\mathscr{H}(u)$ with finite linear combinations of kernel functions, that is, with functions of the form

$$\sum_{1 \leqslant j \leqslant n} c_j k_{\lambda_j},$$

where $c_1, \ldots, c_n \in \mathbb{C}$ and $\lambda_1, \ldots, \lambda_n \in \mathbb{D}$. Define a norm on these finite linear combinations in a way that makes $k_\lambda(z)$ the reproducing kernel for $\mathscr{H}(u)$. One does this by defining

$$\left\| \sum_{1 \leqslant j \leqslant n} c_j k_{\lambda_j} \right\|_{\mathscr{H}(u)}^2 = \left\langle \sum_{1 \leqslant j \leqslant n} c_j k_{\lambda_j}, \sum_{1 \leqslant \ell \leqslant n} c_\ell k_{\lambda_\ell} \right\rangle_{\mathscr{H}(u)}$$

$$:= \sum_{1 \leqslant \ell, j \leqslant n} c_j \overline{c_\ell} k_{\lambda_j}(\lambda_\ell).$$

This defines a norm and the corresponding inner product space is a pre-Hilbert space. The de Branges–Rovnyak space $\mathscr{H}(u)$ is the completion of this pre-Hilbert space. There is an alternate, perhaps more standard, way of defining $\mathscr{H}(u)$ in terms of Toeplitz operators [165].

Comparing the kernel above with the Cauchy kernel, one can prove that $\mathscr{H}(u)$ is contractively contained in H^2, that is to say,

$$\|f\|_{H^2} \leqslant \|f\|_{\mathscr{H}(u)}, \qquad f \in \mathscr{H}(u).$$

When $\|u\|_\infty < 1$, we have $\mathscr{H}(u) = H^2$ with an equivalent norm. On the other extreme, when u is inner, then $\mathscr{H}(u) = \mathcal{K}_u$, and both spaces are endowed with the usual H^2 norm.

5.10.6 Other connections

In [58] de Branges and Rovnyak introduced a notion of a "complementary space" which is a generalization of the classical notion of an orthogonal complement. This concept became useful in defining de Branges–Rovnyak spaces which, as we have seen, are generalizations of model spaces. In [57] de Branges discussed an important class of entire functions that have close connections to model spaces of the upper half plane. Other connections along these lines are discussed in [105, 106].

5.11 For further exploration

5.11.1 The projection P_u

By Proposition 5.14,

$$P_u f = f - u P(\overline{u} f), \qquad f \in H^2.$$

Prove the formula

$$P_u f = P f - u P(\overline{u} f), \qquad f \in L^2. \tag{5.26}$$

5.11.2 TMW basis again

For a Blaschke product u, we know from Proposition 5.25 that $\{v_n\}_{n \geqslant 1}$, defined in (5.22), is an orthonormal basis for $\mathcal{K}_u$. Show that

$$\{u^m v_n : n \geqslant 1, m \geqslant 0\}$$

is an orthonormal basis for H^2.

5.11.3 The backward shift on the Bergman and Dirichlet spaces

For the Bergman space L_a^2 (see (3.28)) and Dirichlet space $\mathcal{D}$ (see (3.29)), one can define the backward shift operator

$$Bf = \frac{f - f(0)}{z}.$$

One can ask, as was done for H^2: What are the B-invariant subspaces? What are the cyclic vectors for B? One can even ask these two questions for a very broad class of Hilbert spaces of analytic functions on $\mathbb{D}$. Some references for this are [14, 15, 154].

5.11.4 Fermat's Last Theorem for model spaces

Dyakonov recently asked whether or not "Fermat's Last Theorem" is true for model spaces [68]. In other words, for fixed $n \geqslant 3$, does there exist $f, g, h \in \mathcal{K}_u$ such that dim $\bigvee\{f, g, h\} > 1$, and such that

$$f^n + g^n = h^n?$$

The Mason–Stothers Theorem tells us that if a, b, and c are relatively prime polynomials, not all constants, such that $a + b = c$; then

$$\max\{\deg a, \deg b, \deg c\} < N(abc),$$

where $N(abc)$ denotes the number of distinct zeros of abc in $\mathbb{C}$ [139, 183]. As a consequence, one can show that the Fermat equation $f^n + g^n = h^n$ has no non-trivial polynomial solutions (that is, no solutions in any space $\mathcal{K}_{z^m}$) [67]. Because of this, (5.17) tells us that the Fermat equation has no non-trivial solutions in $\mathcal{K}_u$ when u is a finite Blaschke product. The situation for general inner functions appears largely open.

6

Operators between model spaces

Having discussed the basics of model spaces in the previous chapter, we now introduce a few operators that act on, or between, them. We discuss composition operators, the Crofoot transform, and the multipliers between model spaces. Other operators, such as the compressed shift and truncated Toeplitz operators, are more involved and will be covered later in Chapters 9 and 13.

6.1 Littlewood Subordination Principle

An *analytic self-map* of the unit disk $\mathbb{D}$ is an analytic function $\varphi : \mathbb{D} \to \mathbb{D}$. If $\mathcal{O}(\mathbb{D})$ denotes the set of all analytic functions on $\mathbb{D}$, then for each analytic self-map φ of $\mathbb{D}$, the linear transformation

$$C_\varphi : \mathcal{O}(\mathbb{D}) \to \mathcal{O}(\mathbb{D}), \qquad C_\varphi f = f \circ \varphi,$$

is called the *composition operator* with symbol φ.

For two Hilbert spaces $\mathcal{X}$ and $\mathcal{Y}$ of analytic functions on $\mathbb{D}$, when is

$$C_\varphi : \mathcal{X} \to \mathcal{Y}$$

well-defined and bounded? In this section, we study this question when

$$\mathcal{X} = \mathcal{Y} = H^2,$$

and in the next section when

$$\mathcal{X} = \mathcal{K}_u, \quad \mathcal{Y} = \mathcal{K}_v,$$

where u and v are inner functions.

The discussion of composition operators on H^2 begins with the Littlewood Subordination Principle, which says that if $f \in H^2$ and $\varphi : \mathbb{D} \to \mathbb{D}$ is analytic with $\varphi(0) = 0$, then $f \circ \varphi \in H^2$ and $\|f \circ \varphi\| \leqslant \|f\|$. This classical result has several proofs [53, 63, 174]. We will give a somewhat different proof that is better suited for our upcoming discussion of composition operators between model spaces.

Lemma 6.1 *For $\varphi : \mathbb{D} \to \mathbb{D}$ analytic with $\varphi(0) = 0$, the composition operator $C_\varphi : H^2 \to H^2$ is bounded with $\|C_\varphi\| = 1$.*

Proof   For $f = \sum_{n \geq 0} a_n z^n \in H^2$, we have

$$(S^{*k}f)(z) = \sum_{n \geq 0} a_{n+k} z^n, \qquad k \geq 0,$$

which implies that

$$(S^{*k}f)(0) = a_k. \tag{6.1}$$

The formula

$$S^*f = \frac{f - f(0)}{z}$$

can be equivalently rewritten as

$$f(z) = f(0) + z(S^*f)(z), \qquad z \in \mathbb{D}.$$

Since φ maps $\mathbb{D}$ into itself and $\varphi(0) = 0$, we can replace z with $\varphi(z)$ in the preceding identity to get

$$f(\varphi(z)) = f(0) + \varphi(z)(S^*f)(\varphi(z)).$$

In the language of operator theory, this identity can be written as

$$C_\varphi f = f(0) + T_\varphi C_\varphi S^* f, \tag{6.2}$$

where T_φ is an analytic Toeplitz operator (see (4.7)) and C_φ is the composition operator. One may question the validity of this identity since we do not yet know if C_φ maps H^2 into itself. To take this into account, we will initially apply (6.2) to polynomials f, in which case everything is well-defined.

The assumption that $\varphi(0) = 0$, along with (6.2), implies that

$$(T_\varphi C_\varphi S^* f)(0) = 0,$$

and thus $T_\varphi C_\varphi S^* f$ is orthogonal to the constant functions. Hence by (6.2),

$$\begin{aligned}
\|C_\varphi f\|^2 &= |f(0)|^2 + \|T_\varphi C_\varphi S^* f\|^2 \\
&\leq |f(0)|^2 + \|T_\varphi\|^2 \|C_\varphi S^* f\|^2 \\
&\leq |f(0)|^2 + \|C_\varphi S^* f\|^2. \qquad \text{(Proposition 4.12)}
\end{aligned}$$

We can replace f with $S^{*k}f$ in the previous estimate and use (6.1) to get

$$\|C_\varphi S^{*k}f\|^2 \leqslant |a_k|^2 + \|C_\varphi S^{*(k+1)}f\|^2, \qquad k \geqslant 0. \tag{6.3}$$

Sum both sides of these inequalities for $0 \leqslant k \leqslant n = \deg f$ and use the fact that $S^{*(n+1)}f \equiv 0$ to get

$$\sum_{0\leqslant k\leqslant n} \|C_\varphi S^{*k}f\|^2 \leqslant \sum_{0\leqslant k\leqslant n} |a_k|^2 + \sum_{0\leqslant k\leqslant n} \|C_\varphi S^{*(k+1)}f\|^2,$$

which, by a telescoping series argument, implies that

$$\|C_\varphi f\|^2 \leqslant \sum_{0\leqslant k\leqslant n} |a_k|^2 = \|f\|^2. \tag{6.4}$$

Hence the Littlewood Subordination Principle holds for polynomials.

To establish the Littlewood Subordination Principle for H^2, let $f \in H^2$ with $f = \sum_{k\geqslant 0} a_k z^k$ and let

$$f_n = \sum_{0\leqslant k\leqslant n} a_k z^k, \qquad n \geqslant 1.$$

By (3.6) and Proposition 3.4, the sequence of polynomials $\{f_n\}_{n\geqslant 1}$ converges uniformly on compact subsets of $\mathbb{D}$ to f. For a fixed $r \in (0,1)$ we observe that

$$\int_{\mathbb{T}} |(f_n \circ \varphi)(r\zeta)|^2 dm(\zeta) \leqslant \|f_n \circ \varphi\|^2$$
$$\leqslant \|f_n\|^2 \qquad \text{(by (6.4))}$$
$$\leqslant \|f\|^2. \qquad \text{(Parseval's Theorem)}$$

Since $r\mathbb{T}$ is a compact subset of $\mathbb{D}$, we can let $n \to \infty$ in the previous inequality to obtain

$$\int_{\mathbb{T}} |(f \circ \varphi)(r\zeta)|^2 dm(\zeta) \leqslant \|f\|^2.$$

Since this estimate holds uniformly in r, we conclude that $f \circ \varphi \in H^2$ and $\|f \circ \varphi\| \leqslant \|f\|$ (Proposition 3.9). In other words, $\|C_\varphi\| \leqslant 1$. Since $C_\varphi 1 = 1$, we have $\|C_\varphi\| = 1$. $\qquad\square$

When $\varphi(0) \neq 0$ we prove the continuity of the composition operator C_φ on H^2 by combining the estimates above with some properties of disk automorphisms. For each fixed $w \in \mathbb{D}$, recall from (2.1) that

$$\tau_{1,w}(z) = \frac{w-z}{1-\overline{w}z}, \qquad z \in \mathbb{D},$$

is an automorphism of $\mathbb{D}$ such that $(\tau_{1,w} \circ \tau_{1,w})(z) = z$ for every $z \in \mathbb{D}$.

Lemma 6.2 *For each $w \in \mathbb{D}$, the composition operator $C_{\tau_{1,w}} : H^2 \to H^2$ is bounded with*

$$1 \leqslant \|C_{\tau_{1,w}}\| < \left(\frac{1 + |w|}{1 - |w|}\right)^{1/2}.$$

Proof   Since $C_{\tau_{1,w}} 1 = 1$, we have $\|C_{\tau_{1,w}}\| \geqslant 1$. To verify the upper bound, fix any $f \in H^2$ and use the change of variable formula

$$e^{is} = \tau_{1,w}(e^{it}) \iff e^{it} = \tau_{1,w}(e^{is}) \qquad \text{and} \qquad dt = \frac{1 - |w|^2}{|1 - \overline{w}e^{is}|^2} \, ds$$

to obtain

$$\begin{aligned}
\|f \circ \tau_{1,w}\|^2 &= \frac{1}{2\pi} \int_0^{2\pi} |f(\tau_{1,w}(e^{it}))|^2 \, dt \\
&= \frac{1}{2\pi} \int_0^{2\pi} |f(e^{is})|^2 \frac{1 - |w|^2}{|1 - \overline{w}e^{is}|^2} \, ds \\
&\leqslant \frac{1 - |w|^2}{(1 - |w|)^2} \frac{1}{2\pi} \int_0^{2\pi} |f(e^{is})|^2 \, ds \\
&= \frac{1 + |w|}{1 - |w|} \|f\|^2.
\end{aligned}$$

The result now follows.   □

Combining Lemmas 6.1 and 6.2, we obtain the following.

Theorem 6.3 *For each analytic self-map of $\mathbb{D}$, the composition operator $C_\varphi : H^2 \to H^2$ is well-defined and satisfies the estimate*

$$1 \leqslant \|C_\varphi\| \leqslant \left(\frac{1 + |\varphi(0)|}{1 - |\varphi(0)|}\right)^{1/2}.$$

Proof   Since $C_\varphi 1 = 1$, we have the lower bound $\|C_\varphi\| \geqslant 1$. To verify the upper bound, let $w = \varphi(0)$ and $\psi = \tau_{1,w} \circ \varphi$. Then ψ is also an analytic self-map satisfying the extra property $\psi(0) = 0$. Thus, by Lemma 6.1, $\|C_\psi\| = 1$.

The identity $\psi = \tau_{1,w} \circ \varphi$ is equivalent to $\varphi = \tau_{1,w} \circ \psi$; and the latter implies that $C_\varphi = C_\psi C_{\tau_{1,w}}$. Hence by Lemmas 6.1 and 6.2, we have

$$\|C_\varphi\| \leqslant \|C_{\tau_{1,w}}\| \, \|C_\psi\| \leqslant \left(\frac{1 + |w|}{1 - |w|}\right)^{1/2} = \left(\frac{1 + |\varphi(0)|}{1 - |\varphi(0)|}\right)^{1/2}. \qquad □$$

6.2 Composition operators on model spaces

We will discuss when a vector is cyclic for S^* in Chapter 7. For now, we bring in the following fact: given an *inner* function φ, it is known [61, Thm. 2.4.4] that f is non-cyclic for S^* (that is, $\bigvee\{S^{*n}f : n \geqslant 0\} = \mathcal{K}_u$ for some non-constant inner u) if and only if $f \circ \varphi$ is non-cyclic for S^*. In other words, if we set

$$\mathcal{K} = \bigcup_{u \text{ inner}} \mathcal{K}_u,$$

then, for any inner function φ, the restricted mappings $C_\varphi : \mathcal{K} \to \mathcal{K}$ and $C_\varphi : H^2 \setminus \mathcal{K} \longrightarrow H^2 \setminus \mathcal{K}$ are both well-defined. The next result from [135] (see also [138]) provides a precise refinement of the first mapping.

Theorem 6.4 *Let φ and u be inner and define*

$$v(z) = \begin{cases} (u \circ \varphi)(z) & \text{if } u(0) \neq 0 \text{ and } \varphi(0) = 0, \\ z(u \circ \varphi)(z) & \text{if } u(0) \neq 0 \text{ and } \varphi(0) \neq 0, \\ z\dfrac{u(\varphi(z))}{\varphi(z)} & \text{if } u(0) = 0. \end{cases} \tag{6.5}$$

Then v is inner and the composition operator

$$C_\varphi : \mathcal{K}_u \longrightarrow \mathcal{K}_v$$

is well-defined and bounded. Moreover, $\mathcal{K}_v$ is the smallest closed S^*-invariant subspace of H^2 that contains $C_\varphi(\mathcal{K}_u)$.

The alert reader has probably noticed there is a little detail to deal with here, namely, that v is actually an inner function! In other words, we need to prove that the composition of two inner functions is another inner function. The more casual reader might think this detail is obvious. Indeed, φ is inner and so $\varphi(\zeta) \in \mathbb{T}$ for almost every $\zeta \in \mathbb{T}$. Moreover, since u is also inner, we have $u(\xi) \in \mathbb{T}$ for almost every $\xi \in \mathbb{T}$. This means that $u(\varphi(\zeta)) \in \mathbb{T}$ for almost every $\zeta \in \mathbb{T}$. One of the subtle problems with this analysis is that when we write $u(\xi)$ we mean

$$u(\xi) = \angle \lim_{z \to \xi} u(z),$$

the non-tangential limit, and, *prima facie*, it is not not immediately clear that if z approaches ζ within a Stolz domain, that $\varphi(z)$ approaches $\varphi(\zeta) \in \mathbb{T}$ within a Stolz domain. This is important when considering the value of $u(\varphi(\zeta))$, which is defined by the non-tangential limit

$$\angle \lim_{z \to \zeta} u(\varphi(z)).$$

As it turns out, thanks to some classical theorems of Lindelöf and Riesz, things all work out fine in the end.

Proposition 6.5 *If u and v are inner functions, then u ∘ v is also an inner function.*

Proof The result depends on the following classical result of Lindelöf [50, p. 19]: Suppose $f \in H^\infty$ and $f(z) \to A$ as z moves along a continuous path $L \subset \mathbb{D}$ terminating at $\zeta \in \mathbb{T}$. Then $f(z) \to A$ uniformly as $z \to \zeta$ in any Stolz domain $\Gamma_\alpha(\zeta)$. We also need the following classical result of Riesz [42, p. 215]: If $E \subset \mathbb{T}$ has (Lebesgue) measure zero and u is an inner function, then $u^{-1}(E)$ also has measure zero.

We apply these two results as follows. Let

$$F = F_1 \cup F_2 \cup F_3,$$

where F_1 is the subset of $\mathbb{T}$ on which $u \circ v$ does not have a radial limit, F_2 is the subset of $\mathbb{T}$ on which v does not have radial limit, $F_3 = v^{-1}(F_4)$, where F_4 is the subset of $\mathbb{T}$ on which u fails to have a radial limit. The set

$$E = \mathbb{T} \setminus F$$

has full measure in $\mathbb{T}$.

For $\zeta \in E$, fix a ray R from 0 to ζ and let $L = v(R)$. Notice that L is a continuous path in $\mathbb{D}$ that terminates at $v(\zeta) \in \mathbb{T}$ (see Figure 6.1).

As $v(z)$ moves along L to the boundary point $v(\zeta)$, we see that $u(v(z))$ approaches $u(v(\zeta))$. By Lindelof's theorem, $(u \circ v)(r\zeta)$ tends to $u(v(\zeta)) \in \mathbb{T}$ as $r \to 1$. □

Proof of Theorem 6.4 By Theorem 6.3, the composition operator $C_\varphi : H^2 \to H^2$ is well-defined and bounded. Therefore, to prove the first part of the theorem (the boundedness of C_φ from $\mathcal{K}_u$ to $\mathcal{K}_v$), it suffices to show that C_φ maps $\mathcal{K}_u$ into $\mathcal{K}_v$. Since the linear span of the reproducing kernels for $\mathcal{K}_u$ are dense in $\mathcal{K}_u$ (Proposition 5.20), it is enough to show that $C_\varphi k_\lambda^u \in \mathcal{K}_v$ for all $\lambda \in \mathbb{D}$.

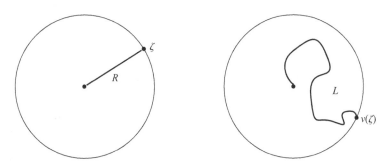

Figure 6.1 The ray R from the origin to ζ (left) and the path $L = v(R)$ terminating at $v(\zeta)$ (right).

Recall that $\mathcal{K}_v = H^2 \cap v\overline{z H^2}$ (Proposition 5.4). Hence, we need to verify that

$$C_\varphi k_\lambda^u \in H^2 \quad \text{and} \quad v\overline{C_\varphi k_\lambda^u} \in zH^2.$$

The first containment is automatic since C_φ maps H^2 to itself (Theorem 6.3). For the second containment, we recall from Proposition 6.5 that $u \circ \varphi$ is an inner function since both φ and u are inner. Moreover, we use the fact that $\varphi\overline{\varphi} = 1$ and $u(\varphi)\overline{u(\varphi)} = 1$ a.c. on $\mathbb{T}$ to get, for every $\lambda \in \mathbb{D}$ and almost every $z \in \mathbb{T}$,

$$
\begin{aligned}
\overline{C_\varphi k_\lambda^u(z)} &= \frac{1 - \overline{u(\lambda)}u(\varphi(z))}{1 - \overline{\lambda}\varphi(z)} \\
&= \frac{1 - \overline{u(\lambda)}u(\varphi(z))}{1 - \overline{\lambda}\varphi(z)} \cdot \frac{\varphi(z)}{\varphi(z)} \cdot \frac{u(\varphi(z))}{u(\varphi(z))} \\
&= \frac{u(\varphi(z)) - u(\lambda)}{\varphi(z) - \lambda} \cdot \frac{\varphi(z)}{u(\varphi(z))}.
\end{aligned}
$$

Using the difference quotient operator Q_λ from Lemma 4.28, we rewrite the first quotient in the identity above as

$$\frac{u(\varphi(z)) - u(\lambda)}{\varphi(z) - \lambda} = C_\varphi Q_\lambda u(z).$$

Since C_φ and Q_λ are both bounded operators on H^2, the representation above reveals that

$$\frac{u \circ \varphi - u(\lambda)}{\varphi - \lambda} = C_\varphi Q_\lambda u \in H^2$$

and

$$\overline{C_\varphi k_\lambda^u} = C_\varphi Q_\lambda u \cdot \frac{\varphi}{u \circ \varphi}. \tag{6.6}$$

To deal with the term $\varphi/(u \circ \varphi)$ in (6.6), we need to consider three cases. In fact, the distinction between these different cases is already reflected in the definition of v in (6.5). We will just deal with the first case, that is, when $u(0) \neq 0$ and $\varphi(0) = 0$. The analysis for the other cases is similar.

In this case, the inner function v is defined by $v = u \circ \varphi$. By (6.6) we have

$$v\overline{C_\varphi k_\lambda^u} = (C_\varphi Q_\lambda u)\varphi \in zH^2.$$

Note the use of $\varphi(0) = 0$. Hence, $C_\varphi k_\lambda^u \in \mathcal{K}_v$ for every $\lambda \in \mathbb{D}$. Therefore, $C_\varphi : \mathcal{K}_u \to \mathcal{K}_v$ is well-defined and bounded.

It is more delicate to prove the second part of Theorem 6.4, namely that the smallest (closed) S^*-invariant subspace of H^2 that contains the image $C_\varphi(\mathcal{K}_u)$ (call this subspace $\mathcal{M}$) is precisely $\mathcal{K}_v$. Based on the above discussion, since $C_\varphi(\mathcal{K}_u) \subset \mathcal{K}_v$, we see that $\mathcal{M} \subset \mathcal{K}_v$. To establish $\mathcal{K}_v \subset \mathcal{M}$, we will prove that $\mathcal{M}^\perp \subset vH^2 = \mathcal{K}_v^\perp$ and then use orthogonality.

So suppose $g \perp \mathcal{M}$. This, together with the fact that $\mathcal{M}$ is S^*-invariant imply that

$$\langle g, S^{*n} C_\varphi k_\lambda^u \rangle = 0$$

for all $\lambda \in \mathbb{D}$ and $n \geqslant 0$. From here we obtain

$$0 = \langle g, S^{*n} C_\varphi k_\lambda^u \rangle = \langle S^n g, C_\varphi k_\lambda^u \rangle = \langle z^n, \bar{g} \, C_\varphi k_\lambda^u \rangle$$

and so

$$\overline{C_\varphi k_\lambda^u} \, g \in z H^2,$$

or equivalently, by (6.6), that

$$\frac{\varphi}{u \circ \varphi} \cdot \frac{u \circ \varphi - u(\lambda)}{\varphi - \lambda} \cdot g \in z H^2, \qquad \lambda \in \mathbb{D}. \tag{6.7}$$

From this relation, we want to prove that

$$\frac{g}{u \circ \varphi} = \frac{g}{v} \in H^2.$$

The key to doing this is our freedom to choose the right λ in (6.7). By Remark 2.23, and the Open Mapping Theorem, we can pick a $\lambda \in \mathbb{D}$ such that

$$B_1 = \frac{u \circ \varphi - u(\lambda)}{1 - \overline{u(\lambda)}(u \circ \varphi)} \qquad \text{and} \qquad B_2 = \frac{\varphi - \lambda}{1 - \bar{\lambda} \varphi}$$

are both Blaschke products with simple zeros. Moreover, we can further restrict λ so that

$$u(\lambda) \neq u(0) \qquad \text{and} \qquad u(\lambda) \neq 0.$$

The zeros of B_1 come from the solutions to the equation $u(\varphi(z)) = u(\lambda)$ while the zeros of B_2 come from the solutions to the equation $\varphi(z) = \lambda$. Hence, the zeros of B_1 contain the zeros of B_2. Therefore, $B = B_1/B_2$ is also a Blaschke product. Note the use of the simplicity of the zeros of B_1 and B_2.

The condition $u(\lambda) \neq u(0)$ ensures that $B_1(0) \neq 0$. Therefore,

$$\frac{B_1}{B_2} = \frac{u \circ \varphi - u(\lambda)}{\varphi - \lambda} = B \frac{1 - \bar{\lambda} \varphi}{1 - \overline{u(\lambda)}(u \circ \varphi)} = Bh,$$

where h is an outer function (see Corollary 3.24) that satisfies $\delta \leqslant |h| \leqslant c$ on $\mathbb{D}$ for some positive constants δ and c, and B is a Blaschke product with $B(0) \neq 0$. Therefore, from (6.7), we deduce that

$$\frac{\varphi}{u \circ \varphi} B g \in z H^2.$$

We also picked λ such that $u(\lambda) \neq 0$. The effect of this choice is that B and $u \circ \varphi$ have no common inner factor. Thus, we can go one step further and say that

$$\frac{\varphi}{u \circ \varphi} g \in z H^2. \tag{6.8}$$

For the final step, we also claim that the inner functions φ and $u \circ \varphi$ have no common inner factor. To see this, recall that $u(0) \neq 0$. Hence, in the first place, φ and $u \circ \varphi$ cannot have a common Blaschke factor. Second, suppose that φ has a singular inner factor. For the corresponding singular measure, we can choose a non-empty carrier $A \subset \mathbb{T}$ of the associated singular measure (see Theorem 2.12) such that for each $\zeta \in A$ we have

$$\lim_{r \to 1} \varphi(r\zeta) = 0.$$

Then, at any such point, we would have

$$\lim_{r \to 1} (u \circ \varphi)(r\zeta) = u(0) \neq 0.$$

Hence, if $u \circ \varphi$ has a singular inner factor, a carrier of its corresponding singular measure can be taken to be disjoint from A. Thus φ and $u \circ \varphi$ have no non-constant inner divisors in common. By (6.8) and the fact that the inner-outer factorization of H^2 functions is unique up to a unimodular multiplicative constant (Theorem 3.20), we conclude that $u \circ \varphi$ must divide g. In other words, $g \in (u \circ \varphi) H^2 = v H^2$.   □

In the hierarchy of inner functions, for a given φ and u, the largest inner function v among the three possible situations in Theorem 6.4 (that is, the one that divides the other two) is $z(u \circ \varphi)$, and this gives the largest model subspace $\mathcal{K}_{z(u \circ \varphi)}$ among the three possible cases. Hence, if we are not concerned with the smallest possible model subspace that contains the image of C_φ, we obtain the following result.

Corollary 6.6 *For inner functions φ and u, the composition operator*

$$C_\varphi : \mathcal{K}_u \to \mathcal{K}_{z(u \circ \varphi)}$$

is well-defined and bounded.

6.3 Unitary maps between model spaces

We now consider several standard maps that transform one model space into another. It is helpful to have this information here in one place since these maps tend to appear unannounced in the course of various technical arguments. Moreover, in our experience with the literature, the fact that these maps have the desired properties is often assumed and the uninitiated reader may have difficulty locating proofs of these results. One source is [45].

Theorem 6.7 *For an inner function u and $w \in \mathbb{D}$, the Crofoot transform*

$$J_w f = \frac{\sqrt{1 - |w|^2}}{1 - \overline{w}u} f \qquad (6.9)$$

defines a unitary operator from $\mathcal{K}_u$ onto $\mathcal{K}_{u_w}$, where

$$u_w = \frac{u - w}{1 - \overline{w}u}.$$

Proof Following the ideas of [166], we first claim that $J_w \mathcal{K}_u \subset \mathcal{K}_{u_w}$. To show this, we must verify, for any $f \in \mathcal{K}_u$, that $J_w f$ is orthogonal to every function of the form $u_w g$ where $g \in H^2$. This follows from the computation

$$\langle J_w f, u_w g \rangle = \left\langle \frac{\sqrt{1 - |w|^2}}{1 - \overline{w}u} f, u_w g \right\rangle = \sqrt{1 - |w|^2} \left\langle f, \frac{u_w g}{1 - \overline{w}u} \right\rangle$$

$$= \sqrt{1 - |w|^2} \left\langle f, \frac{u u_w g}{u - w} \right\rangle = \sqrt{1 - |w|^2} \left\langle f, \frac{u g}{1 - \overline{w}u} \right\rangle$$

$$= 0$$

since $ug(1 - \overline{w}u)^{-1} \in uH^2$. On the other hand, since

$$u = \frac{u_w + w}{1 + \overline{w}u_w},$$

an analogous argument shows that the range of the map

$$g \mapsto \frac{\sqrt{1 - |w|^2}}{1 + \overline{w}u_w} g$$

on $\mathcal{K}_{u_w}$ is contained in $\mathcal{K}_u$. However, this map is the inverse of J_w since

$$(1 - \overline{w}u)(1 + \overline{w}u_w) = (1 - \overline{w}u)\left[1 + \overline{w}\left(\frac{u - w}{1 - \overline{w}u}\right)\right]$$

$$= (1 - \overline{w}u)\left[\frac{1 - \overline{w}u + \overline{w}u - |w|^2}{1 - \overline{w}u}\right]$$

$$= 1 - |w|^2.$$

Hence $J_w \mathcal{K}_u = \mathcal{K}_{u_w}$.

We now show that $J_w : \mathcal{K}_u \to \mathcal{K}_{u_w}$ is an isometry. To this end, we first note that if $f, g \in \mathcal{K}_u$, then

$$\langle u^j f, u^k g \rangle = \begin{cases} \langle f, g \rangle & \text{if } j = k, \\ 0 & \text{if } j \neq k. \end{cases}$$

Indeed, for $j < k$ we see that $\langle u^j f, u^k g \rangle = \langle f, u^{k-j} g \rangle = 0$ since $f \in \mathcal{K}_u$. Similarly, $\langle u^j f, u^k g \rangle = 0$ if $j > k$. Putting this all together, it follows that

$$\begin{aligned} \langle J_w f, J_w g \rangle &= (1 - |w|^2) \left\langle \frac{f}{1 - \overline{w}u}, \frac{g}{1 - \overline{w}u} \right\rangle \\ &= (1 - |w|^2) \sum_{j,k \geq 0} \overline{w}^j w^k \langle u^j f, u^k g \rangle \\ &= (1 - |w|^2) \sum_{j \geq 0} |w|^{2j} \langle f, g \rangle \\ &= (1 - |w|^2)(1 - |w|^2)^{-1} \langle f, g \rangle \\ &= \langle f, g \rangle. \end{aligned}$$
$\square$

For the next unitary transformation, we need to recall from (2.1) that every disk automorphism takes the form

$$\tau_{\zeta,a}(z) = \zeta \frac{a - z}{1 - \overline{a}z}, \quad \zeta \in \mathbb{T}, a \in \mathbb{D}. \tag{6.10}$$

Furthermore,

$$\tau'_{\zeta,a}(z) = \zeta \frac{1 - |a|^2}{(1 - \overline{a}z)^2}$$

is non-vanishing on $\mathbb{D}$ and so, for each $f \in H^2$, the function

$$\sqrt{\tau'_{\zeta,a}}(f \circ \tau_{\zeta,a})$$

is an analytic function on $\mathbb{D}$ and belongs to H^2 (Lemma 6.2).

Theorem 6.8 *If u is an inner function and $\varphi = \tau_{\zeta,a}$ is a disk automorphism, then*

$$f \mapsto \sqrt{\varphi'}(f \circ \varphi)$$

defines a unitary operator from $\mathcal{K}_u$ onto $\mathcal{K}_{u \circ \varphi}$.

Proof Let φ be a disk automorphism of the form (6.10) and define

$$U : H^2 \to H^2, \quad Uf = \sqrt{\varphi'}(f \circ \varphi).$$

Performing a change of variables similar to the computation in the proof of Lemma 6.2 shows that U is a unitary operator whose inverse is given by

$$U^{-1}f = \sqrt{(\varphi^{-1})'}(f \circ \varphi^{-1}).$$

Next observe that if $f \in \mathcal{K}_u$, then

$$\langle Uf, (u \circ \varphi)h \rangle = \langle f, U^*((u \circ \varphi)h) \rangle = \langle f, u\sqrt{(\varphi^{-1})'}(h \circ \varphi^{-1}) \rangle = 0$$

for all $h \in H^2$. Hence $U\mathcal{K}_u \subset \mathcal{K}_{u \circ \varphi}$. The same argument applied to U^{-1} shows that $U^{-1}\mathcal{K}_{u \circ \varphi} \subset \mathcal{K}_u$. Thus $U\mathcal{K}_u = \mathcal{K}_{u \circ \varphi}$ and hence U restricts to a unitary map from $\mathcal{K}_u$ onto $\mathcal{K}_{u \circ \varphi}$. □

For $g \in H^2$ define

$$g^{\#}(z) := \overline{g(\overline{z})}, \quad z \in \mathbb{D},$$

and note that $g^{\#} \in H^2$.

Theorem 6.9 *If u is an inner function then, in terms of boundary functions, the map*

$$f(\zeta) \mapsto \overline{\zeta}f(\overline{\zeta})u^{\#}(\zeta) \tag{6.11}$$

defines a unitary operator from $\mathcal{K}_u$ onto $\mathcal{K}_{u^{\#}}$.

Proof Since the map defined by (6.11) is clearly linear and isometric, we need only show that it maps $\mathcal{K}_u$ onto $\mathcal{K}_{u^{\#}}$. In fact, it is easier to view this map as a composition of two conjugate-linear operators. First note that Proposition 5.4 ensures that the map $f \mapsto \overline{fz}u$ is a conjugate-linear isometric bijection from $\mathcal{K}_u$ to itself (see Chapter 8 for more on this). A change of variables will show that the #-operator is a conjugate linear transformation that maps $\mathcal{K}_u$ bijectively and isometrically onto $\mathcal{K}_{u^{\#}}$. Hence the composition of these two conjugate-linear maps yields a linear isomorphic bijection, given by (6.11), from $\mathcal{K}_u$ onto $\mathcal{K}_{u^{\#}}$. □

6.4 Multipliers of $\mathcal{K}_u$

If $\mathcal{H}$ is a Hilbert space of analytic functions on $\mathbb{D}$, we say that an analytic function φ on $\mathbb{D}$ is a *multiplier* for $\mathcal{H}$ if

$$\varphi\mathcal{H} \subset \mathcal{H}.$$

The set of multipliers of $\mathcal{H}$ is denoted by $\mathcal{M}(\mathcal{H})$.

Proposition 6.10 $\mathcal{M}(H^2) = H^\infty$.

Proof   Notice that $\varphi H^2 \subset H^2$ for $\varphi \in H^\infty$. In fact, the map $f \mapsto \varphi f$ on H^2 is just the analytic Toeplitz operator T_φ discussed in Chapter 4.

Now suppose that φ is analytic on $\mathbb{D}$ and $\varphi H^2 \subset H^2$. Let us first show that the operator $M_\varphi : H^2 \rightarrow H^2$ defined by $M_\varphi f = \varphi f$ is bounded. This is done by using the Closed Graph Theorem (Theorem 1.28). Indeed, if $f_n \rightarrow f$ and $\varphi f_n \rightarrow g$ in the norm of H^2, then by Proposition 3.4, both of these sequences converge pointwise on $\mathbb{D}$ and so $g(z) = \varphi(z)f(z)$ for all $z \in \mathbb{D}$. This means that the graph $\{(f, M_\varphi f) : f \in H^2\}$ of M_φ is closed and thus M_φ is bounded.

To show that $\varphi \in H^\infty$, let $\lambda \in \mathbb{D}$ and $n \geq 1$, and note that

$$|\varphi(\lambda)|^n = |\varphi(\lambda)^n| = |(M_\varphi^n 1)(\lambda)|$$

$$\leq \frac{1}{\sqrt{1 - |\lambda|^2}} \|M_\varphi^n 1\| \qquad \text{(Proposition 3.4)}$$

$$\leq \frac{1}{\sqrt{1 - |\lambda|^2}} \|M_\varphi\|^n \|1\|. \qquad \text{(by (1.26))}$$

Taking nth roots and passing to the limit we obtain

$$|\varphi(\lambda)| \leq \lim_{n \to \infty} \left(\frac{1}{\sqrt{1 - |\lambda|^2}} \right)^{1/n} \|M_\varphi\| = \|M_\varphi\|$$

for every $\lambda \in \mathbb{D}$ and so $\varphi \in H^\infty$. $\square$

Remark 6.11   An alternate proof of the fact that a multiplier φ of H^2 must belong to H^∞ is as follows: One shows, as in the proof of Proposition 6.10, that M_φ is a bounded operator. Now prove the identity $M_\varphi^* c_\lambda = \overline{\varphi(\lambda)} c_\lambda$ (same proof as Proposition 4.19). Thus $\overline{\varphi(\mathbb{D})} \subset \sigma_p(M_\varphi) \subset \sigma(M_\varphi)$, which must be a bounded set (Proposition 1.30).

The following result shows that the set of multipliers of $\mathcal{K}_u$ is the smallest possible collection.

Theorem 6.12   *Let u be a non-constant inner function. Then $\mathcal{M}(\mathcal{K}_u) = \mathbb{C}1$.*

Proof   The inclusion $\mathbb{C}1 \subset \mathcal{M}(\mathcal{K}_u)$ is automatic. To prove the converse, we show that $\varphi S^* u \notin \mathcal{K}_u$ for every non-constant $\varphi \in H^\infty$.

Suppose towards a contradiction that $\varphi S^* u \in \mathcal{K}_u$ and $\varphi \in H^\infty \setminus \mathbb{C}$. Since $uz^n \perp \mathcal{K}_u$ for all $n \geq 0$, we have

$$
\begin{aligned}
0 = \langle \varphi S^* u, uz^n \rangle &= \langle \varphi \frac{u - u(0)}{z}, uz^n \rangle = \langle \varphi(u - u(0)), uz^{n+1} \rangle \\
&= \langle \varphi u, uz^{n+1} \rangle - u(0) \langle \varphi, uz^{n+1} \rangle = \langle \varphi, z^{n+1} \rangle - u(0) \langle \overline{u}\varphi, z^{n+1} \rangle \\
&= \langle \varphi, z^{n+1} \rangle - u(0) \langle P(\overline{u}\varphi), z^{n+1} \rangle \\
&= \langle \varphi - u(0) T_{\overline{u}}\varphi, z^{n+1} \rangle.
\end{aligned}
$$

Therefore,

$$
\varphi - u(0) T_{\overline{u}}\varphi \perp z^{n+1}, \qquad n \geq 0.
$$

Since $\varphi - u(0) T_{\overline{u}}\varphi \in H^2$, this function must be a constant. Thus

$$
S^*(\varphi - u(0) T_{\overline{u}}\varphi) = 0,
$$

which we rewrite as

$$
S^*\varphi = u(0) S^* T_{\overline{u}}\varphi = u(0) T_{\overline{z}} T_{\overline{u}}\varphi = u(0) T_{\overline{u}} S^* \varphi.
$$

Note the use of Theorem 4.22 in the previous line. In particular, using the identity $\|T_{\overline{u}}\| = \|T_u\| = \|u\|_\infty = 1$ (Proposition 4.12), we deduce that

$$
\|S^* \varphi\| \leq |u(0)| \, \|S^* \varphi\|.
$$

However, since φ is not a constant function, $\|S^* \varphi\| \neq 0$. Therefore, $|u(0)| \geq 1$, which, by the Maximum Modulus Theorem (Theorem 2.4), forces u to be a unimodular constant. This is a contradiction. $\qquad \square$

6.5 Multipliers between two model spaces

Although the multipliers of $\mathcal{K}_u$ are not interesting, the analytic functions φ on $\mathbb{D}$ that multiply one model space $\mathcal{K}_u$ to *another* model space $\mathcal{K}_v$ do form an interesting, and not completely understood, class of functions. For example, the Crofoot transform yields a non-constant multiplier from $\mathcal{K}_u$ to $\mathcal{K}_{u_w}$ (Theorem 6.7). What are the multipliers from one model space to another? To make our exposition more manageable, we make the simplifying assumption that $u(0) = 0$.

Theorem 6.13 *Suppose that u, v are inner, $u(0) = 0$, and $\varphi \in H^\infty$. Then the following are equivalent:*

(i) $\varphi \mathcal{K}_u \subset \mathcal{K}_v$;
(ii) $u\varphi \in \mathcal{K}_{zv}$;
(iii) $\varphi \in \ker T_{\overline{zvu}}$.

Proof   A key fact in our proof is the fact that for an inner function w, $T_{\bar{w}}f = 0$ if and only if $f \in \mathcal{K}_w$ (Proposition 5.8).

(i) $\implies$ (ii): Suppose $\varphi \mathcal{K}_u \subset \mathcal{K}_v$. Since $S^*u \in \mathcal{K}_u$ (Proposition 5.15), it follows that $\varphi S^*u \in \mathcal{K}_v$. Using the fact just mentioned above, we have

$$\varphi S^*u \in \mathcal{K}_v \implies T_{\bar{v}}(\varphi S^*u) = 0 \qquad \text{(Proposition 5.8)}$$
$$\implies T_{\bar{v}}(\bar{z}u\varphi) = 0 \qquad \text{(since } u(0) = 0)$$
$$\implies T_{\overline{zv}}(\varphi u) = 0 \qquad \text{(Theorem 4.22)}$$
$$\implies \varphi u \in \mathcal{K}_{zv}. \qquad \text{(Proposition 5.8)}$$

(ii) $\iff$ (iii): Observe the equivalences

$$u\varphi \in \mathcal{K}_{zv} \iff T_{\overline{zv}}(u\varphi) = 0 \qquad \text{(Proposition 5.8)}$$
$$\iff T_{\overline{zv}}T_u\varphi = 0$$
$$\iff T_{\overline{zvu}}\varphi = 0 \qquad \text{(Theorem 4.22)}$$
$$\iff \varphi \in \ker T_{\overline{zvu}}. \qquad \text{(Proposition 5.8)}$$

(ii) $\implies$ (i): We have

$$u\varphi \in \mathcal{K}_{zv} = \mathcal{K}_z \oplus z\mathcal{K}_v = \mathbb{C} \oplus z\mathcal{K}_v$$

(Lemma 5.10). Since $u(0) = 0$, this implies that $u\varphi \in z\mathcal{K}_v$ from which we conclude

$$\frac{u\varphi}{z} \in \mathcal{K}_v. \tag{6.12}$$

Using the fact that we can always divide out the inner factor of a function in $\mathcal{K}_v$ and still remain in $\mathcal{K}_v$ (Proposition 5.6), we see that

$$\varphi \in \mathcal{K}_v. \tag{6.13}$$

This fact will be important in a moment. Also notice that

$$u\varphi \in z\mathcal{K}_v \implies S^{*n}(u\varphi) \in \mathcal{K}_v, \quad n \geqslant 1.$$

Indeed, by (6.12), and the fact that $u(0) = 0$, we see that $S^*(u\varphi) \in \mathcal{K}_v$ and thus, since $\mathcal{K}_v$ is S^*-invariant,

$$S^{*n}(u\varphi) \in \mathcal{K}_v, \quad \forall n \geqslant 1. \tag{6.14}$$

By our "product rule" formula from (4.19), we have

$$S^*(u\varphi) = \varphi S^*u + u(0)S^*\varphi = \varphi S^*u. \tag{6.15}$$

We conclude that

$$\varphi S^*u \in \mathcal{K}_v. \tag{6.16}$$

Take S^* of both sides of (6.15) to get

$$S^{*2}(u\varphi) = S^*(\varphi S^* u) = \varphi S^{*2} u + (S^* u)(0) S^* \varphi. \tag{6.17}$$

Now use (6.13) and (6.14) to deduce that

$$\varphi S^{*2} u \in \mathcal{K}_v. \tag{6.18}$$

Take S^* of both sides of (6.17) to get

$$S^*(\varphi S^{*2} u + (S^* u)(0) S^* \varphi) = \varphi S^{*3} u + (S^{*2} u)(0) S^* \varphi + (S^* u)(0) S^{*2} \varphi.$$

Again, by (6.14), $S^{*3}(u\varphi)$, along with $S^* \varphi, S^{*2} \varphi$ belong to $\mathcal{K}_v$ and so

$$\varphi S^{*3} u \in \mathcal{K}_v. \tag{6.19}$$

Combine (6.16), (6.18), and (6.19) and continue in this manner to prove that

$$\varphi S^{*n} u \in \mathcal{K}_v \qquad \forall n \geqslant 1. \tag{6.20}$$

From Proposition 5.15, we see that any $f \in \mathcal{K}_v$ can be written as

$$f = \sum_{n \geqslant 1} c_n S^{*n} u,$$

where the sum converges in the H^2 norm. Since $\varphi \in H^\infty$ is a multiplier of H^2 (Proposition 6.10), we see, from (6.20), that $\varphi f \in \mathcal{K}_v$. □

6.6 Notes

6.6.1 Littlewood Subordination Principle

The classical Littlewood Subordination Principle takes a different and more general form than the one stated in this chapter. Indeed, suppose that f, g are analytic on $\mathbb{D}$ and there exists an analytic function φ on $\mathbb{D}$ satisfying the conditions $|\varphi(z)| \leqslant |z|$ and $f = g \circ \varphi$ on $\mathbb{D}$. When this happens, we say that f is "subordinate" to g. A 1925 theorem of Littlewood says that if f is subordinate to g then, for each $0 < p \leqslant \infty$,

$$\int_{\mathbb{T}} |f(r\zeta)|^p dm(\zeta) \leqslant \int_{\mathbb{T}} |g(r\zeta)|^p dm(\zeta), \quad 0 < r < 1.$$

A nice treatment of this is found in [63]. Using the Littlewood Subordination Principle, one can show that for any analytic self map φ of $\mathbb{D}$, the composition operator C_φ is well-defined and continuous on all of the Hardy spaces H^p.

6.6.2 Crofoot transforms

The map (6.9) is called the *Crofoot transform* [54]. Part of its significance lies in the fact that if one deals with $\mathcal{K}_u$ where u has a Blaschke factor, one can often make the simplifying assumption that $u(0) = 0$. Also of great importance is the fact that Crofoot transforms intertwine the conjugations (see Chapter 8) on the corresponding model spaces [166, Lem. 3.1]. A detailed treatment of the Crofoot transform can be found in Sarason's article [166, Sec. 13]. We followed his approach in our presentation.

6.6.3 Multipliers

It turns out that $\mathcal{M}(\mathcal{H}) \subset H^\infty$ for many standard Hilbert spaces of analytic functions on $\mathbb{D}$. For example, $\mathcal{M}(H^2) = H^\infty$ (Proposition 6.10), and a similar argument will show that $\mathcal{M}(L_a^2) = H^\infty$, where L_a^2 is the Bergman space (see (3.28)). For the Dirichlet space $\mathcal{D}$ (see (3.29)), the inclusion $\mathcal{M}(\mathcal{D}) \subset H^\infty$ is strict and the multipliers of $\mathcal{D}$ are not completely understood [70, 182].

6.6.4 Kernels of Toeplitz operators

Theorem 6.13 involves the kernel of a Toeplitz operator. This is much-explored territory [66, 86, 107, 108, 111, 163, 164].

6.7 For further exploration

6.7.1 Multipliers between model spaces

The multipliers from $\mathcal{K}_u$ to itself are not very interesting since they are just the constant functions (Theorem 6.12). Crofoot [54] explored the multipliers from $\mathcal{K}_u$ onto $\mathcal{K}_v$. In particular, he showed that if $\varphi \mathcal{K}_u = \mathcal{K}_v$, then φ must be outer. Must φ be bounded? Crofoot also examined the *isometric multipliers* from $\mathcal{K}_u$ onto $\mathcal{K}_v$, that is to say, $\varphi \mathcal{K}_u = \mathcal{K}_v$ and $\|\varphi f\| = \|f\|$ for all $f \in \mathcal{K}_u$. He proved there exists an isometric multiplier from $\mathcal{K}_u$ onto $\mathcal{K}_v$ if and only if $v = \tau_{\zeta,a} \circ u$ for some disk automorphism $\tau_{\zeta,a}$.

What happens when we relax the *onto* condition in Crofoot's analysis? For what φ do we have $\varphi \mathcal{K}_u \subset \mathcal{K}_v$? Must such φ be bounded? When does there exist an isometric multiplier φ from $\mathcal{K}_u$ *into* (not necessarily onto) $\mathcal{K}_v$? Can we classify all such isometric multipliers?

6.7.2 Multipliers and kernels of Toeplitz operators

In Theorem 6.13, we saw a connection between $\ker T_{\overline{z}\overline{v}u}$ and the existence of a multiplier from $\mathcal{K}_u$ to $\mathcal{K}_v$ (if $\varphi \in \ker T_{\overline{z}\overline{v}u}$ is a bounded function, then φ is a multiplier from $\mathcal{K}_u$ into $\mathcal{K}_v$). When $\ker T_{\overline{z}\overline{v}u} \neq \{0\}$, does it always contain a non-zero bounded function? In general, kernels of Toeplitz operators need not contain (non-zero) bounded functions. But do they for the special symbols $\overline{z}\overline{v}u$ for inner u and v? When $\ker T_{\overline{z}\overline{v}u} \neq \{0\}$, will it contain an isometric multiplier from $\mathcal{K}_u$ to $\mathcal{K}_v$?

6.7.3 Other properties of composition operators

In this chapter, we explored composition operators between model spaces, and, in certain cases, proved they were well-defined and continuous. What else can be said? When are these operators compact? What are the norms of these operators?

7

Boundary behavior

In this chapter we discuss two types of boundary phenomena for model spaces. The first is that for any model space $\mathcal{K}_u$, each function in $\mathcal{K}_u$ has a "continuation" across $\mathbb{T}$ to a meromorphic function on the exterior disk. The second is that for certain model spaces $\mathcal{K}_u$, it is possible for all functions in $\mathcal{K}_u$ to have a finite non-tangential limit at a fixed boundary point.

7.1 Pseudocontinuation

Functions in model spaces enjoy a certain type of continuation across $\mathbb{T}$. The type of generalized analytic continuation that is relevant to model spaces is called a *pseudocontinuation* and was first explored by H. S. Shapiro [172, 173]. In short, for each $f \in \mathcal{K}_u$ there is an analytic function F_u, defined on the extended exterior disk

$$\mathbb{D}_e := \{|z| > 1\} \cup \{\infty\},$$

such that the non-tangential limits of f/u (from $\mathbb{D}$) are equal to the non-tangential limits of F_u (from $\mathbb{D}_e$) almost everywhere on $\mathbb{T}$, that is to say,

$$\angle \lim_{\substack{z \to \zeta \\ z \in \mathbb{D}}} \frac{f}{u}(z) = \angle \lim_{\substack{z \to \zeta \\ z \in \mathbb{D}_e}} F_u(z) \quad \text{a.e. } \zeta \in \mathbb{T}. \tag{7.1}$$

In a sense, F_u can be regarded as a "continuation" of f/u.

To add some precision to this, recall the definitions of the Hardy space H^2, the set of functions of bounded type $\mathfrak{N}$, and the Smirnov class N^+ from Chapter 3.

Definition 7.1 For each of these spaces $\mathfrak{X} = H^2$, $\mathfrak{N}$, or N^+ of analytic (resp. meromorphic) functions on $\mathbb{D}$, define its corresponding space $\mathfrak{X}(\mathbb{D}_e)$ of analytic (resp. meromorphic) functions on $\mathbb{D}_e$ by

$$\mathfrak{X}(\mathbb{D}_e) := \{f(1/z) : f \in \mathfrak{X}\}. \tag{7.2}$$

From Chapter 1 recall, for fixed $\zeta \in \mathbb{T}$, the Stolz domains

$$\Gamma_\alpha(\zeta) := \{z \in \mathbb{D} : |z - \zeta| < \alpha(1 - |z|)\}, \quad \alpha > 1,$$

anchored at ζ, and, for a meromorphic function f on $\mathbb{D}$ and fixed $\zeta \in \mathbb{T}$, recall that

$$L = \angle \lim_{\substack{z \to \zeta \\ z \in \mathbb{D}}} f(z)$$

means $f(z) \to L$ as $z \to \zeta$ within any fixed Stolz domain $\Gamma_\alpha(\zeta)$ anchored at ζ. We extend this definition to meromorphic functions F on $\mathbb{D}_e$ and fixed $\zeta \in \mathbb{T}$ by saying that

$$M = \angle \lim_{\substack{z \to \zeta \\ z \in \mathbb{D}_e}} F(z)$$

if $F(z) \to M$ as $z \to \zeta$ within any fixed exterior Stolz domain

$$\{z \in \mathbb{D}_e : 1/\overline{z} \in \Gamma_\alpha(\zeta)\}.$$

Definition 7.2

(i) Let f and $\widetilde{f}$ be meromorphic functions on $\mathbb{D}$ and $\mathbb{D}_e$, respectively. If

$$\angle \lim_{\substack{z \to \zeta \\ z \in \mathbb{D}}} f(z) = \angle \lim_{\substack{z \to \zeta \\ z \in \mathbb{D}_e}} \widetilde{f}(z) \quad \text{a.e. } \zeta \in \mathbb{T},$$

we say that f and $\widetilde{f}$ are *pseudocontinuations* of one another.

(ii) A meromorphic function f on $\mathbb{D}$ is *pseudocontinuable of bounded type* (*PCBT*) if f has a pseudocontinuation $\widetilde{f}$ that belongs to $\mathfrak{N}(\mathbb{D}_e)$.

Remark 7.3

(i) By Privalov's Uniqueness Theorem (see the end notes of Chapter 2), pseudocontinuations are unique in the sense that if F and G (meromorphic functions on $\mathbb{D}_e$) are both pseudocontinuations of the same meromorphic function f on $\mathbb{D}$, then $F = G$.

(ii) Another application of Privalov's Theorem shows that pseudocontinuations are compatible with analytic continuations in the sense that if a meromorphic function on $\mathbb{D}$ has both a pseudocontinuation to $\mathbb{D}_e$ and an analytic continuation across some neighborhood of a point on $\mathbb{T}$, then these two extensions must agree on their common domain.

(iii) The definition of a pseudocontinuation is stated in terms of *non-tangential limits* as rather than *radial limits*, since Privalov's Uniqueness Theorem is no longer valid for radial limits [21]. However, this will not be

an issue for us in our presentation since the functions we deal with
are of bounded type and consequently have non-tangential limits almost
everywhere (Proposition 3.27).

Most of the following examples are discussed in [154].

Example 7.4   Each inner function u has a pseudocontinuation to $\mathbb{D}_e$ defined by

$$\widetilde{u}(z) := \frac{1}{\overline{u(1/\overline{z})}}, \quad z \in \mathbb{D}_e \setminus \{u(1/\overline{z}) = 0\}, \tag{7.3}$$

the Schwarz reflection of u with respect to $\mathbb{T}$. Moreover, u is PCBT since the
right side of (7.3) is a quotient of two bounded analytic functions on $\mathbb{D}_e$. If u is
a Blaschke product whose zeros accumulate on all of $\mathbb{T}$, then u has a pseudocontinuation (as above) but not an analytic continuation across any arc of $\mathbb{T}$.

Example 7.5   If f is a rational function whose poles lie in $\mathbb{D}_e$, then f is PCBT.
Indeed, if $f = p/q$, where p and q are polynomials of whose degrees are at
most n, then

$$\widetilde{f}(z) = \frac{z^{-n}p(z)}{z^{-n}q(z)}$$

is a quotient of meromorphic functions that are bounded on $\mathbb{D}_e$. Thus $\widetilde{f}$ is a
pseudocontonuation (of bounded type) of f.

For examples of functions without pseudocontinuations, we need two
definitions. The first is a proper definition of a winding point.

Definition 7.6   We say an analytic function f on $\mathbb{D}$ has an *isolated winding
point* at $\zeta \in \mathbb{T}$ if there is a punctured disk $G_\zeta = \{z : 0 < |z - \zeta| < a\}$ such that
(i) f is analytically continuable (without singularities) along every path $\gamma \subset G_\zeta$
originating in $\mathbb{D}$ and (ii) there is some closed path $\gamma \subset G_\zeta$, originating in $\mathbb{D}$
along which the continuation of f does not return to its original branch. For
example, $f(z) = \log(1 - z)$ has an isolated winding point at $z = 1$.

The second is a localization of the concept of a pseudocontinuation.

Definition 7.7   For an arc $I \subset \mathbb{T}$ and an analytic function f on $\mathbb{D}$, we say that
f *has a pseudocontinuation across* $I \subset \mathbb{T}$ if there is a meromorphic function $\widetilde{f}$
defined on a domain $\Omega \subset \mathbb{D}_e$ such that $I \subset \Omega^-$ (see Figure 7.1) and

$$\angle \lim_{\substack{z \to \zeta \\ z \in \mathbb{D}}} f(z) = \angle \lim_{\substack{z \to \zeta \\ z \in \Omega}} \widetilde{f}(z) \quad \text{a.e. } \zeta \in I.$$

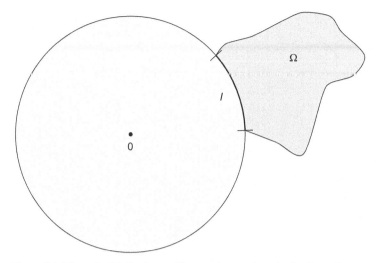

Figure 7.1 The unit disk $\mathbb{D}$ along with a contiguous domain Ω whose closures meet on an arc $I \subset \mathbb{T}$.

Proposition 7.8 *If an analytic function f on $\mathbb{D}$ has an isolated winding point $\zeta \in \mathbb{T}$, then f does not have a pseudocontinuation across any arc of $\mathbb{T}$ which contains ζ.*

Proof Let I be any open arc of $\mathbb{T}$ containing ζ. For $t > 0$, consider the annular domain

$$R = \{z : t < |z - \zeta| < 3t\}.$$

We assume t is so small that f extends analytically throughout the outer disk $\{z : |z - \zeta| < 3t\}$ (as a multiple-valued function), except for the winding point at ζ.

Suppose that f has a pseudocontinuation $\widetilde{f}$ (meromorphic in $\mathbb{D}_e$) across I to to contiguous domain $\Omega \subset \mathbb{D}_e$ with $\Omega^- \cap \mathbb{T} = I$. We can assume that $\Omega \subset \{z : |z - \zeta| < 3t\}$. Let γ be the circle

$$\gamma = \{z : |z - \zeta| = 2t\}$$

and suppose that z_0 and w_0 are points on γ with $z_0 \in \mathbb{D}$ and $w_0 \in \mathbb{D}_e$ (see Figure 7.2). If we now apply the compatibility of analytic continuation with pseudocontinuation (Remark 7.3) to the two possible analytic continuations of f from z_0 to w_0 along arcs of γ, the values of both these continuations at w_0 are required to be equal to the same number, namely $\widetilde{f}(w_0)$. This contradicts the assumed branching of f. $\qquad\square$

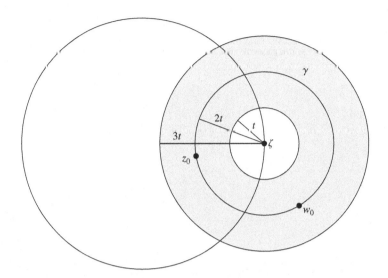

Figure 7.2 The region R (shaded) along with the circle γ.

Example 7.9 By Proposition 7.8, analytic functions on $\mathbb{D}$ having iso-lated branch points on $\mathbb{T}$, such as $\sqrt{1-z}$ and $\log(1-z)$, do not possess pseudocontinuations across $\mathbb{T}$.

Example 7.10 Although the function e^z is analytically continuable to $\mathbb{C}$, its continuation is not meromorphic on $\mathbb{D}_e$ due to its essential singularity at ∞. Thus e^z does not have a pseudocontinuation to $\mathbb{D}_e$.

Example 7.11 The classical gap theorems of Hadamard [101] and Fabry [72] show that analytic functions on $\mathbb{D}$ with lacunary power series, for example,

$$\sum_{n \geqslant 0} z^{2^n}$$

do not have analytic continuations across any point of the unit circle. Using modern techniques, one can show that these functions do not have pseudocon-tinuations either [1, 11, 173].

Example 7.12 It can be shown that the function

$$f(z) = \sum_{n \geqslant 1} \frac{2^{-n}}{z - (1 + \frac{1}{n})}$$

belongs to H^2 (in fact, $|f| \leqslant 2$ on $\mathbb{D}$) and has an analytic continuation to $\mathbb{C} \backslash \{1 + \frac{1}{n} : n \geqslant 1\}$ that vanishes at ∞. Although f has a pseudocontinuation to $\mathbb{D}_e$,

f is not PCBT because $f(1/z)$ is not a function of bounded type on $\mathbb{D}$ since the zeros of $f(1/z)$ in $\mathbb{D}$, namely $\{1 - \frac{1}{n}\}_{n\geqslant 1}$, do not form a Blaschke sequence (Corollary 3.16).

The following seminal result from [61] demonstrates the connection between membership in a model space and pseudocontinuability.

Theorem 7.13 (Douglas, Shapiro, Shields) *Let u be inner. A function $f \in H^2$ belongs to $\mathcal{K}_u$ if and only if f/u has a pseudocontinuation $\widetilde{f/u} \in H^2(\mathbb{D}_e)$ satisfying $\widetilde{f/u}(\infty) = 0$.*

Proof Suppose that $f \in \mathcal{K}_u$. By Proposition 5.4, there exists a $g \in H^2$ such that $f = \overline{gz}u$ almost everywhere on $\mathbb{T}$. Define the function $\widetilde{f/u} \in H^2(\mathbb{D}_e)$ by

$$\widetilde{f/u}(z) := \frac{1}{z}\overline{g\left(\frac{1}{\overline{z}}\right)} \tag{7.4}$$

and note that $\widetilde{f/u}(\infty) = 0$. Since

$$\widetilde{f/u}(\zeta) = \overline{\zeta g(\zeta)} = f(\zeta)/u(\zeta) \quad \text{a.e. } \zeta \in \mathbb{T},$$

it follows that $\widetilde{f/u}$ is a pseudocontinuation of f/u.

For the other direction, suppose that f/u has a pseudocontinuation $\widetilde{f/u} \in H'(\mathbb{D}_e)$ with $\widetilde{f/u}(\infty) = 0$. By (7.2) we have

$$f(\zeta)/u(\zeta) = \widetilde{f/u}(\zeta) = \frac{1}{\zeta}\overline{h\left(\frac{1}{\overline{\zeta}}\right)} \quad \text{a.e. } \zeta \in \mathbb{T},$$

for some $h \in H^2$. The function $g(z) = \overline{h(\overline{z})}$ belongs to H^2 and satisfies $f = \overline{gz}u$ a.e. on $\mathbb{T}$. By Proposition 5.4, $f \in \mathcal{K}_u$. $\qquad\square$

The result above says that each $f \in \mathcal{K}_u$ enjoys a pseudocontinuation $\widetilde{f}$ to $\mathbb{D}_e$ of bounded type. For $\lambda \in \mathbb{D}_e \setminus \{1/\overline{z} : u(z) = 0\}$, the identities in (7.4) and (7.3) imply that

$$\frac{\widetilde{f}(\lambda)}{\overline{\widetilde{u}(\lambda)}} = \widetilde{f/u}(\lambda) = \frac{1}{\lambda}\overline{g\left(\frac{1}{\overline{\lambda}}\right)},$$

where $f = \overline{gz}u$. Thus

$$\widetilde{f}(\lambda) = \frac{1}{\overline{u(1/\overline{\lambda})}}\frac{1}{\lambda}\overline{g\left(\frac{1}{\overline{\lambda}}\right)};$$

and so

$$|\widetilde{f}(\lambda)| \leqslant C_\lambda|g(1/\overline{\lambda})| \leqslant D_\lambda\|g\| = D_\lambda\|f\|,$$

where C_λ and D_λ are positive constants depending only on λ. The inequality above shows that the evaluation functional

$$f \mapsto \widetilde{f}(\lambda)$$

is bounded on $\mathcal{K}_u$. Hence, by the Riesz Representation Theorem (Theorem 1.26),

$$\widetilde{f}(\lambda) = \langle f, v_\lambda \rangle$$

for some $v_\lambda \in \mathcal{K}_u$. The next proposition identifies v_λ.

Proposition 7.14 *For each $\lambda \in \mathbb{D}_e \setminus \{1/\bar{z} : u(z) = 0\}$, the function*

$$v_\lambda(z) := \frac{1 - \overline{\widetilde{u}(\lambda)}u(z)}{1 - \bar{\lambda}z}$$

belongs to $\mathcal{K}_u$ and satisfies

$$\widetilde{f}(\lambda) = \langle f, v_\lambda \rangle$$

for all $f \in \mathcal{K}_u$.

Proof Writing $f = \overline{g}\overline{z}u$, where $g \in H^2$, we see from (7.4) that

$$\frac{\widetilde{f}(\lambda)}{\widetilde{u}(\lambda)} = \frac{1}{\lambda}\overline{g\left(\frac{1}{\bar{\lambda}}\right)}$$

for $\lambda \in \mathbb{D}_e \setminus \{1/\bar{z} : u(z) = 0\}$. Since $k_{1/\bar{\lambda}} \in \mathcal{K}_u$, it follows that g, as well as $\overline{uzk_{1/\bar{\lambda}}}$, belong to $\mathcal{K}_u$ (Proposition 5.4). It now follows that

$$\widetilde{f}(\lambda) = \frac{\widetilde{u}(\lambda)}{\lambda}\overline{g\left(\frac{1}{\bar{\lambda}}\right)}$$

$$= \frac{\widetilde{u}(\lambda)}{\lambda}\langle k_{1/\bar{\lambda}}, g\rangle \qquad\qquad (g \in \mathcal{K}_u)$$

$$= \frac{\widetilde{u}(\lambda)}{\lambda}\langle \overline{g}\overline{z}u, \overline{zk_{1/\bar{\lambda}}}u\rangle$$

$$= \widetilde{u}(\lambda)\Big\langle f, \frac{u(z) - u(1/\bar{\lambda})}{\bar{\lambda}(z - 1/\bar{\lambda})}\Big\rangle$$

$$= \Big\langle f, \frac{\overline{\widetilde{u}(\lambda)}u(z) - \overline{\widetilde{u}(\lambda)}u(1/\bar{\lambda})}{\bar{\lambda}z - 1}\Big\rangle$$

$$= \Big\langle f, \frac{1 - \overline{\widetilde{u}(\lambda)}u(z)}{1 - \bar{\lambda}z}\Big\rangle \qquad\qquad (\overline{\widetilde{u}(\lambda)}u(1/\bar{\lambda}) = 1)$$

$$= \langle f, v_\lambda \rangle.$$

Following the equalities above, notice that v_λ is a constant times $\overline{zk_{1/\bar\lambda}u}$ which, as we have just discussed, belongs to $\mathcal{K}_u$.   □

7.2 Cyclicity via pseudocontinuation

It turns out that the lack of a pseudocontinuation is what characterizes the cyclic vectors for S^*. Recall that $f \in H^2$ is *cyclic* for S^* if

$$\bigvee \{S^{*n}f : n \geqslant 0\} = H^2.$$

If f is not cyclic for S^*, then $\bigvee\{S^{*n}f : n \geqslant 0\}$ is a proper S^*-invariant subspace of H^2 and hence must be equal to $\mathcal{K}_u$ for some inner function u.

Proposition 7.15   *A function $f \in H^2$ is not cyclic for S^* if and only if f is PCBT.*

Proof   If f is non-cyclic for S^*, then f must belong to some model space $\mathcal{K}_u$. By Theorem 7.13, f/u has a pseudocontinuation $\widetilde{f/u}$ (see (7.4)) that belongs to $H^2(\mathbb{D}_e)$. However, since u already has a natural pseudocontinuation $\widetilde{u}$ given by (7.3), we see that f has a pseudocontinuation $\widetilde{u}\widetilde{f/u}$ that is of bounded type.

For the other direction, suppose there exists an $F \in \mathfrak{N}(\mathbb{D}_e)$ for which $F(\zeta) = f(\zeta)$ almost everywhere. The definition of $\mathfrak{N}(\mathbb{D}_e)$ says that

$$F(z) = \frac{u_1(1/z)G_1(1/z)}{u_2(1/z)G_2(1/z)},$$

where u_1, u_2 are inner functions on $\mathbb{D}$ and G_1, G_2 are bounded outer functions on $\mathbb{D}$. But $f = F$ almost everywhere on $\mathbb{T}$ and so it follows that for almost every $\zeta \in \mathbb{T}$,

$$f(\zeta) = \overline{u_2(\bar\zeta)}u_1(\bar\zeta)\frac{\overline{G_1(\bar\zeta)}}{\overline{G_2(\bar\zeta)}}$$

$$= \left(\zeta\overline{u_2(\bar\zeta)}\right) \cdot \left(\overline{\bar\zeta u_1(\bar\zeta)\frac{G_1(\bar\zeta)}{G_2(\bar\zeta)}}\right).$$

The first factor in the previous line is the boundary function for an inner function u, while the second factor is the complex conjugate of a boundary function for a $g \in N^+$ which vanishes at the origin. Since g has L^2 boundary values, we see that $g \in H^2$ (Theorem 3.29). Now use Proposition 5.4 to see that $f \in \mathcal{K}_u$, which means that f is non-cyclic.   □

In light of the preceding proposition, Examples 7.4 and 7.5 show that inner functions and rational functions are non-cyclic for S^*. On the other hand, the functions discussed in Examples 7.9, 7.10, 7.11, and 7.12 are cyclic vectors for S^*.

Proposition 7.15 implies the following result concerning algebraic operations with cyclic and non-cyclic vectors.

Corollary 7.16 *Let $f, g \in H^2$ be non-cyclic and $h \in H^2$ be cyclic for S^*. Then:*

(i) *$f + g$ is non-cyclic;*
(ii) *$f + h$ is cyclic;*
(iii) *If $fg \in H^2$, then fg is non-cyclic;*
(iv) *If $f/g \in H^2$, then f/g is non-cyclic;*
(v) *If $fh \in H^2$, then fh is cyclic;*
(vi) *If $f/h \in H^2$, then f/h is cyclic.*

7.3 Analytic continuation

Every $f \in \mathcal{K}_u$ has a pseudocontinuation to a meromorphic function on $\mathbb{D}_e$ of bounded type. The situation for analytic continuation is more complicated, since the existence of an analytic continuation across an arc of $\mathbb{T}$ is closely tied to the behavior of the inner function u near that arc.

Recall from Theorem 2.14 that an inner function u can be written as

$$u(z) = e^{i\gamma} z^n \underbrace{\left(\prod_{n \geq 1} \frac{\overline{a_n}}{|a_n|} \frac{a_n - z}{1 - \overline{a_n}z} \right)}_{B} \underbrace{\exp\left(-\int_{\mathbb{T}} \frac{\zeta + z}{\zeta - z} d\mu(\zeta) \right)}_{s_\mu}, \qquad (7.5)$$

where $\gamma \in [0, 2\pi)$, the first factor B is a Blaschke product with zero set $\{a_n\}_{n \geq 1}$ (taking multiplicities into account), and the second factor s_μ is a singular inner function with corresponding positive singular measure μ. Initially, the function u is defined only on $\mathbb{D}$. However, it turns out that u might have an analytic continuation across certain portions of $\mathbb{T}$.

Definition 7.17 For a non-constant inner function $u = B s_\mu$ as in (7.5), the *spectrum* of u is the set

$$\sigma(u) = \{a_n\}_{n \geq 1}^- \cup \operatorname{supp} \mu. \qquad (7.6)$$

The term *spectrum* arises from the fact that $\sigma(u)$ is the spectrum of the *compressed shift*, an important operator defined on $\mathcal{K}_u$ that we will discuss in Chapter 9. Since the support of μ and the closure of the zero set $\{a_n\}_{n \geq 1}$ are both closed, it follows that as long as u is not a constant function, the spectrum $\sigma(u)$ is a non-empty compact subset of $\mathbb{D}^-$.

Theorem 7.18 *If u is an inner function and $\sigma(u) \cap \mathbb{T} \neq \mathbb{T}$, then:*

(i) u has an analytic continuation across $\mathbb{T} \setminus \sigma(u)$ to

$$\widehat{\mathbb{C}} \setminus \{1/\bar{z} : z \in \sigma(u)\}; \tag{7.7}$$

(ii) $|u(\zeta)| = 1$ for all $\zeta \in \mathbb{T} \setminus \sigma(u)$;
(iii) The analytic continuation $\widetilde{u}$ of u satisfies the formula

$$\widetilde{u}(z) = \frac{1}{\overline{u(1/\bar{z})}}, \quad z \in \mathbb{D}_e \setminus \{1/\bar{z} : z \in \sigma(u)\};$$

(iv) If $\lambda \in \mathbb{D}$ is a zero of u having order n, then $1/\bar{\lambda} \in \mathbb{D}_e$ is a pole of $\widetilde{u}$ of order n.

Proof   Throughout the following, we assume that u is of the form (7.5). For the proof of (i) and (ii), we consider the Blaschke factor B and the singular inner factor s_μ of u separately.

That s_μ has an analytic continuation to the set W defined by (7.7) follows from the definition of s_μ. Indeed, we only need to concern ourselves with what happens on $\mathbb{T}$. If $z \in W \cap \mathbb{T}$, then z does not belong to the support of μ and hence the integral in the definition of s_μ (from (7.5)) may be differentiated under the integral sign near z.

We now consider the Blaschke product B where, without loss of generality, we assume that $0 \notin \{a_n\}_{n \geq 1}$ (the zeros of B). Thus $|a_n| \geq C > 0$ for some constant C (the existence of such a constant follows from the fact that the zeros of B must tend to the boundary $\mathbb{T}$). Now observe that if K is a compact subset of W, then there exists a second constant C_K such that

$$|1 - \bar{a}_n z| \geq C_K > 0 \quad \forall n \geq 1, z \in K.$$

Letting

$$b_{a_n}(z) = \frac{\bar{a}_n}{|a_n|} \frac{a_n - z}{1 - \bar{a}_n z}$$

denote the Blaschke factor with zero $a_n \neq 0$, we find that for all $z \in K$,

$$|1 - b_{a_n}(z)| = \left| 1 - \left(\frac{\bar{a}_n}{|a_n|} \frac{a_n - z}{1 - \bar{a}_n z} \right) \right|$$

$$= \left| \frac{1}{|a_n|} \left(1 - \frac{|a_n|^2 - \bar{a}_n z}{1 - \bar{a}_n z} \right) + \left(1 - \frac{1}{|a_n|} \right) \right|$$

$$= \left| \frac{1 - |a_n|}{|a_n|} \left(\frac{1 + |a_n|}{1 - \overline{a}_n z} - 1 \right) \right|$$

$$\leqslant \frac{1 - |a_n|}{C} \left(1 + \frac{2}{C_K} \right).$$

Since $\sum_{n \geqslant 1}(1 - |a_n|) < \infty$, we see that $\sum_{n \geqslant 1} |1 - b_{a_n}(z)|$ converges uniformly on K. It follows [158, Ch. 15] that the product defining B converges uniformly on each compact subset K of W and hence defines an analytic function on W. This completes the proof of (i).

(ii) Let $\zeta \in \mathbb{T} \setminus \sigma(u)$ and observe how the proof of (i) shows that the Blaschke product B converges uniformly in a neighborhood of ζ. Since each of the partial products has modulus one at ζ, so does B. For the singular inner factor s_μ, note that since ζ does not belong to the support of the singular measure μ, it follows that $(D\mu)(\zeta)$, the symmetric derivative (see Definition 1.1) of μ at ζ, equals zero. Since s_μ is analytic in a neighborhood of ζ, we see from Fatou's Theorem (Theorem 1.10) that

$$|s_\mu(\zeta)| = \lim_{r \to 1^-} \exp(-\mathscr{P}(\mu)(r\zeta)) = \exp(-(D\mu)(\zeta)) = 1.$$

This establishes (ii).

For (iii), note that $\widetilde{u}$ is the Schwarz reflection of u across $\mathbb{T}$. Moreover, by (ii), $u(\zeta) = \widetilde{u}(\zeta)$ for every $\zeta \in \mathbb{T} \setminus \sigma(u)$. By the Schwarz Reflection Principle [158, p. 230] (or Privalov's uniqueness theorem), $\widetilde{u}$ is a formula for the analytic continuation of u.

The proof of (iv) follows from the formula for $\widetilde{u}$ from (iii).   □

The following theorem provides a second characterization of $\sigma(u)$. Our initial approach (Theorem 7.18) provides us with a convenient description of $\sigma(u)$ in terms of the canonical factorization of u. The following result provides us with a useful characterization of $\sigma(u)$ in terms of the so-called *liminf zero set* of u.

Proposition 7.19   *For an inner function u we have*

$$\sigma(u) = \left\{ \lambda \in \mathbb{D}^- : \varliminf_{z \to \lambda} |u(z)| = 0 \right\}. \tag{7.8}$$

Proof   Let L denote the right-hand side of (7.8) (that is, the liminf zero set of u). Suppose that λ does not belong to the spectrum $\sigma(u)$ of u, as defined by (7.6). If $\lambda \in \mathbb{D}$, then $u(\lambda) \neq 0$ since λ is not a zero of the Blaschke factor of u. By continuity, it follows that $\lambda \notin L$. If $\lambda \in \mathbb{T}$, then Theorem 7.18 tells us that $|u(\lambda)| = 1$. Since both cases yield that $\lambda \notin L$, we conclude that $L \subset \sigma(u)$.

For the reverse inclusion, suppose that $\lambda \in \sigma(u)$. If $\lambda \in \mathbb{D}$, then λ is a zero of u. By continuity, $\lambda \in L$. If $\lambda \in \mathbb{T} \cap \sigma(u)$, then the definition (7.6) of $\sigma(u)$ tells us that either λ is an accumulation point of the zeros of u, in which case $\lambda \in L$ by the definition of liminf, or λ belongs to the support of the singular measure μ associated with the singular inner factor of u.

Suppose $\lambda \in \text{supp}\,\mu$. Since $\mu \perp m$, we know from Theorem 1.2 that μ is carried by the set

$$E := \{w : (D\mu)(w) = \infty\},$$

by which we mean $\mu(\mathbb{T} \setminus E) = 0$. If $\xi \in E$, and $\zeta \in \mathbb{T}$ with $|\zeta - \xi| < 1 - r$, we have

$$|\zeta - r\xi| \leqslant |\zeta - \xi| + |\xi - r\xi| \leqslant 2(1 - r).$$

From here we get

$$P_{r\xi}(\zeta) = \frac{1 - r^2}{|\zeta - r\xi|^2} \geqslant \frac{1 - r^2}{4(1 - r)^2} \geqslant \frac{1}{4(1 - r)}.$$

Therefore,

$$\int_{\mathbb{T}} P_{r\xi}(\zeta)d\mu(\zeta) \geqslant \int_{\{|\zeta - \xi| < 1 - r\}} P_{r\xi}(\zeta)d\mu(\zeta)$$

$$\geqslant \int_{\{|\zeta - \xi| < 1 - r\}} \frac{1}{4(1 - r)}d\mu(\zeta)$$

$$= \frac{1}{4(1 - r)}\mu(\{|\zeta - \xi| < 1 - r\}),$$

from which it follows that

$$\lim_{r \to 1^-} \int_{\mathbb{T}} P_{r\xi}(\zeta)d\mu(\zeta) = \infty \quad \forall \xi \in E.$$

Thus

$$\lim_{r \to 1^-} |s_\mu(r\xi)| = \lim_{r \to 1^-} \exp(-\mathscr{P}(\mu)(r\xi)) = 0,$$

and so $\xi \in \sigma(u)$. If $\xi \in \text{supp}\,\mu \setminus E$, then ξ must be an accumulation point of E, in which case $\xi \in \sigma(u)$ as well. $\qquad\square$

By Theorem 7.18, we know that u can be analytically continued across $\mathbb{T} \setminus \sigma(u)$. To complete our characterization, we need to show that u cannot be analytically continued across any point of $\mathbb{T} \cap \sigma(u)$.

Proposition 7.20 *An inner function u does not have an analytic continuation across any point of $\mathbb{T} \cap \sigma(u)$.*

Proof Suppose that $\zeta \in \sigma(u) \cap \mathbb{T}$ and u is analytic in a neighborhood of ζ. By the liminf description of $\sigma(u)$ (Proposition 7.19), and the assumed continuity of u near ζ, we see that $u(\zeta) = 0$. However, u is analytic near ζ and so, by the Identity Theorem for analytic functions, u is zero free on some punctured disk $\{z : 0 < |z - \zeta| < \delta\}$. At each point $\xi \in \{z : 0 < |z - \zeta| < \delta\} \cap \mathbb{T}$, we use Proposition 7.19 once again to see that $|u(\xi)| = 1$. By the continuity of u near ζ, we see that $|u(\zeta)| = 1$, which is a contradiction. $\square$

The relevance of $\sigma(u)$ to the function theoretic properties of $\mathcal{K}_u$ lies in the following observation [41, p. 84; 61; 141, p. 65].

Proposition 7.21 *Every $f \in \mathcal{K}_u$ has an analytic continuation across $\mathbb{T} \setminus \sigma(u)$.*

Proof   Suppose $\lambda_0 \in \mathbb{T} \setminus \sigma(u)$. By Theorem 7.18 there is an $r_0 > 0$ so that u is analytic in an open neighborhood of the closure $D(\lambda_0, r_0)^-$ of the disk $D(\lambda_0, r_0) = \{|z - \lambda_0| < r_0\}$. Let

$$J = \mathbb{T} \cap D(\lambda_0, r_0)^-$$

and note that any $f \in \mathcal{K}_u$ can be written, as a function of $\lambda \in \mathbb{D}$, by means of the formula

$$f(\lambda) = \langle f, k_\lambda \rangle$$

$$= \int_{\mathbb{T}} f(\zeta) \left(\frac{1 - \overline{u(\zeta)}u(\lambda)}{1 - \overline{\zeta}\lambda} \right) dm(\zeta)$$

$$= \int_{\mathbb{T} \setminus J} \frac{\zeta f(\zeta)}{u(\zeta)} \left(\frac{u(\zeta) - u(\lambda)}{\zeta - \lambda} \right) dm(\zeta) + \int_J \frac{\zeta f(\zeta)}{u(\zeta)} \left(\frac{u(\zeta) - u(\lambda)}{\zeta - \lambda} \right) dm(\zeta).$$

$$(7.9)$$

Note that the first integral in (7.9),

$$\int_{\mathbb{T} \setminus J} \frac{\zeta f(\zeta)}{u(\zeta)} \left(\frac{u(\zeta) - u(\lambda)}{\zeta - \lambda} \right) dm(\zeta),$$

is an analytic function of λ on $D(\lambda_0, r_0)$. To deal with the second integral, observe that

$$u(z) = \sum_{n \geq 0} a_n (z - \lambda_0)^n, \quad z \in D(\lambda_0, r_0);$$

and so for $\zeta \in J$,

$$u(\lambda) - u(\zeta) = \sum_{n \geq 0} a_n \left[(\lambda - \lambda_0)^n - (\zeta - \lambda_0)^n \right]$$

$$= \sum_{n \geq 1} a_n (\lambda - \zeta) \cdot \sum_{0 \leq m \leq n-1} (\lambda - \lambda_0)^m (\zeta - \lambda_0)^{n-m-1}$$

$$= (\lambda - \zeta) \sum_{m \geq 0} \left[\sum_{1 \leq n \leq m+1} a_n (\zeta - \lambda_0)^{n-m-1} \right] (\lambda - \lambda_0)^m.$$

Hence for $\lambda \in D(\lambda_0, r_0)$, the second integral in (7.9) becomes

$$\int_J \frac{\zeta f(\zeta)}{u(\zeta)} \left(\frac{u(\zeta) - u(\lambda)}{\zeta - \lambda} \right) dm(\zeta)$$

$$= \int_J \frac{\zeta f(\zeta)}{u(\zeta)} \cdot \sum_{m \geqslant 0} \left[\sum_{1 \leqslant n \leqslant m+1} a_n (\zeta - \lambda_0)^{n-m-1} \right] \cdot (\lambda - \lambda_0)^m dm(\zeta)$$

$$= \sum_{m \geqslant 0} \left[\sum_{1 \leqslant n \leqslant m+1} \int_J \frac{\zeta f(\zeta)}{u(\zeta)} (\zeta - \lambda_0)^{n-m-1} dm(\zeta) \right] \cdot (\lambda - \lambda_0)^m.$$

Notice how the above defines an analytic function of λ on $D(\lambda_0, r_0)$. Thus f can be extended to be an analytic function on $D(\lambda_0, r_0)$. $\square$

Remark 7.22   There is an alternate proof of this result using Morera's Theorem (see [41, 61]).

We conclude this section with the following observation that is most likely known to experts in the field, although we have not been able to find a reference for it. Although initially somewhat surprising, when one thinks in terms of the Aleksandrov Density Theorem (Theorem 5.24), which states that the functions in $\mathcal{K}_u$ that are continuous on $\mathbb{D}^-$ are dense in $\mathcal{K}_u$, this next result is not so implausible.

Proposition 7.23   *If u is an inner function, then there exists a non-constant function in $\mathcal{K}_u$ that can be analytically continued across some non-empty open arc of $\mathbb{T}$.*

Proof   If u is a finite Blaschke product, the result follows from the fact that all of the functions in $\mathcal{K}_u$ are rational functions whose poles lie in $|z| > 1$ (Corollary 5.18). Now recall that if v is an inner function that divides u, then $\mathcal{K}_v \subset \mathcal{K}_u$ (Corollary 5.9). If u contains a Blaschke factor, then the preceding remarks ensure that $\mathcal{K}_u$ contains a function that can be analytically contin- ued across $\mathbb{T}$ itself. It therefore suffices to consider the case where $u = s_\mu$ is a singular inner function. Moreover, we may also assume that $\operatorname{supp} \mu = \mathbb{T}$ since otherwise $\mathbb{T} \setminus \operatorname{supp} \mu$ contains a non-empty open arc J across which every function in $\mathcal{K}_u$ is analytically continuable (Proposition 7.21).

Since μ can have only countably many atoms (else μ would not be a finite measure), we may find a non-empty open arc J with $0 < m(J) < 1$ so that the endpoints of J are not atoms of μ. Define a singular measure v on $\mathbb{T}$ by setting $v(E) = \mu(E \cap J)$ for every Borel set E and let $w = s_v$. Using the fact that each function in $\mathcal{K}_w$ can be analytically continued across $\mathbb{T} \setminus J^-$ (Proposition 7.21), along with the inclusion $\mathcal{K}_w \subset \mathcal{K}_u$ (since w divides u (Corollary 5.9)), the desired result follows.   □

7.4 Boundary limits

Functions in H^2 possess non-tangential limits almost everywhere on $\mathbb{T}$. However, for each *fixed* ζ in $\mathbb{T}$, it is easy to produce an H^2 function that does not have a finite non-tangential limit at ζ (for example, $f(z) = \log(\zeta - z)$). In contrast to this, for a given model space $\mathcal{K}_u$, there may exist a point ζ in $\mathbb{T}$ such that every function in $\mathcal{K}_u$ possesses a non-tangential limit at ζ. For example, if $\zeta \in \mathbb{T} \setminus \sigma(u)$, then *every* $f \in \mathcal{K}_u$ has an analytic continuation to a neighborhood of ζ (Proposition 7.21) and so certainly each f has a non-tangential limit at ζ. On the other hand, it is sometimes possible for ζ to belong to $\sigma(u) \cap \mathbb{T}$, which rules out the possibility of every $f \in \mathcal{K}$ having an analytic continuation to a neighborhood of ζ (Proposition 7.20), yet every $f \in \mathcal{K}_u$ has a non-tangential limit at ζ. References for this are [4, 5, 165].

Theorem 7.24 *Let u be an inner function, $\zeta \in \mathbb{T}$, and define*

$$c = \lim_{z \to \zeta} \frac{1 - |u(z)|}{1 - |z|}.$$

Then the following are equivalent:

 (i) $c < \infty$;
 (ii) *There exists a $\lambda \in \mathbb{T}$ such that*

$$\frac{u - \lambda}{z - \zeta} \in \mathcal{K}_u;$$

(iii) *Each $f \in \mathcal{K}_u$ has a non-tangential limit at ζ;*
(iv) *u has an angular derivative in the sense of Carathéodory at ζ.*

Moreover, under the equivalent conditions above, u has the following properties:

(a) $\lambda = u(\zeta)$ *and* $u'(\zeta) = \overline{\zeta} u(\zeta) |u'(\zeta)|$;
(b) *For each $f \in \mathcal{K}_u$, we have*

$$f(\zeta) = \langle f, k_\zeta \rangle,$$

   where

$$k_\zeta = \frac{1 - \overline{u(\zeta)} u}{1 - \overline{\zeta} z} \in \mathcal{K}_u;$$

(c) $c > 0$ and

$$c = \angle \lim_{z \to \zeta} \frac{1 - |u(z)|}{1 - |z|} = |u'(\zeta)| = k_\zeta(\zeta) = \|k_\zeta\|^2;$$

(d)

$$\angle \lim_{z \to \zeta} \|k_z - k_\zeta\| = 0.$$

Before starting the proof, let us make an important observation which is at the heart of our approach. We claim that for every $\zeta \in \mathbb{T}$,

$$\lim_{z \to \zeta} \frac{1 - |u(z)|}{1 - |z|} = \lim_{z \to \zeta} \frac{1 - |u(z)|^2}{1 - |z|^2}. \tag{7.10}$$

Denote the right-hand side by c'.

Suppose that $c = \infty$. Since

$$\frac{1 - |u(z)|}{1 - |z|} = \frac{1 - |u(z)|^2}{1 - |z|^2} \frac{1 + |z|}{1 + |u(z)|} \leqslant 2 \frac{1 - |u(z)|^2}{1 - |z|^2},$$

we must also have $c' = \infty$. A similar argument shows that if $c' = \infty$ then $c = \infty$.

If $c < \infty$, there is a sequence $\{z_n\}_{n \geqslant 1} \subset \mathbb{D}$ such that $z_n \to \zeta$ and

$$\lim_{n \to \infty} \frac{1 - |u(z_n)|}{1 - |z_n|} = c.$$

This implies $\lim_{n \to \infty} |u(z_n)| = 1$, and thus

$$\lim_{n \to \infty} \frac{1 + |u(z_n)|}{1 + |z_n|} = 1.$$

Consequently,

$$c' = \lim_{z \to \zeta} \frac{1 - |u(z)|^2}{1 - |z|^2} \leqslant \lim_{n \to \infty} \frac{1 - |u(z_n)|^2}{1 - |z_n|^2} = \lim_{n \to \infty} \frac{1 - |u(z_n)|}{1 - |z_n|} = c.$$

Now pick a sequence $\{w_n\}_{n \geqslant 1} \subset \mathbb{D}$ with $w_n \to \zeta$, such that

$$c' = \lim_{n \to \infty} \frac{1 - |u(w_n)|^2}{1 - |w_n|^2}.$$

A similar argument shows that $|u(w_n)| \to 1$ and thus

$$c' = \lim_{n \to \infty} \frac{1 - |u(w_n)|}{1 - |w_n|} \geqslant c.$$

This verifies (7.10).

The main advantage of (7.10) is that we can rewrite it as

$$c = \lim_{z \to \zeta} \|k_z\|^2, \tag{7.11}$$

which places the problem in the language of model spaces and reproducing kernels.

Proof of Theorem 7.24   We will show that (i) $\implies$ (ii) $\implies$ (iii) $\implies$ (i) and then (i), (ii), (iii) $\implies$ (iv) $\implies$ (i). The properties (a) – (d) will be established at different steps of the proof.

(i) $\implies$ (ii): By (7.11) there exists a sequence $\{z_n\}_{n \geqslant 1} \subset \mathbb{D}$ such that

$$\lim_{n \to \infty} z_n = \zeta \quad \text{and} \quad c = \lim_{n \to \infty} \|k_{z_n}\|^2 < \infty.$$

By the Banach–Alaoglu Theorem (Theorem 1.25), $\{k_{z_n}\}_{n \geqslant 1}$ has a weakly convergent subsequence with a limit $k \in \mathcal{K}_u$. Since $\{u(z_n)\}_{n \geqslant 1}$ is a bounded sequence of complex numbers, it also has a convergent subsequence with a limit in $\mathbb{D}^-$. Thus, by passing to subsequences, we have a $\lambda \in \mathbb{D}^-$ and a $k \in \mathcal{K}_u$ such that

$$k_{z_n} \to k \text{ weakly and } u(z_n) \to \lambda.$$

The reproducing kernels of $\mathcal{K}_u$ enable us to obtain more information about λ and k. Indeed, for each $z \in \mathbb{D}$, we have

$$
\begin{aligned}
k(z) = \langle k, k_z \rangle &= \lim_{n \to \infty} \langle k_{z_n}, k_z \rangle = \lim_{n \to \infty} k_{z_n}(z) \\
&= \lim_{n \to \infty} \frac{1 - \overline{u(z_n)}\, u(z)}{1 - \bar{z}_n z} \\
&= \frac{1 - \bar{\lambda}\, u(z)}{1 - \bar{\zeta} z}.
\end{aligned}
$$

Since $k \in H^2 \setminus \{0\}$ and $1/(1 - \bar{\zeta} z) \notin H^2$ (note that $(1 - \bar{\zeta} z)^{-1}$ does not have square summable Taylor coefficients), it must be the case that $|\lambda| = 1$. However, remembering that k_{z_n} converges weakly to k, we get

$$\|k\|^2 \leqslant \lim_{n \to \infty} \|k_{z_n}\|^2 = c. \tag{7.12}$$

The above stems from the inequality

$$\|k_{z_n}\| \, \|k\| \geqslant |\langle k_{z_n}, k \rangle| \to \langle k, k \rangle = \|k\|^2.$$

As a byproduct of this, we conclude that $c > 0$. We also have the identity

$$\lambda \bar{\zeta} k(z) = \frac{u(z) - \lambda}{z - \zeta}, \qquad z \in \mathbb{D}, \tag{7.13}$$

which will be needed below when we discuss $u'(\zeta)$, the angular derivative of u at ζ. Moreover, (7.13) shows that the difference quotient on the right side is a function belonging to $\mathcal{K}_u$. This completes the proof of (ii).

(ii) $\implies$ (iii): Define k using the identity in (7.13) and note that our assumption implies that $k \in \mathcal{K}_u$. Hence,

$$u(z) = \lambda + \lambda\overline{\zeta}\,(z - \zeta)k(z), \quad z \in \mathbb{D},$$

which implies that

$$|u(z) - \lambda| \leqslant |z - \zeta||k(z)| = |z - \zeta|\,|\langle k, k_z\rangle| \leqslant |z - \zeta|\,\|k\|\,\|k_z\|$$

$$= |z - \zeta|\,\|k\| \left(\frac{1 - |u(z)|^2}{1 - |z|^2}\right)^{1/2}$$

$$\leqslant \|k\|\frac{|z - \zeta|}{(1 - |z|^2)^{1/2}}.$$

If we restrict z to be inside the Stolz domain $\Gamma_\alpha(\zeta)$, then

$$|u(z) - \lambda| \leqslant \alpha\|k\|(1 - |z|)^{1/2}.$$

The right-hand side of the previous inequality tends to zero as z tends to ζ from within $\Gamma_\alpha(\zeta)$, and thus u has a non-tangential limit $u(\zeta)$ at ζ. Moreover, this limit is precisely the constant λ. From now on, we will adhere to the following convention:

$$u(\zeta) := \lambda, \qquad k := k_\zeta. \tag{7.14}$$

These definitions are designed so that

$$k_\zeta(z) = \langle k_\zeta, k_z\rangle, \qquad z \in \mathbb{D}.$$

We now want to explore how k_z approaches k_ζ as z approaches ζ. This requires some further estimates.

First, by the Cauchy–Schwarz Inequality, we have

$$|k_\zeta(z)| = |\langle k_\zeta, k_z\rangle| \leqslant \|k_\zeta\|\,\|k_z\|. \tag{7.15}$$

Second,

$$|k_\zeta(z)| = \frac{|1 - \overline{u(\zeta)}\,u(z)|}{|1 - \overline{\zeta}\,z|} \geqslant \frac{1 - |u(z)|}{|z - \zeta|}.$$

But since

$$1 - |u(z)| = \frac{(1 - |z|^2)\,\|k_z\|^2}{(1 + |u(z)|)},$$

we have

$$|k_\zeta(z)| \geqslant \frac{(1 - |z|^2)\|k_z\|^2}{(1 + |u(z)|)|z - \zeta|}. \tag{7.16}$$

The inequalities (7.15) and (7.16) together imply that

$$\|k_z\| \leqslant \|k_\zeta\| \frac{1 + |u(z)|}{1 + |z|} \frac{|z - \zeta|}{1 - |z|}.$$

Hence, in each Stoltz domain $\Gamma_\alpha(\zeta)$,

$$\|k_z\| \leqslant 2\alpha\|k_\zeta\|, \qquad z \in \Gamma_\alpha(\zeta). \tag{7.17}$$

In other words, $\|k_z\|$ remains bounded as z approaches ζ inside $\Gamma_\alpha(\zeta)$.

Fix $w \in \mathbb{D}$. The definition of $u(\zeta)$ from (7.14) implies that

$$\angle \lim_{z \to \zeta} \frac{1 - \overline{u(z)}u(w)}{1 - \bar{z}w} = \frac{1 - \overline{u(\zeta)}u(w)}{1 - \bar\zeta w}.$$

Write the identity above as

$$\angle \lim_{z \to \zeta} k_z(w) = k_\zeta(w),$$

and then interpret it in the following way:

$$\angle \lim_{z \to \zeta} \langle k_z, k_w \rangle = \langle k_\zeta, k_w \rangle.$$

Therefore,

$$\angle \lim_{z \to \zeta} \langle k_z, f \rangle = \langle k_\zeta, f \rangle \tag{7.18}$$

whenever $f \in \mathcal{K}_u$ is any element of the form

$$f = \alpha_1 k_{w_1} + \cdots + \alpha_n k_{w_n}, \qquad \alpha_j \in \mathbb{C}, \ w_j \in \mathbb{D}.$$

On one hand, this collection is dense in $\mathcal{K}_u$ (Proposition 5.20). On the other hand, we can use (7.17) to see that as z approaches ζ from within $\Gamma_\alpha(\zeta)$, the family k_z is uniformly (norm) bounded. This implies that the identity (7.18) holds for all $f \in \mathcal{K}_u$, and furthermore,

$$f(\zeta) = \angle \lim_{z \to \zeta} f(z) = \langle f, k_\zeta \rangle, \qquad f \in \mathcal{K}_u.$$

This completes the proof of (iii). In particular, taking $f = k_\zeta$, we obtain $k_\zeta(\zeta) = \|k_\zeta\|^2$. Also note that the relation (7.18) says that k_z tends weakly to k_ζ as z tends non-tangentially to ζ.

(iii) $\implies$ (i): Fix any Stoltz domain $\Gamma_\alpha(\zeta)$. The relation $f(z) = \langle f, k_z \rangle$ and our assumption (existence of a non-tangential limit for every $f \in \mathcal{K}_u$) imply that

$$\sup_{z \in \Gamma_\alpha(\zeta)} |\langle f, k_z \rangle| = c_f < \infty.$$

Therefore, by the Principle of Uniform Boundedness, we must have

$$c' = \sup_{z \in \Gamma_\alpha(\zeta)} \|k_z\| < \infty. \tag{7.19}$$

Take any sequence inside $\Gamma_\alpha(\zeta)$ that approaches ζ, for example, $z_n = (1 - 1/n)\zeta$, $n \geqslant 1$, and observe that

$$\frac{1 - |u(z_n)|^2}{1 - |z_n|^2} = \|k_{z_n}\|^2 \leqslant c'^2, \qquad n \geqslant 1.$$

This implies that $\lim_{n \to \infty} |u(z_n)| = 1$. Moreover,

$$c \leqslant \varliminf_{n \to \infty} \frac{1 - |u(z_n)|^2}{1 - |z_n|^2} = \lim_{n \to \infty} \|k_{z_n}\|^2 \leqslant c'^2.$$

(i), (ii), (iii) $\implies$ (iv): Since $k_\zeta \in \mathcal{K}_u$ we can write

$$\frac{u(z) - u(\zeta)}{z - \zeta} = \overline{\zeta} u(\zeta) k_\zeta(z) = \overline{\zeta} u(\zeta) \langle k_\zeta, k_z \rangle, \qquad z \in \mathbb{D}.$$

In the course of the proof, we saw that our assumptions imply that k_z converges weakly to k_ζ as z tends non-tangentially to ζ. Hence,

$$\angle \lim_{z \to \zeta} \frac{u(z) - u(\zeta)}{z - \zeta} = \overline{\zeta} u(\zeta) \|k_\zeta\|^2,$$

which, by Theorem 2.19, means that u has an angular derivative at ζ in the sense of Carathéodory and

$$u'(\zeta) = \overline{\zeta} u(\zeta) \|k_\zeta\|^2. \tag{7.20}$$

As the first byproduct of (7.20), note that $|u'(\zeta)| = \|k_\zeta\|^2$. To obtain further information, we proceed to show that

$$\angle \lim_{z \to \zeta} \|k_z\| = \|k_\zeta\|. \tag{7.21}$$

This fact, which we will prove momentarily, has at least three consequences:

1. By (7.12), $c \geqslant \|k_\zeta\|^2$. Using (7.11) we also see the reverse inequality $c \leqslant \|k_\zeta\|^2$. Thus $c = \|k_\zeta\|^2$.

2. In light of the fact that k_z converges weakly to k_ζ as z tends non-tangentially to ζ, along with the identity

$$\|k_z - k_\zeta\|^2 = \|k_z\|^2 - 2\operatorname{Re}\langle k_z, k_\zeta\rangle + \|k_\zeta\|^2,$$

and the facts that $\|k_z\| \to \|k_\zeta\|$ (by (7.21)) and $\langle k_z, k_\zeta\rangle \to \|k_\zeta\|^2$, we obtain

$$\angle \lim_{z\to\zeta} \|k_z - k_\zeta\| = 0.$$

3. By (7.11),

$$c = \angle \lim_{z\to\zeta} \|k_z\|^2 = \|k_\zeta\|^2 = \angle \lim_{z\to\zeta} \frac{1 - |u(z)|^2}{1 - |z|^2} = \lim_{z\to\zeta} \frac{1 - |u(z)|}{1 - |z|}.$$

To prove that $\|k_z\| \to \|k_\zeta\|$, as z tends non-tangentially to ζ, let

$$g(z) = \frac{u(z) - u(\zeta)}{z - \zeta} - u'(\zeta), \qquad z \in \mathbb{D}.$$

Thus,

$$u(z) = u(\zeta) + u'(\zeta)(z - \zeta) + (z - \zeta)g(z), \qquad z \in \mathbb{D};$$

and, by (7.20),

$$|u(z)|^2 = 1 - 2\|k_\zeta\|^2 \operatorname{Re}(1 - \bar{\zeta}z) + h(z), \qquad z \in \mathbb{D}, \tag{7.22}$$

where

$$h(z) = (|u'(\zeta)|^2 + |g(z)|^2)|z - \zeta|^2 + 2\operatorname{Re}\left(g(z)(z - \zeta)\overline{(u(\zeta) + u'(\zeta)(z - \zeta))}\right).$$

Proposition 3.25 gives us

$$\angle \lim_{z\to\zeta} \frac{|h(z)|}{1 - |z|} = 0. \tag{7.23}$$

Another elementary fact is

$$\frac{\operatorname{Re}(1 - \bar{\zeta}z)}{1 - |z|^2} = \frac{1}{2} + \frac{1}{2}\frac{|z - \zeta|^2}{1 - |z|^2},$$

and from this identity we have

$$\angle \lim_{z\to\zeta} \frac{\operatorname{Re}(1 - \bar{\zeta}z)}{1 - |z|^2} = \frac{1}{2}. \tag{7.24}$$

Therefore by (7.22), (7.23), and (7.24),

$$\angle \lim_{z\to\zeta} \|k_z\|^2 = \angle \lim_{z\to\zeta} \frac{1 - |u(z)|^2}{1 - |z|^2} = \|k_\zeta\|^2.$$

(iv) $\implies$ (i): The inequality

$$\frac{1 - |u(r\zeta)|}{1 - r} \leqslant \left| \frac{u(r\zeta) - u(\zeta)}{r\zeta - \zeta} \right|$$

yields

$$c = \lim_{z \to \zeta} \frac{1 - |u(z)|}{1 - |z|} \leqslant \lim_{r \to 1} \left| \frac{u(r\zeta) - u(\zeta)}{r\zeta - \zeta} \right| = |u'(\zeta)| < \infty. \qquad \square$$

Recall the explicit condition from Theorem 2.21 which says that $u = Bs_\mu$ has a finite angular derivative at $\zeta \in \mathbb{T}$ if and only if

$$\sum_{n \geqslant 1} \frac{1 - |a_n|}{|\zeta - a_n|^2} + \int \frac{1}{|\zeta - \xi|^2} d\mu(\xi) < \infty.$$

Here are two examples from [102] of when this condition is satisfied, thus implying that every $f \in \mathcal{K}_u$ has a non-tangential limit at ζ. In each of these examples, notice that $\zeta \in \sigma(u)$ (where analytic continuation does not always occur – Proposition 7.20).

Example 7.25 Suppose $a_n = r_n e^{i\theta_n}$ with

$$r_n = 1 - x_n \theta_n^2, \quad \theta_n \downarrow 0, \quad x_n > 0, \quad \sum_{n \geqslant 1} x_n < \infty. \tag{7.25}$$

In a moment, we will show this sequence approaches 1 *tangentially* and will eventually lie outside any given Stolz domain $\Gamma_\alpha(1)$ (see Figure 7.3). For now, notice that $\theta_n \downarrow 0$ and so

$$\sum_{n \geqslant 1}(1 - |a_n|) = \sum_{n \geqslant 1}(1 - r_n) = \sum_{n \geqslant 1} \theta_n^2 x_n \leqslant \theta_1^2 \sum_{n \geqslant 1} x_n < \infty.$$

Thus $\{a_n\}_{n \geqslant 1}$ is indeed a Blaschke sequence.

We will need the well-known Pythagorean type result (see Figure 7.4): if $a = re^{i\theta}$ and $r \in (0, 1)$, then

$$|1 - a|^2 \asymp (1 - r)^2 + \theta^2 \asymp [(1 - r) + |\theta|]^2, \quad r \approx 1, \theta \approx 0. \tag{7.26}$$

Observe that by using (7.26) and the inequality $0 < x_n \theta_n^2 \leqslant \theta_n$, we get

$$|1 - a_n| \asymp (1 - r_n) + \theta_n \asymp x_n \theta_n^2 + \theta_n \asymp \theta_n.$$

Hence

$$\frac{|1 - a_n|}{1 - |a_n|} \asymp \frac{\theta_n}{1 - r_n} = \frac{1}{\theta_n r_n} \to \infty, \quad n \to \infty. \tag{7.27}$$

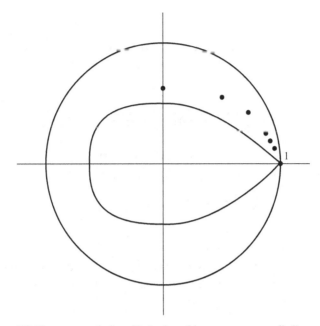

Figure 7.3 The sequence $\{a_n\}_{n\geqslant 1}$. Notice how this sequence eventually lies outside any given Stolz domain $\Gamma_\alpha(1)$.

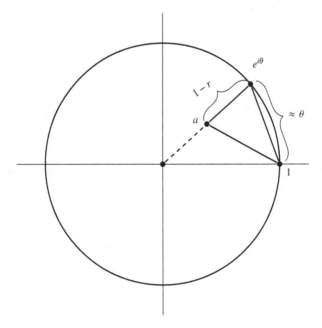

Figure 7.4 The estimate $|1 - a| \asymp (1 - r) + |\theta|$ as $r \approx 1$, $\theta \approx 0$.

This means that for any Stolz domain $\Gamma_\alpha(1)$ anchored at 1, there exists an N_α such that $a_n \notin \Gamma_\alpha(1)$ for all $n > N_\alpha$. The estimate in (7.27) also gives us

$$\sum_{n\geqslant 1} \frac{1 - |a_n|}{|1 - a_n|} \asymp \sum_{n\geqslant 1} \frac{1 - r_n}{\theta_n} = \sum_{n\geqslant 1} \theta_n x_n \leqslant \theta_1 \sum_{n\geqslant 1} x_n < \infty$$

and so the Frostman condition in Theorem 2.17 is satisfied. This ensures that if B is a Blaschke product whose zeros are $\{a_n\}_{n\geqslant 1}$, then

$$\angle \lim_{z \to 1} B(z) = \eta \in \mathbb{T}.$$

In a similar way,

$$\sum_{n\geqslant 1} \frac{1 - |a_n|}{|1 - a_n|^2} \asymp \sum_{n\geqslant 1} x_n < \infty.$$

In light of Theorem 2.21, B has a finite angular derivative at 1, even though $1 \in \{a_n\}_{n\geqslant 1}^- = \sigma(B)$.

Example 7.26   Suppose that μ is the discrete measure

$$d\mu = \sum_{n\geqslant 1} \alpha_n \delta_{e^{i\theta_n}},$$

where $\alpha_n > 0$, $\sum_{n\geqslant 1} \alpha_n < \infty$ (so μ is a finite measure), and $\theta_n \downarrow 0$. To determine whether or not the associated singular inner function s_μ has a finite angular derivative at $\zeta = 1$, we need to determine whether or not

$$\int_{\mathbb{T}} \frac{1}{|1 - \xi|^2} d\mu(\xi) = \sum_{n\geqslant 1} \alpha_n \frac{1}{|1 - e^{i\theta_n}|^2} < \infty.$$

Observe that $|1 - e^{i\theta_n}|^2 = 2(1 - \cos\theta_n) \asymp \theta_n^2$ and so

$$\sum_{n\geqslant 1} \alpha_n \frac{1}{|1 - e^{i\theta_n}|^2} \asymp \sum_{n\geqslant 1} \frac{\alpha_n}{\theta_n^2}.$$

For example, setting $\alpha_n = 1/n^4$ and $\theta_n = 1/n$ produces a measure μ so that s_μ has a finite angular derivative at $\zeta = 1 \in \{\alpha_n\}_{n\geqslant 1}^- = \operatorname{supp}\mu = \sigma(s_\mu)$.

7.5 Notes

7.5.1 Cyclic vectors for S^*

At least in principle, the cyclic vector problem for S^* is solved (f is non-cyclic for S^* if and only if f is PCBT – Proposition 7.15). However, this solution is not as explicit as the solution to the cyclic vector problem for S (f is cyclic for S if and only if f is outer – Corollary 4.5). Outer functions are, in a sense, readily identifiable via Proposition 3.22. On the other hand, PCBT functions are not so easily recognized. We refer the reader to some papers which partially characterize the non-cyclic vectors for S^* by means of such conditions as the growth of its Taylor coefficients [119, 171], the modulus of the function [61, 119], and gaps in its Taylor coefficients [1, 11, 61]. In the next chapter, we will parameterize the non-cyclic vectors in terms of a conjugation operator on a model space.

7.5.2 Pseudocontinuation and cyclic vectors

The non-cyclic vectors for the backward shift on H^2 have pseudocontinuations of bounded type (Proposition 7.15). This phenomenon is not isolated to H^2 and appears in other Hilbert spaces of analytic functions. For example, in the Bergman space L^2_a (see (3.28)), and other closely related spaces, the non-cyclic vectors for the backward shift also have pseudocontinuations of bounded type, though this is not the full description of the non-cyclic vectors. For the Dirichlet space $\mathcal{D}$ (see (3.29)), the situation is much different in that there are non-cyclic vectors for the backward shift which do not have pseudocontinuations across any arc of $\mathbb{T}$. See [14, 15, 154] for more details

7.5.3 Generalized analytic continuation

There are a number of meaningful ways that one can consider two meromorphic functions, one on $\mathbb{D}$ and another on $\mathbb{D}_e$, to be "extensions" of one another. For example, if $f = \sum_{n \geqslant 0} a_n z^n$ and $g = \sum_{n \geqslant 0} b_n z^n$ are analytic on $\mathbb{D}$ while $F = \sum_{n \geqslant 0} \frac{c_n}{z^n}$ is analytic on $\mathbb{D}_e$, then, under very mild technical conditions, one can consider F to be an extension of f/g if $f(z) = g(z)F(z)$ by formal multiplication of power series. A Fourier series argument will show how this occurs with f/u and F_u as discussed in (7.1). We refer the reader to [154, 156] for other examples of generalized analytic continuation and how they appear in many unexpected areas of analysis.

7.5.4 Model spaces and tangential limits

One can also discuss *tangential* limits of functions from model spaces [28, 36]. In fact, there is an alternate proof of Theorem 7.24 in [104] which can be used to prove a bit more. Indeed, suppose that u is inner, $\zeta \in \mathbb{T}$, and Ω_ζ is an approach region with vertex at ζ. By the term "approach region," we mean an open subset of $\mathbb{D}$ whose closure intersects $\mathbb{T}$ only at ζ. For example, an approach region might be the interior of a triangle whose vertex is at ζ (a Stolz approach region), or the interior of an internally tangent circle with contact point ζ (an *oricyclic* approach region), or perhaps one might want to consider other, even more tangential, approach regions. Whatever the approach region Ω_ζ, the theorem here is that every $f \in \mathcal{K}_u$ has a limit as $z \to \zeta$ (with $z \in \Omega_\zeta$) if and only if $\sup\{\|k_z\| : z \in \Omega_\zeta\} < \infty$. Is there a condition, similar to Frostman's Theorem (Theorem 2.21), that one could test to determine whether or not $\sup\{\|k_z\| : z \in \Omega_\zeta\} < \infty$?

7.5.5 When Ahern–Clark fails

There are some results from [103] which control the radial growth of functions from $\mathcal{K}_u$ when the conditions of Theorem 7.24 fail.

8

Conjugation

Each model space $\mathcal{K}_u$ comes equipped with a conjugation, a certain conjugate-linear operator on $\mathcal{K}_u$ that generalizes complex conjugation $z \mapsto \bar{z}$ on $\mathbb{C}$. Not only does this conjugation cast a new light on pseudocontinuations, it also interacts with a number of important linear operators that act on model spaces.

8.1 Abstract conjugations

We begin by introducing the following simple, but surprisingly fruitful, concept.

Definition 8.1 A *conjugation* on a complex separable Hilbert space $\mathcal{H}$ is a mapping $C : \mathcal{H} \to \mathcal{H}$ that is:

(i) conjugate linear: $C(\alpha\mathbf{x} + \beta\mathbf{y}) = \bar{\alpha}C\mathbf{x} + \bar{\beta}C\mathbf{y}$ for all $\mathbf{x}, \mathbf{y}$ in $\mathcal{H}$ and $\alpha, \beta \in \mathbb{C}$;
(ii) involutive: $C^2 = I$;
(iii) isometric: $\|C\mathbf{x}\| = \|\mathbf{x}\|$ for all $\mathbf{x}$ in $\mathcal{H}$.

In the literature, some authors use the term *anti-linear* instead of *conjugate-linear*. From this perspective, a mapping that satisfies the first and third conditions listed above is called an *anti-unitary* operator. Thus a conjugation is simply an anti-unitary operator that is involutive. Let us also remark that in light of the well-known polarization identity

$$4\langle \mathbf{x}, \mathbf{y} \rangle = \|\mathbf{x} + \mathbf{y}\|^2 - \|\mathbf{x} - \mathbf{y}\|^2 + i\|\mathbf{x} + i\mathbf{y}\|^2 - i\|\mathbf{x} - i\mathbf{y}\|^2, \quad \mathbf{x}, \mathbf{y} \in \mathcal{H},$$

the isometric condition is equivalent to the seemingly stronger assertion that

$$\langle C\mathbf{x}, C\mathbf{y} \rangle = \langle \mathbf{y}, \mathbf{x} \rangle, \quad \mathbf{x}, \mathbf{y} \in \mathcal{H}.$$

170

Notice how this condition, along with the fact that $C^2 = I$, implies the useful identity

$$\langle C\mathbf{x}, \mathbf{y}\rangle = \langle C\mathbf{y}, \mathbf{x}\rangle, \qquad \mathbf{x}, \mathbf{y} \in \mathcal{H}. \tag{8.1}$$

Although our interest in conjugations stems primarily from a single, specific example in the context of model spaces, we first consider a number of apparently unrelated examples that will turn out to be relevant later on.

Example 8.2 For a measure space (X, μ), the *canonical conjugation* on $L^2(X, \mu)$ is pointwise complex conjugation:

$$[Cf](x) = \overline{f(x)}.$$

Particular instances include the canonical conjugation

$$C(z_1, z_2, \ldots, z_n) = (\overline{z_1}, \overline{z_2}, \ldots, \overline{z_n}) \tag{8.2}$$

on $\mathbb{C}^n$ (which can be regarded as the L^2 space corresponding to a measure consisting of n distinct unit point masses) and the canonical conjugation

$$C(z_1, z_2, z_3, \ldots) = (\overline{z_1}, \overline{z_2}, \overline{z_3}, \ldots) \tag{8.3}$$

on the space $\ell^2(\mathbb{N})$ of all square-summable sequences of complex numbers.

Example 8.3 The *Toeplitz conjugation* on $\mathbb{C}^n$ is defined by

$$C(z_1, z_2, \ldots, z_n) = (\overline{z_n}, \overline{z_{n-1}}, \ldots, \overline{z_1}). \tag{8.4}$$

As its name suggests, the Toeplitz conjugation is intimately related to the study of Toeplitz matrices. In light of its appearance in the *Szegő recurrence* from the theory of orthogonal polynomials on the unit circle [179, eq. 1.1.7], one might also choose to refer to (8.4) as the *Szegő conjugation*.

Example 8.4 Along similar lines, if one has a measure space (X, μ) that possesses a certain amount of symmetry, one can sometimes form a corresponding conjugation. For instance, the conjugation

$$[Cf](x) = \overline{f(1 - x)}$$

on $L^2[0, 1]$ arises in the study of the Volterra integration operator.

Example 8.5 Another example along the lines of the previous one is the conjugation C on $L^2(\mathbb{T}, m)$ defined by

$$(Cf)(\zeta) = \overline{f(\overline{\zeta})}.$$

Note that C is conjugate linear and involutive. The fact that C is isometric comes from Parseval's Theorem. Using (8.1), this conjugation yields the useful integral identity

$$\int_{\mathbb{T}} f(\zeta)\overline{g(\zeta)}\,dm(\zeta) = \int_{\mathbb{T}} f(\overline{\zeta})\overline{g(\overline{\zeta})}\,dm(\zeta), \quad f, g \in L^2, \tag{8.5}$$

which will be used later on.

Although varied in appearance, it turns out that the structure of conjugations is remarkably simple. The following observation is from [80, Lem. 1], although no doubt it must have been discovered numerous times before.

Lemma 8.6 *If C is a conjugation on $\mathcal{H}$, then there exists an orthonormal basis $\{e_n\}_{n \geqslant 1}$ of $\mathcal{H}$ such that $Ce_n = e_n$ for all n.*

Proof   Consider the set $\mathcal{K} = (I + C)\mathcal{H}$, noting that each vector in $\mathcal{K}$ is fixed by C and that $\mathcal{K}$ is a closed subset of $\mathcal{H}$ that is invariant under addition and multiplication by *real* scalars. Since

$$\langle \mathbf{x}, \mathbf{y} \rangle = \langle C\mathbf{y}, C\mathbf{x} \rangle = \langle \mathbf{y}, \mathbf{x} \rangle = \overline{\langle \mathbf{x}, \mathbf{y} \rangle}, \qquad \mathbf{x}, \mathbf{y} \in \mathcal{K},$$

we see that $\mathcal{K}$ is a real Hilbert space (that is, a Hilbert space whose field of scalars is $\mathbb{R}$) when endowed with the inner product $\langle \cdot, \cdot \rangle$. Let $\{e_n\}_{n \geqslant 1}$ be an orthonormal basis for $\mathcal{K}$. In light of the fact that

$$2\mathbf{x} = (\mathbf{x} + C\mathbf{x}) - i(i\mathbf{x} + C(i\mathbf{x}))$$

holds for all $\mathbf{x}$ in $\mathcal{H}$, it follows that $\{e_n\}_{n \geqslant 1}$ is also an orthonormal basis for the complex Hilbert space $\mathcal{H} = \mathcal{K} + i\mathcal{K}$ as well.   □

A basis of the type described in Lemma 8.6 is called a *C-real* orthonormal basis of $\mathcal{H}$. Such bases will be important to us when we continue our study of various classes of operators defined on model spaces. At present, however, we require an additional result about general conjugations before arriving at the specific conjugation relevant to model spaces.

Lemma 8.7 *Every conjugation is unitarily equivalent to the canonical conjugation on an ℓ^2-space of the appropriate dimension.*

Proof   If $\{e_n\}_{n \geqslant 1}$ is a C-real orthonormal basis for $\mathcal{H}$, then

$$C\Big(\sum_{n \geqslant 1} \alpha_n \mathbf{e}_n \Big) = \sum_{n \geqslant 1} \overline{\alpha_n} \mathbf{e}_n,$$

since C is conjugate linear and fixes each $\mathbf{e}_n$. Now observe that the coordinate map $U : \mathcal{H} \to \ell^2$ defined by $U\mathbf{x} = \{\langle \mathbf{x}, \mathbf{e}_n \rangle\}_{n \geqslant 1}$ is unitary (being an invertible isometry) and satisfies $JU = UC$ where J denotes the canonical conjugation on the appropriate ℓ^2-space (see (8.3)). $\qquad\qquad\qquad\qquad$ □

8.2 Conjugation on $\mathcal{K}_u$

Being a complex Hilbert space, each model space $\mathcal{K}_u$ possesses many conjugations. Indeed, given any orthonormal basis $\{\mathbf{e}_n\}_{n \geqslant 1}$ of $\mathcal{K}_u$, we may define a conjugation C on $\mathcal{K}_u$ by setting $C\mathbf{e}_n = \mathbf{e}_n$ and extending to the rest of $\mathcal{K}_u$ by conjugate-linearity. Of course, this observation is of limited usefulness since it makes no particular use of the function theoretic structure of $\mathcal{K}_u$. It turns out, however, that $\mathcal{K}_u$ comes pre-equipped with a natural conjugation that respects, to a broad degree, the function theoretic properties of $\mathcal{K}_u$. In order to understand this conjugation, we first remind the reader that

$$\mathcal{K}_u = H^2 \cap u\overline{zH^2}$$

when regarded as a space of functions on $\mathbb{T}$ via non-tangential boundary values (Proposition 5.4).

Proposition 8.8 *The conjugate linear operator* $C : \mathcal{K}_u \to \mathcal{K}_u$, *defined in terms of boundary functions on $\mathbb{T}$ by*

$$Cf = \overline{f}zu, \qquad\qquad (8.6)$$

is a conjugation. In particular, $|f| = |Cf|$ almost everywhere on $\mathbb{T}$ so that f and Cf share the same outer factor.

Proof Since $u\overline{u} = 1$ a.e. on $\mathbb{T}$, it follows that C is conjugate-linear, isometric, and involutive. We need only prove that C maps $\mathcal{K}_u$ into $\mathcal{K}_u$. If $f \in H^2$ is orthogonal to uH^2, it follows that

$$\langle Cf, \overline{zh} \rangle = \langle \overline{f}zu, \overline{zh} \rangle = \langle uzh, fz \rangle = \langle uh, f \rangle = 0 \quad \forall h \in H^2.$$

In other words, $Cf \in H^2$. Similarly,

$$\langle Cf, uh \rangle = \langle \overline{f}zu, uh \rangle = \langle \overline{fz}, h \rangle = 0 \quad \forall h \in H^2,$$

from which we obtain $Cf \in \mathcal{K}_u$. $\qquad\qquad\qquad\qquad\qquad\qquad\qquad$ □

Remark 8.9 The conjugation C is defined by the inner function u. However, $\mathcal{K}_u = \mathcal{K}_{\xi u}$ for any unimodular constant ξ. This equality will result in a unimodular multiple of C when defining the conjugation on $\mathcal{K}_{\xi u}$. Normally this unimodular constant is unimportant. However, it does come into play when one wants to normalize the inner function in a certain way (see Section 8.5).

The conjugation on $\mathcal{K}_u$ defined by (8.6) is closely related to the notion of pseudocontinuation introduced in Section 7.1. We now make this connection more explicit. Letting $g^{\#}(z) = \overline{g(\bar{z})}$ denote the function obtained by conjugating the Taylor coefficients of g, the equation $f = \overline{gz}u$ tells us that the functions $f(z)/u(z)$ on $\mathbb{D}$ and $\frac{1}{z}g^{\#}(\frac{1}{z})$ on $\mathbb{D}_e$ (the extended exterior disk) have matching non-tangential limits almost everywhere on $\mathbb{T}$. Thus the difference between conjugation and pseudocontinuation is mostly one of interpretation. Our present approach is to think of the complementary function g as Cf, a function belonging to $\mathcal{K}_u$, and hence with domain $\mathbb{D}$, as opposed to a function defined on $\mathbb{D}_e$.

Example 8.10 If $u = z^n$, then $\mathcal{K}_u = \bigvee\{1, z, \ldots, z^{n-1}\}$ and the conjugation C assumes the form

$$C(a_0 + a_1 z + \cdots a_{n-1} z^{n-1}) = \overline{a_{n-1}} + \overline{a_{n-2}} z + \cdots + \overline{a_0} z^{n-1}. \qquad (8.7)$$

Indeed, working on $\mathbb{T}$, one observes that

$$C(z^j) = \overline{z^j} z z^n = z^{n-1-j}$$

for $0 \leqslant j \leqslant n - 1$. The conjugation (8.7) is thus unitarily equivalent to the Toeplitz conjugation (8.4) defined on $\mathbb{C}^n$.

Example 8.11 More generally, suppose that

$$u = \prod_{1 \leqslant i \leqslant n} \frac{z - \lambda_i}{1 - \overline{\lambda_i} z}$$

is a finite Blaschke product having zeros $\lambda_1, \lambda_2, \ldots, \lambda_n$, repeated according to multiplicity. Corollary 5.18 says that every $f \in \mathcal{K}_u$ takes the form

$$f = \frac{a_0 + a_1 z + \cdots + a_{n-1} z^{n-1}}{\prod_{1 \leqslant i \leqslant n} (1 - \overline{\lambda_i} z)}.$$

Using this representation for $f \in \mathcal{K}_u$, and thinking of the computations below on $\mathbb{T}$, the conjugation becomes

$$Cf = \overline{zf}u$$

$$= \bar{z} \cdot \frac{\overline{a_0} + \overline{a_1}\bar{z} + \cdots + \overline{a_{n-1}}\bar{z}^{n-1}}{\prod_{1 \leqslant i \leqslant n}(1 - \lambda_j \bar{z})} \cdot \prod_{1 \leqslant i \leqslant n} \frac{z - \lambda_i}{1 - \overline{\lambda_i} z}$$

$$= \bar{z} \cdot \frac{\bar{z}^{n-1}(\overline{a_{n-1}} + \overline{a_{n-2}}z + \cdots + \overline{a_0}z^{n-1})}{\bar{z}^n \prod_{1 \leqslant i \leqslant n}(z - \lambda_j)} \cdot \prod_{1 \leqslant i \leqslant n} \frac{z - \lambda_i}{1 - \overline{\lambda_i}z}$$

$$= \frac{\overline{a_{n-1}} + \overline{a_{n-2}}z + \cdots + \overline{a_0}z^{n-1}}{\prod_{1 \leqslant i \leqslant n}(1 - \overline{\lambda_i}z)}.$$

In other words,

$$C\left(\frac{a_0 + a_1 z + \cdots + a_{n-1}z^{n-1}}{\prod_{1 \leqslant i \leqslant n}(1 - \overline{\lambda_i}z)}\right) = \frac{\overline{a_{n-1}} + \overline{a_{n-2}}z + \cdots + \overline{a_0}z^{n-1}}{\prod_{1 \leqslant i \leqslant n}(1 - \overline{\lambda_i}z)}.$$

Example 8.12 Suppose that $f = \sum_{n \geqslant 0} a_n z^n$ and $g = \sum_{n \geqslant 0} b_n z^n$ are conjugate to each other in $\mathcal{K}_u$ (that is, $g = Cf$ and vice-versa). From Proposition 5.15 and its proof, it follows that

$$f = \sum_{n \geqslant 0} \overline{b_n} S^{*(n+1)} u, \qquad g = \sum_{n \geqslant 0} \overline{a_n} S^{*(n+1)} u,$$

where

$$\|f\|^2 = \sum_{n \geqslant 0} |a_n|^2 = \sum_{n \geqslant 0} |b_n|^2 = \|g\|^2.$$

Example 8.13 A straightforward computation reveals that the conjugation operator on $\mathcal{K}_u$ sends reproducing kernels to difference quotients:

$$(Ck_\lambda)(z) = \overline{\left(\frac{1 - \overline{u(\lambda)}u(z)}{1 - \overline{\lambda}z}\right)} \cdot \bar{z}u(z) = \frac{1 - u(\lambda)\overline{u(z)}}{1 - \lambda\bar{z}} \cdot \frac{u(z)}{z}$$

$$= \frac{u(z) - u(\lambda)}{z - \lambda} = (Q_\lambda u)(z),$$

where, as usual, we consider all of the functions involved as functions on $\mathbb{T}$ (so that $z\bar{z} = 1$ and $u(z)\overline{u(z)} = 1$ a.e. on $\mathbb{T}$). For each λ, the kernel k_λ is outer since it is the quotient of the two outer functions $1 - \overline{u(\lambda)}u(z)$ and $1 - \overline{\lambda}z$ (Corollary 3.24). In light of Proposition 8.8, we expect that the difference quotient Ck_λ is simply an inner multiple of k_λ. This is indeed the case as the following computation shows:

$$\frac{(Ck_\lambda)(z)}{k_\lambda(z)} = \frac{u(z) - u(\lambda)}{z - \lambda} \cdot \frac{1 - \overline{\lambda}z}{1 - \overline{u(\lambda)}u(z)}$$

$$= \frac{u(z) - u(\lambda)}{1 - \overline{u(\lambda)}u(z)} \cdot \frac{1 - \overline{\lambda}z}{z - \lambda}$$

$$= \frac{b_{u(\lambda)}(u(z))}{b_\lambda(z)},$$

where $b_w(z) = (z - w)/(1 - \overline{w}z)$. When $\lambda = 0$, we obtain a specific identity that arises frequently in the study of Clark unitary operators (see Chapter 11), namely

$$C(1 - \overline{u(0)}u) = S^*u. \qquad (8.8)$$

In particular, if $u(0) = 0$, then $C1 = u/z$ and vice-versa.

Example 8.14 If u has a finite angular derivative in the sense of Carathéodory at $\zeta \in \mathbb{T}$, then $u(\zeta)$ exists and is unimodular. Moreover, the boundary kernel

$$k_\zeta = \frac{1 - \overline{u(\zeta)}u}{1 - \overline{\zeta}z}$$

belongs to $\mathcal{K}_u$ (Theorem 7.24). In this case, something interesting occurs. Since ζ and $u(\zeta)$ are of unit modulus, we obtain

$$Ck_\zeta = \frac{u - u(\zeta)}{z - \zeta} = \frac{u(\zeta)}{\zeta} \cdot \frac{1 - \overline{u(\zeta)}u}{1 - \overline{\zeta}z} = \overline{\zeta}u(\zeta)k_\zeta.$$

For either branch of the square root, it follows that the function

$$(\overline{\zeta}u(\zeta))^{\frac{1}{2}}k_\zeta$$

belongs to $\mathcal{K}_u$ and is fixed by C.

Under certain circumstances, it is possible to construct C-real (orthonormal) bases for $\mathcal{K}_u$ using boundary kernels. This is relatively straightforward for finite-dimensional model spaces, as the following example demonstrates.

Example 8.15 Let u be a finite Blaschke product with n zeros

$$\lambda_1, \lambda_2, \ldots, \lambda_n \in \mathbb{D},$$

repeated according to multiplicity. For the sake of simplicity, suppose that $u(0) = 0$. Fix $\alpha \in \mathbb{T}$ and observe that the equation $u(\zeta) = \alpha$ has precisely n distinct solutions

$$\zeta_1, \zeta_2, \ldots, \zeta_n \in \mathbb{T}.$$

To see this, first observe that the equation $u(z) = \alpha$ has at most n solutions in $\mathbb{C}$ since it can be rewritten as a polynomial equation of degree n. Since $|u(z)| = 1$ if and only if $z \in \mathbb{T}$, we see that these solutions must all lie on $\mathbb{T}$. Evaluate the identity

$$u'(z) = u(z) \sum_{1 \leq j \leq n} \frac{1 - |\lambda_j|^2}{(1 - \overline{\lambda_j}z)(z - \lambda_j)}$$

(see (2.21)) at $z = \zeta \in \mathbb{T}$ to get

$$u'(\zeta) = u(\zeta)\overline{\zeta} \sum_{1 \leqslant j \leqslant n} \frac{1 - |\lambda_j|^2}{|1 - \overline{\lambda_j}\zeta|^2}$$

so that $|u'(\zeta)| > 0$ for all $\zeta \in \mathbb{T}$. Thus every root of the equation $u(z) - \alpha = 0$ must be simple. Putting this all together, it follows that $u(\zeta) = \alpha$ has precisely n distinct solutions on $\mathbb{T}$ and hence the functions

$$\mathbf{e}_j(z) = \frac{e^{\frac{i}{2}(\arg \alpha - \arg \zeta_j)}}{\sqrt{|u'(\zeta_j)|}} \frac{1 - \overline{\alpha}u(z)}{1 - \overline{\zeta_j}z}, \quad 1 \leqslant j \leqslant n,$$

form a C-real basis of $\mathcal{K}_u$. It will turn out (see Theorem 11.4) that vectors $\mathbf{e}_j$ are the eigenvectors of a certain certain unitary operator on $\mathcal{K}_u$ and thus form an orthonormal basis for $\mathcal{K}_u$.

8.3 Inner functions in $\mathcal{K}_u$

Although H^2 contains all of the inner functions, it is not clear whether or not $\mathcal{K}_u$ contains any. The following result answers this question and is another nice application of conjugation on $\mathcal{K}_u$.

Theorem 8.16 *The model space $\mathcal{K}_u$ contains an inner function if and only if $u(0) = 0$. In this case, the inner functions belonging to $\mathcal{K}_u$ are precisely u/z and all of its inner divisors.*

Proof If $u(0) = 0$, then

$$\frac{u}{z} = S^*u \in \mathcal{K}_u.$$

By Proposition 5.6, $\mathcal{K}_u$ is closed under the operation of removing inner divisors and so it follows that all of the inner divisors of u/z belong to $\mathcal{K}_u$.

Conversely, if $f \in \mathcal{K}_u$ is inner, then $Cf = \overline{f}zu$ is also inner. Hence $u = zfCf$ and so $u(0) = 0$. Moreover, if f is any inner function that divides u/z we may write $u = zfg$ where g is inner. Since $f = \overline{gz}u$ a.e. on $\mathbb{T}$, we conclude that $f \in \mathcal{K}_u$ (Proposition 5.4). □

Example 8.17 If $u = z^n$, then $\mathcal{K}_u$ is the space of polynomials of degree at most $n - 1$ (Example 5.17). Among this collection, the only inner functions are ξz^j, where $0 \leqslant j \leqslant n - 1$ and ξ is a unimodular constant (Example 2.15).

8.4 Generators of $\mathcal{K}_u$

Fix an inner function u and let $f = I_1 F$ denote the inner–outer factorization of an $f \in \mathcal{K}_u \setminus \{0\}$. Proposition 8.8 ensures that $g = Cf$ has the same outer factor as f, whence we may write $g = I_2 F$ for some inner function I_2. Since $g = \overline{f} zu$ a.e. on $\mathbb{T}$ it follows that $I_2 F = \overline{I_1 F} zu$ and so

$$I_1 I_2 = \frac{\overline{F} zu}{F} = \frac{CF}{F}. \tag{8.9}$$

Following [86] we make the following definition.

Definition 8.18 Let $F \in \mathcal{K}_u$ be outer. The inner function

$$\mathcal{I}_F = \frac{CF}{F} = \frac{\overline{F} zu}{F}$$

is called the *associated inner function* of F with respect to u.

Before proceeding, we remark that some papers refer to the inner function u itself as the associated inner function of F [61, Rem. 3.1.6].

Example 8.19 As pointed out in Example 8.13, we know that all of the kernel functions k_λ are outer. Setting $F = k_\lambda$ we get

$$\mathcal{I}_F = \frac{Ck_\lambda}{k_\lambda} = \frac{\frac{u-u(\lambda)}{z-\lambda}}{\frac{1-\overline{u(\lambda)}u}{1-\overline{\lambda}z}} = \frac{\frac{u-u(\lambda)}{1-\overline{u(\lambda)}u}}{\frac{z-\lambda}{1-\overline{\lambda}z}},$$

which is a certain Frostman shift of u with its zero at $z = \lambda$ divided out with a single Blaschke factor.

When $u(0) = 0$, then $k_0 = 1$ is outer and

$$\mathcal{I}_1 = \frac{u}{z}. \tag{8.10}$$

If v_1 and v_2 are inner functions such that $v_1 v_2 = \mathcal{I}_F$, then the definition of $\mathcal{I}_F$ implies that

$$v_1 F = \overline{v_2 F} zu$$

almost everywhere on $\mathbb{T}$. Hence the functions $f = v_1 F$ and $g = v_2 F$ satisfy $f = \overline{g} zu$ a.e. on $\mathbb{T}$. By Proposition 5.4, f and g belong to $\mathcal{K}_u$ and satisfy $Cf = g$. Furthermore,

$$C(v_1 F) = v_2 F = \frac{\mathcal{I}_F}{v_1} F. \tag{8.11}$$

Putting this all together, we obtain the following result from [86].

Proposition 8.20 *Let $F \in \mathcal{K}_u$ be outer. The set of all functions in $\mathcal{K}_u$ having outer factor F is precisely*

$$\{vF : v \text{ inner and } v | \mathcal{I}_F\}. \tag{8.12}$$

We define a partial order $\preceq$ on (8.12) by declaring that $v_1 F \preceq v_2 F$ if and only if $v_1 | v_2$. With respect to this ordering, the functions F and $\mathcal{I}_F F$ are, up to unimodular constants, the unique minimal and maximal elements, respectively. Moreover, C restricts to an order-reversing bijection from the set (8.12) onto itself.

Example 8.21 If u is inner with $u(0) = 0$, then $F = k_0 = 1$ is outer. Hence the functions in $\mathcal{K}_u$ with this outer factor are the inner functions in $\mathcal{K}_u$. By the previous proposition, these are the inner divisors of $\mathcal{I}_1 = u/z$ (see (8.10)). Notice how this reproduces Theorem 8.16.

Recall that $f \in \mathcal{K}_u$ *generates* $\mathcal{K}_u$ if

$$\bigvee \{S^{*n}f : n \geqslant 0\} = \mathcal{K}_u. \tag{8.13}$$

The following is part of [61, Thm. 3.1.5].

Proposition 8.22 *Suppose that $f, g \in H^2$ and u is an inner function such that $f = \overline{gz}u$ on $\mathbb{T}$. If u and the inner factor of g are relatively prime, then f generates $\mathcal{K}_u$.*

Proof From Proposition 5.4, we see that $f \in \mathcal{K}_u$ and so

$$\mathcal{M} = \bigvee \{S^{*n}f : n \geqslant 0\} \subset \mathcal{K}_u$$

since $\mathcal{K}_u$ is S^*-invariant. Because $\mathcal{M}$ is a non-trivial S^*-invariant subspace of H^2, it must be of the form $\mathcal{K}_v$ for some inner function v. Moreover, since $\mathcal{K}_v \subset \mathcal{K}_u$ we have $v | u$ (Corollary 5.9). Using the fact that $f \in \mathcal{K}_v$, we may write $f = \overline{hz}v$ a.e. on $\mathbb{T}$ and so $\overline{hz}v = \overline{gz}u$ for some $h \in \mathcal{K}_u$. This implies that $|g| = |h|$ a.e. on $\mathbb{T}$ so that the outer factors of g and h are identical up to a unimodular constant multiple (which we absorb in to the inner factors). Let v_g and v_h denote the inner factors of g and h, respectively. It follows that $\overline{v_h}v = \overline{v_g}u$, from which we obtain the identity $v_g v = v_h u$. Since v_g and u are relatively prime, this identity proves that $u/v = v_g/v_h$ is a unimodular constant. Thus $\mathcal{M} = \mathcal{K}_v = \mathcal{K}_u$, as claimed. $\square$

Recognizing that $g = Cf$ in this setting immediately yields the following [86, Prop. 4.3].

Proposition 8.23 *If $f \in \mathcal{K}_u$ and the inner factor of Cf is relatively prime to u, then f generates $\mathcal{K}_u$.*

Example 8.24 If u is a singular inner function (that is, u is not divisible by any Blaschke product), then Frostman's Theorem (Theorem 2.22) asserts that k_λ generates $\mathcal{K}_u$ for all $\lambda \in \mathbb{D}$ that do not lie in some exceptional set. Indeed, by Example 8.13, the inner factor of Ck_λ is precisely

$$\frac{b_{u(\lambda)} \circ u}{b_\lambda},$$

which is a Blaschke product whenever $u(\lambda)$ is non-zero and does not lie in the exceptional set for u.

Corollary 8.25 *If $f \in \mathcal{K}_u$ is outer, then Cf generates $\mathcal{K}_u$. In particular, any self-conjugate outer function in $\mathcal{K}_u$ generates $\mathcal{K}_u$.*

Example 8.26 Each difference quotient

$$Q_\lambda u = \frac{u - u(\lambda)}{z - \lambda} = Ck_\lambda$$

generates $\mathcal{K}_u$ since its conjugate $C(Q_\lambda u) = k_\lambda$ is outer (Corollary 3.24).

Example 8.27 Any boundary kernel $k_\zeta \in \mathcal{K}_u$ generates $\mathcal{K}_u$. Indeed, Example 8.14 shows that any such function is a constant multiple of a self-conjugate outer function.

Example 8.28 Every outer function $F \in \mathcal{K}_u$ is the sum of two generators since

$$F = \tfrac{1}{2}(1 + \mathcal{I}_F)F + i\tfrac{1}{2i}(1 - \mathcal{I}_F)F.$$

Moreover, by (8.11), $CF = \mathcal{I}_F F$ and thus

$$C(\tfrac{1}{2}(1 + \mathcal{I}_F)F) = \tfrac{1}{2}(1 + \mathcal{I}_F)F,$$
$$C(\tfrac{1}{2i}(1 - \mathcal{I}_F)F) = \tfrac{1}{2i}(1 - \mathcal{I}_F)F.$$

8.5 Cartesian decomposition

Each $f \in \mathcal{K}_u$ enjoys a *Cartesian decomposition* $f = a + ib$ where the functions

$$a = \tfrac{1}{2}(f + Cf), \qquad b = \tfrac{1}{2i}(f - Cf),$$

are both fixed by C. With respect to this decomposition, the conjugation C on $\mathcal{K}_u$ assumes the form $Cf = a - ib$. To explicitly describe the functions belonging to $\mathcal{K}_u$, it suffices to describe those functions that are *C-real* (that is, $Cf = f$).

If $f = Cf$, then

$$f = \overline{f}zu.$$

Since u has unit modulus a.e. on $\mathbb{T}$, by replacing u with a suitable unimodular constant multiple of u (which does not change the model space $\mathcal{K}_u$), we may assume $u(\zeta) = \zeta$ for some $\zeta \in \mathbb{T}$. We can define k_ζ by

$$k_\zeta(z) = \frac{1 - \overline{u(\zeta)}u(z)}{1 - \overline{\zeta}z} = \frac{1 - \overline{\zeta}u(z)}{1 - \overline{\zeta}z}$$

which resides in the Smirnov class N^+ (see Definition 3.26) since the numerator is a bounded function while the denominator is a bounded outer function (Corollary 3.24). If u has a finite angular derivative in the sense of Carathéodory at ζ then k_ζ will belong to $\mathcal{K}_u$, but at this point, we do not assume this extra condition.

A computation yields

$$k_\zeta/\overline{k_\zeta} = \frac{1 - \overline{\zeta}u}{1 - \overline{\zeta}z} \cdot \frac{1 - \zeta\overline{z}}{1 - \zeta\overline{u}} = \frac{1 - \overline{\zeta}u}{1 - \zeta\overline{u}} \cdot \frac{1 - \zeta\overline{z}}{1 - \overline{\zeta}z}$$

$$= \overline{\zeta}u\left(\frac{\zeta\overline{u} - 1}{1 - \zeta\overline{u}}\right) \cdot \zeta\overline{z}\left(\frac{\overline{\zeta}z - 1}{1 - \overline{\zeta}z}\right)$$

$$= \zeta\overline{\zeta}\,\overline{z}u = \overline{z}u,$$

which, when substituted into the equation $f = \overline{f}zu$, reveals

$$f/k_\zeta = \overline{f/k_\zeta}$$

almost everywhere on $\mathbb{T}$. Thus a function $f \in \mathcal{K}_u$ satisfies $f = Cf$ if and only if f/k_ζ is real almost everywhere on $\mathbb{T}$.

Definition 8.29 A function $f \in N^+$ is called a *real Smirnov function* if its boundary function is real-valued almost everywhere on $\mathbb{T}$. Denote the set of all real Smirnov functions by $\mathcal{R}^+$.

The following elegant theorem of Helson [110] provides an explicit formula for real Smirnov functions.

Theorem 8.30 *A function $f \in N^+$ belongs to $\mathcal{R}^+$ if and only if there are inner functions ψ_1, ψ_2 such that*

$$f = i\frac{\psi_1 + \psi_2}{\psi_1 - \psi_2} \tag{8.14}$$

and $\psi_1 - \psi_2$ is outer.

Proof One direction follows from direct computation using the fact that $\psi_j \overline{\psi_j} = 1$ almost everywhere on $\mathbb{T}$. For the other direction, note that

$$\tau(z) = i\frac{1 + z}{1 - z}$$

maps $\mathbb{D}$ onto the upper half plane. Thus $\tau^{-1} \circ f$ is of bounded type (that is, a quotient of H^∞ functions) and is unimodular almost everywhere on $\mathbb{T}$. Hence $\tau^{-1} \circ f = \psi_1/\psi_2$ is a quotient of inner functions and so f has the desired form. $\quad\square$

Helson's Theorem and the representation

$$f = (\alpha + i\beta)k_\zeta, \quad \alpha, \beta \in \mathcal{R}^+,$$

permit us to write any function $f \in \mathcal{K}_u$ as a rational expression involving a finite number of inner functions. Since each inner function has a pseudocontinuation to $\mathbb{D}_e$ given explicitly by the Schwarz reflection formula (see Example 7.4), we conclude that the pseudocontinuability of f itself can be explained entirely through the mechanism of Schwarz reflection.

8.6 2×2 inner functions

In at least one instance, it turns out that conjugations on model spaces can yield structural insights into the nature of higher order inner functions.

Definition 8.31 A matrix-valued function $\Theta : \mathbb{D} \to M_2(\mathbb{C})$ is called an *inner function* if all four of its entries belong to H^∞ and $\Theta(\zeta)$ is unitary for a.e. $\zeta \in \mathbb{T}$. In other words, Θ is inner if

$$\sup_{z \in \mathbb{D}} \|\Theta(z)\| < \infty,$$

where $\| \cdot \|$ denotes the operator norm of a 2×2 matrix, and $\Theta^{-1} = \Theta^*$ a.e. on $\mathbb{T}$.

For $\Theta : \mathbb{D} \to M_2(\mathbb{C})$ to be an inner function, it is necessary that $u = \det \Theta$ is a scalar-valued inner function. Indeed, u is a bounded analytic function on $\mathbb{D}$

whose boundary values are of unit modulus a.e. on $\mathbb{T}$ since the determinant of a unitary matrix must have unit modulus (the eigenvalues of a unitary matrix are of unit modulus and the determinant of a matrix is the product of its eigenvalues).

The following result can be found in [87] along with several other related results, although it is stated in [80] without proof. This representation was exploited in [39] to study the characteristic function of a complex symmetric contraction.

Theorem 8.32 *Let u be an inner function and let $a, b, c, d \in H^\infty$. The analytic matrix-valued function*

$$\Theta = \begin{bmatrix} a & -b \\ c & d \end{bmatrix} \tag{8.15}$$

is unitary almost everywhere on $\mathbb{T}$ and satisfies $\det \Theta = u$ if and only if:

(i) a, b, c, d belong to $\mathcal{K}_{zu}$;
(ii) $|a|^2 + |b|^2 = 1$ almost everywhere on $\mathbb{T}$;
(iii) $Ca = d$ and $Cb = c$.

Here $C : \mathcal{K}_{zu} \to \mathcal{K}_{zu}$ is the conjugation $Cf = \overline{f}u$ on $\mathcal{K}_{zu}$.

Proof   If Θ is unitary a.e. on $\mathbb{T}$, then the scalar-valued function $u = ad + bc$ is inner. Comparing entries in the identity $\Theta = (\Theta^*)^{-1}$ tells us that $a = \overline{d}u$ and $b = \overline{c}u$ a.e. on $\mathbb{T}$. By Proposition 5.4, we conclude that $a, b, c, d \in \mathcal{K}_{zu}$, $Ca = d$, and $Cb = c$, proving (i) and (iii). Expanding the identity $\Theta\Theta^* = I$ and looking at the $(1, 1)$ entry yields (ii).

Conversely, if (i), (ii), and (iii) hold, then a.e. on $\mathbb{T}$ we have

$$\Theta\Theta^* = \begin{bmatrix} a & -b \\ \overline{b}u & \overline{a}u \end{bmatrix} \begin{bmatrix} \overline{a} & b\overline{u} \\ -\overline{b} & a\overline{u} \end{bmatrix} = \begin{bmatrix} |a|^2 + |b|^2 & ab\overline{u} - ab\overline{u} \\ \overline{a}bu - \overline{a}bu & |a|^2 + |b|^2 \end{bmatrix} = I$$

so that Θ is unitary a.e. on $\mathbb{T}$. Moreover,

$$\det \Theta = ad + bc = a\overline{a}u + b\overline{b}u = (|a|^2 + |b|^2)u = u,$$

as claimed.   □

We remark that (8.15) is entirely analogous to the representation of quaternions of unit modulus using 2×2 complex matrices. Recall that the quaternions $\mathbb{H}$ are a four-dimensional division algebra over $\mathbb{R}$ whose elements are of the form $a + bi + cj + dk$ where the symbols i, j, k satisfy $i^2 = j^2 = k^2 = -1$ and $ij = k = -ji$. The norm

$$\|a + bi + cj + dk\| = \sqrt{a^2 + b^2 + c^2 + d^2}$$

on $\mathbb{H}$ is multiplicative, in light of Euler's famous four-square identity. It is well known that, as a division algebra, $\mathbb{H}$ is isomorphic to the subset of $M_2(\mathbb{C})$ consisting of all matrices of the form

$$\begin{bmatrix} z & -w \\ \overline{w} & \overline{z} \end{bmatrix}.$$

In this representation, the quaternions of unit modulus are precisely those for which $|z|^2 + |w|^2 = 1$. See the text [69] for further details.

We now turn our attention to a matrix completion problem that originated in systems theory. The simplest version of the *Darlington synthesis problem* asks whether, given $a \in H^\infty$, it is possible to find $b, c, d \in H^\infty$ such that (8.15) is inner. Theorem 8.32 tells us that $\|a\|_\infty \leqslant 1$ and $a \in \mathcal{K}_{zu}$, where the inner function u must equal the determinant of Θ. If $\|a\|_\infty \leqslant 1$, then the following result asserts that we can produce a second function $b \in \mathcal{K}_{zu}$ such that $|a|^2 + |b|^2 = 1$ a.e. $\mathbb{T}$ [86, 87]. This is the key to our treatment of the Darlington synthesis problem.

Lemma 8.33 *If $a \in \mathcal{K}_{zu}$ and $\|a\|_\infty \leqslant 1$, then there exists $b \in \mathcal{K}_{zu}$ such that $|a|^2 + |b|^2 = 1$ a.e. on $\mathbb{T}$.*

Proof   If a is an inner function, then we may let $b = 0$. So suppose that a is not an inner function. Since $Ca = \overline{a}u$, it follows that

$$u - aCa = u(1 - |a|^2)$$

a.e. on $\mathbb{T}$. From here we obtain $u - aCa \in H^\infty \setminus \{0\}$ (note a is not inner) and so by Corollary 3.21,

$$\int_{\mathbb{T}} \log(1 - |a|^2)\, dm = \int_{\mathbb{T}} \log|u - aCa|\, dm > -\infty.$$

Thus there exists an outer function $h \in H^\infty$ such that $|h|^2 = 1 - |a|^2$ a.e. on $\mathbb{T}$. Since the function $u - aCa = u|h|^2$ belongs to H^∞ and has the same modulus as h^2, there exists an inner function v such that $vh^2 = u|h|^2$. Since $\log|h|$ is integrable, h cannot vanish on a set of positive measure. Thus $vh = \overline{h}u$ holds a.e. on $\mathbb{T}$. By Proposition 5.4, the function $b = vh$ belongs to $\mathcal{K}_{zu}$ and satisfies $|a|^2 + |b|^2 = 1$ a.e. on $\mathbb{T}$.   □

We are now ready to prove the following result of Arov [19] and Douglas and Helton [60]:

Theorem 8.34 *The Darlington synthesis problem with data $a \in H^\infty$ has a solution if and only if:*

(i) $\|a\|_\infty \leqslant 1$;
(ii) a is pseudocontinuable of bounded type (PCBT).

Proof If a is PCBT, then a belongs to a model space $\mathcal{K}_u$. Since $\mathcal{K}_u \subset \mathcal{K}_{zu}$ by Corollary 5.9, we conclude that $a \in \mathcal{K}_{zu}$. By Lemma 8.33, there exists a function $b \in \mathcal{K}_{zu}$ such that $|a|^2 + |b|^2 = 1$. By Theorem 8.32, it follows that the matrix-valued function

$$\Theta = \begin{bmatrix} a & -b \\ Cb & Ca \end{bmatrix}$$

is inner.   □

8.7 Notes

8.7.1 Darlington synthesis

The paper [87] contains a detailed discussion of the solutions set to the 2×2 Darlington synthesis problem for various classes of upper-left corners a. Among other things, an algorithm to produce all possible solutions is given in the case where a is a rational function whose poles lie in $\mathbb{D}_e$.

8.7.2 Real outer functions

The Helson representation from (8.14) can be difficult to compute. For instance, it is a somewhat messy calculation to find a Helson representation for the *Köbe function*

$$k(z) = \frac{z}{(1-z)^2} = z + 2z^2 + 3z^3 + \cdots,$$

which maps $\mathbb{D}$ bijectively onto $\mathbb{C} \setminus (-\infty, -\frac{1}{4}]$. We therefore propose a more constructive description of the functions in $\mathcal{R}^+$, which originates in [85, 94]. First note that if $f = I_f F$ is the inner–outer factorization of a function $\mathcal{R}^+$, then

$$I_f F = \left[\frac{-4I_f}{(1 - I_f)^2} \right] \cdot \left[\frac{(1 - I_f)^2 F}{-4} \right],$$

where the first term is in $\mathcal{R}^+$ and has the same inner factor as f and where the second is *outer* (Corollary 3.24) and in $\mathcal{R}^+$. Thus to describe functions in $\mathcal{R}^+$, it suffices to describe *real outer (RO) functions*. An infinite product expansion for real outer functions, in terms of Cayley-like transforms of inner functions, is obtained in [94].

One of the original motivations for the study of real outer functions stems from interest in exposed points of the unit ball $b(H^1)$ of H^1 and the associated problem of characterizing rigid functions in H^1. A function $F \in b(H^1)$ is called *exposed* if there exists a real linear functional that attains its maximum at F and nowhere else on $b(H^1)$. A function $F \in H^1$ is called *rigid* if it is determined by its argument on $\mathbb{T}$, in the sense that if $G \in H^1$ and $\arg F = \arg G$ a.e. on $\mathbb{T}$, then F and G are positive scalar multiples of each other. It is known that $F \in H^1$ is an exposed point of the unit ball of H^1 if and only if $\|F\|_1 = 1$ and F is rigid [193].

If u is an inner function, then

$$u = \frac{1+u}{1+\bar{u}} = \frac{(1+u)^2}{|1+u|^2}$$

a.e. on $\mathbb{T}$ so that u and $(1+u)^2$ have the same argument a.e. on $\mathbb{T}$. Thus the exposed points of $b(H^1)$ must be outer functions. Moreover, an outer function $F \in b(H^1)$ is not exposed if there exists a non-constant inner function u such that $F/(1+u)^2$ belongs to H^1. This occurs if and only if there is a non-constant real outer function that is non-negative a.e. on $\mathbb{T}$ and multiplies F into H^1. Some important references for this line of research are [94, 110, 114, 128, 147, 164, 168].

9

The compressed shift

We now examine the compression of the unilateral shift S to a model space $\mathcal{K}_u$, resulting in the so-called *compressed shift*. We will discuss many of the basic properties one wants to know when introduced to a new operator: its spectrum, invariant subspaces, cyclic vectors, the C^*-algebra generated by the operator, its commutant, etc.

9.1 What is a compression?

We begin with a discussion that will place the compressed shift in a more general setting. Let $\mathcal{H}$ be a separable complex Hilbert space. Given $T \in \mathcal{B}(\mathcal{H})$ and a subspace $\mathcal{M}$ of $\mathcal{H}$, define the operator

$$R : \mathcal{M} \to \mathcal{M}, \quad R = P_\mathcal{M} T|_\mathcal{M},$$

where $P_\mathcal{M}$ is the orthogonal projection of $\mathcal{H}$ onto $\mathcal{M}$. If one decomposes T according to the orthogonal decomposition

$$\mathcal{H} = \mathcal{M} \oplus \mathcal{M}^\perp,$$

then T has the matrix representation

$$[T] = \begin{bmatrix} R & * \\ * & * \end{bmatrix}.$$

In certain situations, we have the additional property that

$$f(R) = P_\mathcal{M} f(T)|_\mathcal{M} \tag{9.1}$$

for all analytic polynomials f.

Definition 9.1 When (9.1) is satisfied for all analytic polynomials f, we say that R is a *compression* of T to the subspace M and that T is a *dilation* of R to $\mathcal{H}$.

Remark 9.2 The operator theory literature is not always in agreement as there are several definitions of the term "compression." Some use the term compression to mean $R = P_M T|_M$ without insisting that $f(R) = P_M f(T)|_M$ for all analytic polynomials f.

The following theorem is due to Sarason [167].

Theorem 9.3 *Let $T \in \mathcal{B}(\mathcal{H})$, M be a subspace of $\mathcal{H}$, and $R = P_M T|_M$. Then the following are equivalent:*

(i) R is a compression of T;

(ii) There is a subspace N of $\mathcal{H}$ such that $N \perp M$, $TN \subset N$ and $T(M{\oplus}N) \subset M \oplus N$;

(iii) There are subspaces N and $\mathcal{K}$ of $\mathcal{H}$ such that the three subspaces M, N and $\mathcal{K}$ are mutually orthogonal, $TN \subset N$, $T(M \oplus N) \subset M \oplus N$, and $\mathcal{H} = M \oplus N \oplus \mathcal{K}$.

Proof (i) $\implies$ (ii): Let

$$M' = \bigvee \{T^n M : n \geqslant 0\}.$$

Clearly, $M \subset M'$ and $TM' \subset M'$. Define $N = M' \ominus M$, which implies that $M \perp N$ and that $M \oplus N = M'$ is invariant under T. To finish, we need to verify that $TN \subset N$. Indeed, for each $\mathbf{x} \in M$ and $n \geqslant 0$ we have

$$P_M T(T^n \mathbf{x}) = P_M T^{n+1} \mathbf{x} = R^{n+1} \mathbf{x} = RR^n \mathbf{x} = RP_M T^n \mathbf{x} = RP_M(T^n \mathbf{x}).$$

According to the definition of M', we obtain the identity

$$P_M T\mathbf{y} = RP_M \mathbf{y}, \qquad \mathbf{y} \in M'.$$

In particular, choosing any $\mathbf{y} \in N$, we get $P_M T\mathbf{y} = RP_M \mathbf{y} = 0$. This implies that $T\mathbf{y} \in M' \ominus M = N$ and thus $TN \subset N$.

(ii) $\implies$ (iii): Let $\mathcal{K} = (M \oplus N)^\perp$. The definition of $\mathcal{K}$ implies that the subspaces M, N, and $\mathcal{K}$ are mutually orthogonal and that $\mathcal{H} = M \oplus N \oplus \mathcal{K}$.

(iii) $\implies$ (i): With respect to the orthogonal decomposition $\mathcal{H} = M{\oplus}N{\oplus}\mathcal{K}$, the operator T has the matrix representation

$$[T] = \begin{bmatrix} R & 0 & * \\ * & * & * \\ 0 & 0 & * \end{bmatrix}.$$

Notice how the assumption $T\mathcal{N} \subset \mathcal{N}$ yields the second column, and the assumption $T(\mathcal{M} \oplus \mathcal{N}) \subset (\mathcal{M} \oplus \mathcal{N})$ yields the third row. The R in the upper left corner comes from definition of $R = P_{\mathcal{M}} T|_{\mathcal{M}}$.

The matrix representation above shows that for each $n \geqslant 1$, the operator T^n has the matrix representation

$$[T^n] = \begin{bmatrix} R^n & 0 & * \\ * & * & * \\ 0 & 0 & * \end{bmatrix}$$

with respect to the orthogonal decomposition $\mathcal{H} = \mathcal{M} \oplus \mathcal{N} \oplus \mathcal{K}$. Thus the operators T and R fulfill the property $P_{\mathcal{M}} T^n|_{\mathcal{M}} = R^n$ for all $n \geqslant 0$. By linearity we see that $f(R) = P_{\mathcal{M}} f(T)|_{\mathcal{M}}$ for any analytic polynomial f. Thus R is the compression of T to the subspace $\mathcal{M}$. □

A particularly useful case of Theorem 9.3 is the following.

Corollary 9.4 *Let* $T \in \mathcal{B}(\mathcal{H})$ *and let* $\mathcal{M}$ *be a subspace of* $\mathcal{H}$ *such that* $T\mathcal{M}^{\perp} \subset \mathcal{M}^{\perp}$. *Then* $R = P_{\mathcal{M}} T|_{\mathcal{M}}$ *is a compression of* T *to* $\mathcal{M}$.

Proof Apply Theorem 9.3 to $\mathcal{N} = \mathcal{M}^{\perp}$ and $\mathcal{K} = \{0\}$. □

Note that

$$R^* = P_{\mathcal{M}} T^*|_{\mathcal{M}}, \tag{9.2}$$

which will be useful later on.

9.2 The compressed shift

Recall from Proposition 5.13 that the orthogonal projection P_u from L^2 onto the model space $\mathcal{K}_u$ is given by the formula $(P_u f)(\lambda) = \langle f, k_\lambda \rangle$, where k_λ is the reproducing kernel for $\mathcal{K}_u$.

Definition 9.5 For an inner function u, the operator

$$S_u : \mathcal{K}_u \to \mathcal{K}_u, \quad S_u f = P_u S f$$

is called the *compressed shift operator*.

When we study the Sz.-Nagy–Foiaş model in Section 9.4, we will see that a contraction that satisfies several simple conditions is unitarily equivalent to a corresponding compressed shift. Let us first prove several facts about S_u. First and foremost is (see Definition 9.1):

Theorem 9.6 S_u *is a compression of* S.

Proof In Corollary 9.4, let $\mathcal{M} = \mathcal{K}_u$ and note that

$$S\,\mathcal{K}_u^\perp = S(uH^2) \subseteq uH^2 = \mathcal{K}_u^\perp.$$

Since S_u is a compression of S, we have the identity

$$S_u^n f = P_u S^n f, \qquad f \in \mathcal{K}_u, \; n \geqslant 0.$$

Let us now prove some other useful facts about S_u.

Proposition 9.7 $\quad S_u^* f = P_u(\overline{\zeta} f) = S^* f$ for all $f \in \mathcal{K}_u$.

Proof By (9.2) and the fact that $\mathcal{K}_u$ is S^*-invariant, we have

$$S_u^* = P_u S^* f = S^* f \quad \forall f \in \mathcal{K}_u.$$

To prove that $S_u^* f = P_u(\overline{\zeta} f)$, we first observe that $\overline{\zeta u} f \in (H^2)^\perp$ for all $f \in \mathcal{K}_u$. Indeed, for all $n \geqslant 0$,

$$\langle \overline{\zeta u} f, \zeta^n \rangle = \langle \overline{u} f, \zeta^{n+1} \rangle = \langle f, u\zeta^{n+1} \rangle = 0.$$

The previous line says that $P(\overline{\zeta u} f) = 0$ and so, by (5.26),

$$P_u(\overline{\zeta} f) = P(\overline{\zeta} f) - u P(\overline{\zeta u} f) = P(\overline{\zeta} f) = T_{\overline{z}} f = S^* f. \qquad \square$$

Next recall the conjugation C on $\mathcal{K}_u$ from (8.6) defined in terms of boundary values on $\mathbb{T}$ by $Cf = \overline{f} z u$. The following result says that the compressed shift S_u is a *complex symmetric operator* [80, 81]. We will explore this topic in greater detail in Chapter 13.

Proposition 9.8 $\quad S_u = C S_u^* C$.

Proof For $f, g \in \mathcal{K}_u$, use the previous proposition and (8.1) to get

$$\langle C S_u^* C f, g \rangle = \langle Cg, S_u^* C f \rangle = \langle Cg, S^* C f \rangle = \langle S Cg, Cf \rangle$$
$$= \langle \zeta u \overline{g}, u \overline{\zeta f} \rangle = \langle \overline{\zeta g}, \overline{f} \rangle = \langle \zeta f, g \rangle$$
$$= \langle S f, g \rangle = \langle S f, P_u g \rangle = \langle P_u S f, g \rangle = \langle S_u f, g \rangle$$

which proves the desired identity. $\qquad \square$

The following operator identities, motivated by the identity

$$I - S S^* = c_0 \otimes c_0$$

on H^2 (where c_λ is the Cauchy kernel), will appear several times in this book.

Lemma 9.9 $\quad I - S_u S_u^* = k_0 \otimes k_0$ and $I - S_u^* S_u = C k_0 \otimes C k_0 = S^* u \otimes S^* u$.

Proof   First note that

$$P_u 1 = k_0. \tag{9.3}$$

To see this, use the identity $u = u(0) + zS^*u$ along with Proposition 5.14 to get

$$P_u 1 = 1 - uP(\overline{u}) = 1 - uP(\overline{u(0)} + \overline{zS^*u}) = 1 - u\overline{u(0)} + 0 = k_0.$$

Next, for each $f \in \mathcal{K}_u$ we have

$$
\begin{aligned}
(I - S_u S_u^*)f &= f - P_u(SS^*f) && \text{(Proposition 9.7)} \\
&= f - P_u(f - f(0)) \\
&= f(0)P_u 1 \\
&= \langle f, k_0 \rangle k_0 && \text{(by (9.3))} \\
&= (k_0 \otimes k_0)f.
\end{aligned}
$$

This proves the operator identity

$$I - S_u S_u^* = k_0 \otimes k_0. \tag{9.4}$$

To verify the second identity, we need the formula

$$C(f \otimes g)C = Cf \otimes Cg, \qquad f, g \in \mathcal{K}_u, \tag{9.5}$$

which can be seen with the computation

$$
\begin{aligned}
[C(f \otimes g)C](h) &= C[(f \otimes g)(Ch)] \\
&= C[\langle Ch, g \rangle f] \\
&= \overline{\langle Ch, g \rangle} Cf \\
&= \langle h, Cg \rangle Cf && \text{(by (8.1))} \\
&= (Cf \otimes Cg)(h).
\end{aligned}
$$

Next we note that

$$
\begin{aligned}
C(I - S_u S_u^*)C &= CC - CS_u S_u^* C \\
&= I - CS_u CC S_u^* C \\
&= I - S_u^* S_u && \text{(Proposition 9.8)}
\end{aligned}
$$

and so

$$
\begin{aligned}
I - S_u^* S_u &= C(I - S_u S_u^*)C \\
&= C(k_0 \otimes k_0)C && \text{(by (9.4))} \\
&= (Ck_0) \otimes (Ck_0). && \text{(by (9.5))}
\end{aligned}
$$

Finally, note from (8.8) that $Ck_0 = S^*u$. $\square$

From Theorem 6.9 recall the conjugate linear map $J : \mathcal{K}_u \to \mathcal{K}_{u^\#}$ defined by $Jf = f^\#$, where $f^\#(z) := \overline{f(\bar{z})}$, and the unitary operator

$$U : \mathcal{K}_u \to K_{u^\#}, \qquad U = JC. \tag{9.6}$$

The following facts can be verified by direct computation. Since we will be dealing with the two different model spaces $\mathcal{K}_u$ and $\mathcal{K}_{u^\#}$, we will let C denote the conjugation on $\mathcal{K}_u$ and $C^\#$ denote the conjugation on $\mathcal{K}_{u^\#}$.

Lemma 9.10 *For an inner function u, we have the following:*

(i) $J^{-1}g = g^\#$ *for all* $g \in \mathcal{K}_{u^\#}$;
(ii) $JC = C^\# J$;
(iii) $(JC)^* = CJ^{-1} = J^{-1}C^\#$.

Proposition 9.11 *For an inner function u, the following identities hold:*

(i) $JS_u J^{-1} = S_{u^\#}$;
(ii) $JS_u^* J^{-1} = S_{u^\#}^*$;
(iii) $US_u U^* = S_{u^\#}^*$.

Proof   First note that for all $f, g \in \mathcal{K}_{u^\#}$ we have

$$\langle JS_u J^{-1}f, g \rangle = \langle J^{-1}g, S_u J^{-1}f \rangle = \langle J^{-1}g, P_u S J^{-1}f \rangle = \langle J^{-1}g, \zeta J^{-1}f \rangle$$

$$= \int_{\mathbb{T}} \overline{g(\bar{\zeta})}\bar{\zeta}f(\bar{\zeta}) \, dm(\zeta)$$

$$= \int_{\mathbb{T}} \zeta f(\zeta)\overline{g(\zeta)} \, dm(\zeta) \quad \text{(by (8.5))}$$

$$= \langle Sf, g \rangle = \langle Sf, P_{u^\#}g \rangle = \langle P_{u^\#}Sf, g \rangle = \langle S_{u^\#}f, g \rangle,$$

which implies that $JS_u J^{-1} = S_{u^\#}$. Now bring in the identities from Lemma 9.10, along with Proposition 9.8, to get

$$US_u U^* = (JC)S_u(JC)^* = C^\# JS_u J^{-1}C^\# = C^\# S_{u^\#}C^\# = S_{u^\#}^*.$$

Finally,

$$JS_u^* J^{-1} = JCS_u CJ^{-1} = US_u U^* = S_{u^\#}^*. \qquad \square$$

We conclude this section with a proof that S_u^n and S_u^{*n} both tend to zero in the strong operator topology (see Chapter 1) as $n \to \infty$. This observation will be important when we consider the Sz.-Nagy–Foiaş model theory (Section 9.4).

Lemma 9.12   *Let u be an inner function. Then for each $f \in \mathcal{K}_u$ we have*

$$\lim_{n \to \infty} \|S_u^{*n} f\| = \lim_{n \to \infty} \|S_u^n f\| = 0.$$

Proof   Since $S_u^{*n} f = S^{*n} f$, the first assertion follows from the fact that for each $f \in \mathcal{K}_u$, we have

$$\|S_u^{*n} f\|^2 = \|S^{*n} f\|^2 = \sum_{k \geqslant n} |\widehat{f}(k)|^2 \to 0$$

as $n \to \infty$ (Theorem 1.8). The second assertion follows from Proposition 9.11 because S_u^n is unitarily equivalent to $S_{u^\#}^{*n}$. □

9.3 Invariant subspaces and cyclic vectors

From Beurling's Theorem (Theorem 4.3), the unilateral shift S is a cyclic operator (any outer function serves as a cyclic vector) and so is the backward shift S^* (any function that does not have a pseudocontinuation of bounded type serves as a cyclic vector (see Proposition 7.15)). We now focus our attention on the cyclic vectors and invariant subspaces of the compressed shift S_u.

Proposition 9.13   *The compressed shift S_u is a cyclic operator with cyclic vector k_0. That is to say,*

$$\bigvee \{S_u^n k_0 : n \geqslant 0\} = \mathcal{K}_u.$$

Proof   From Proposition 5.15 we know that

$$\bigvee \{S^{*n} u : n \geqslant 1\} = \mathcal{K}_u$$

and so

$$
\begin{aligned}
\bigvee \{S_u^n k_0 : n \geqslant 0\} &= \bigvee \{(CS_u^* C)^n k_0 : n \geqslant 0\} && \text{(Proposition 9.8)} \\
&= \bigvee \{CS_u^{*n} C k_0 : n \geqslant 0\} && (C^2 = I) \\
&= C \bigvee \{S^{*(n+1)} u : n \geqslant 0\} && \text{(by (8.8))} \\
&= C\mathcal{K}_u \\
&= \mathcal{K}_u. && \square
\end{aligned}
$$

Proposition 9.14   *If u and v are inner functions and $v|u$, then:*

(i) $vH^2 \cap \mathcal{K}_u$ is an S_u-invariant subspace of $\mathcal{K}_u$;
(ii) $vH^2 \cap \mathcal{K}_u = v\mathcal{K}_{u/v}$.

Moreover, every S_u-invariant subspace of $\mathcal{K}_u$ is of the form $vH^2 \cap \mathcal{K}_u$ where v is an inner function that divides u.

Proof   If M is a subspace of $\mathcal{K}_u$ and $S_u M \subset M$, then $M^\perp = \mathcal{K}_u \ominus M$ satisfies $S_u^* M^\perp \subset M^\perp$. Since $S_u^* = S^*|_{\mathcal{K}_u}$, Proposition 5.2 ensures that $M^\perp = \mathcal{K}_v$ for some inner function v. However, since $\mathcal{K}_v \subset \mathcal{K}_u$, we see that $v|u$ (Corollary 5.9). Putting this all together we get

$$M = \mathcal{K}_u \ominus \mathcal{K}_v = vH^2 \cap \mathcal{K}_u.$$

From Lemma 5.10 we see that

$$\mathcal{K}_u = \mathcal{K}_v \oplus v\mathcal{K}_{u/v}, \tag{9.7}$$

which implies that

$$vH^2 \cap \mathcal{K}_u = v\mathcal{K}_{u/v}.$$

To see that $v\mathcal{K}_{u/v}$ is S_u-invariant, notice that $S_u^* = S^*|_{\mathcal{K}_u}$ and $\mathcal{K}_v$ is S^*-invariant. From (9.7) it follows that orthogonal complement of $\mathcal{K}_v$ in $\mathcal{K}_u$, that is, $v\mathcal{K}_{u/v}$, must be S_u-invariant.   □

Recall the definition of the greatest common divisor (gcd) of two inner functions from (4.5) and Corollary 4.9.

Corollary 9.15   *If $f \in \mathcal{K}_u$ and has inner factor ϑ and $v = \gcd(\vartheta, u)$, then*

$$\bigvee \{S_u^n f : n \geqslant 0\} = vH^2 \cap \mathcal{K}_u.$$

Thus $f \in \mathcal{K}_u$ is a cyclic vector for S_u if and only if the inner factor of f is relatively prime to u. In particular, every outer function in $\mathcal{K}_u$ is cyclic for S_u.

Proof   Set $M = vH^2 \cap \mathcal{K}_u$ and observe that M is S_u-invariant (Proposition 9.14). Note that $\vartheta H^2 \subset vH^2$ (Theorem 4.7) and so $f \in M$. Thus we have

$$\bigvee \{S_u^n f : n \geqslant 0\} \subset M.$$

To show equality, we will use Theorem 1.27. Indeed, let $g \in M$ and $g \perp S_u^n f$ for all $n \geqslant 0$. This implies

$$0 = \langle g, S_u^n f \rangle = \langle g, P_u S^n f \rangle = \langle P_u g, \zeta^n f \rangle = \langle g, \zeta^n f \rangle$$

and so

$$g \perp \bigvee \{S^n f : n \geqslant 0\} = \vartheta H^2 \subset vH^2.$$

However, since $g \in vH^2$, we conclude that $g \equiv 0$.   □

Corollary 9.16   *The compressed shift S_u is irreducible, meaning there are no proper non-trivial subspaces of $\mathcal{K}_u$ that are invariant for both S_u and S_u^*.*

Proof   If $\mathcal{M} \neq \mathcal{K}_u$ is S_u-invariant, then $\mathcal{M} = vH^2 \cap \mathcal{K}_u$ for some (non-constant) inner function v dividing u. If $\mathcal{M}$ were also S_u^*-invariant, then $\mathcal{M}$ would be S^*-invariant. By Proposition 5.5, the outer factor of any function in $\mathcal{M}$ would also belong to $\mathcal{M}$. However, $vH^2 \cap \mathcal{K}_u$ contains no outer functions and so $\mathcal{M} = \{0\}$. $\square$

9.4 The Sz.-Nagy–Foiaş model

One of the main reasons that model spaces are worthy of study in their own right stems from the so-called *model theory* developed by Sz.-Nagy and Foiaş, which shows that a wide range of Hilbert space operators can be realized concretely as restrictions of the backward shift operator to model spaces. These ideas have since been generalized in many directions (for example, de Branges-Rovnyak spaces, vector-valued Hardy spaces, etc.) and we make no attempt to provide an encyclopedic account of the subject, referring the reader instead to the influential texts [27, 141, 143, 153, 186]. Instead, we present few a results to illustrate the connection between operator theory and model spaces.

In the following, we let $\mathcal{H}$ denote a separable complex Hilbert space. If $T \in \mathcal{B}(\mathcal{H})$, then, by rescaling, we may assume that T is a *contraction*. As such, T enjoys a decomposition of the form $T = K \oplus U$ (see [186, p. 8] for more details) where U is a unitary operator and K is a *completely non-unitary (CNU) contraction*, by which we mean there does not exist a non-trivial reducing subspace for K (invariant for both K and K^*) upon which K is unitary.

Since the structure of unitary operators is described by the Spectral Theorem, the study of arbitrary bounded Hilbert space operators can be focused on CNU contractions. With a few additional hypotheses, one can obtain a concrete *functional model* for such operators [186]. The fact that the operator $S_u^* = S^*|_{\mathcal{K}_u}$ (see Proposition 9.7) satisfies conditions (i) and (ii) of the following theorem comes from Lemma 9.12 and Lemma 9.9, respectively.

Theorem 9.17 (Sz.-Nagy–Foiaş)   *If $T \in \mathcal{B}(\mathcal{H})$ is a contraction satisfying:*

(i) $\|T^n x\| \to 0$ *for all $x \in \mathcal{H}$, that is, $T^n \to 0$ in the strong operator topology;*
(ii) $\mathrm{rank}(I - T^*T) = \mathrm{rank}(I - TT^*) = 1$;

then there exists an inner function u such that T is unitarily equivalent to $S^*|_{\mathcal{K}_u}$.

Proof   We let $\cong$ denote the unitary equivalence of Hilbert spaces or their operators. Since the *defect operator* $D = \sqrt{I - T^*T}$ (which is well-defined since T is a contraction and so $I - T^*T$ is a positive operator) has rank 1, we see that $\mathrm{ran}\, D$ (the range of D) is isomorphic to $\mathbb{C}$ so that

$$\widetilde{\mathcal{H}} := \bigoplus_{n \geqslant 1} \mathrm{ran}\, D \cong H^2. \tag{9.8}$$

It follows that for each $n \geqslant 1$ we have

$$
\begin{aligned}
\sum_{0 \leqslant j \leqslant n} \|DT^j\mathbf{x}\|^2 &= \sum_{0 \leqslant j \leqslant n} \langle (I - T^*T)^{\frac{1}{2}}T^j\mathbf{x}, (I - T^*T)^{\frac{1}{2}}T^j\mathbf{x} \rangle \\
&= \sum_{0 \leqslant j \leqslant n} \langle (I - T^*T)T^j\mathbf{x}, T^j\mathbf{x} \rangle \\
&= \sum_{0 \leqslant j \leqslant n} \left(\langle T^j\mathbf{x}, T^j\mathbf{x} \rangle - \langle T^*TT^j\mathbf{x}, T^j\mathbf{x} \rangle \right) \\
&= \sum_{0 \leqslant j \leqslant n} \left(\|T^j\mathbf{x}\|^2 - \|T^{j+1}\mathbf{x}\|^2 \right) \\
&= \|\mathbf{x}\|^2 - \|T^{n+1}\mathbf{x}\|^2.
\end{aligned}
$$

Since, by hypothesis, $\|T^n\mathbf{x}\| \to 0$ for each $\mathbf{x} \in \mathcal{H}$, we conclude that

$$
\sum_{j \geqslant 0} \|DT^j\mathbf{x}\|^2 = \|\mathbf{x}\|^2, \quad \mathbf{x} \in \mathcal{H},
$$

and hence the operator $\Phi : \mathcal{H} \to H^2$ defined by

$$
\Phi\mathbf{x} = (D\mathbf{x}, DT\mathbf{x}, DT^2\mathbf{x}, DT^3\mathbf{x}, \dots)
$$

is an isometric embedding of $\mathcal{H}$ into H^2 (here we have identified a function in H^2 with its sequence of Taylor coefficients). Since Φ is an isometry, its image is closed in H^2 and is clearly S^*-invariant. By Corollary 5.2, $\operatorname{ran}\Phi = \mathcal{K}_u$ for some u (the possibility that $\operatorname{ran}\Phi = H^2$ is ruled out because the argument we are about to give will show that $T \cong S^*$, violating the assumption that $\operatorname{rank}(I - TT^*) = 1$). Now observe that

$$
\Phi T\mathbf{x} = (DT\mathbf{x}, DT^2\mathbf{x}, DT^3\mathbf{x}, \dots) = S^*\Phi\mathbf{x}, \quad \mathbf{x} \in \mathcal{H},
$$

that is to say, the following diagram commutes:

$$
\begin{array}{ccc}
\mathcal{H} & \xrightarrow{\;\;T\;\;} & \mathcal{H} \\
\Big\downarrow{\scriptstyle\Phi} & & \Big\downarrow{\scriptstyle\Phi} \\
H^2 & \xrightarrow{\;\;S^*\;\;} & H^2
\end{array}
$$

Letting $U : \mathcal{H} \to \mathcal{K}_u$ denote the unitary operator obtained from Φ by reducing its codomain from H^2 to $\mathcal{K}_u$, it follows that

$$
UT = (S^*|_{\mathcal{K}_u})U = S_u^*U,
$$

that is to say, the following diagram commutes:

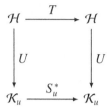

Thus T is unitarily equivalent to the restriction of S^* to $\mathcal{K}_u$. □

Since Proposition 9.11 says that S_u is unitarily equivalent to $S_{u^\#}^*$, the Sz.-Nagy-Foiaş Theorem (Theorem 9.17) can be restated as follows:

Theorem 9.18 (Sz.-Nagy–Foiaş) *If T is a contraction on a Hilbert space which satisfies:*

(i) $\|T^{*n}x\| \to 0$ *for all* $x \in \mathcal{H}$, *that is,* $T^{*n} \to 0$ *as* $n \to \infty$ *in the strong operator topology;*

(ii) $\mathrm{rank}(I - T^*T) = \mathrm{rank}(I - TT^*) = 1;$

then there exists an inner function u such that T is unitarily equivalent to S_u.

9.5 Functional calculus for S_u

For any $T \in \mathcal{B}(\mathcal{H})$ one can define $p(T) \in \mathcal{B}(\mathcal{H})$ for any analytic polynomial

$$p(z) = a_0 + a_1 z + \cdots + a_n z^n$$

by setting

$$p(T) = a_0 I + a_1 T + \cdots + a_n T^n.$$

A much studied problem in operator theory is to determine how to define the operator $f(T)$ for other classes of functions besides polynomials. This line of inquiry falls under the broad heading of *functional calculus*. In particular, for an operator T (perhaps from a certain class of operators), we want to define a map

$$f \mapsto f(T)$$

from a certain class of functions containing the analytic polynomials in such a way that this map is a continuous homomorphism (in some appropriate topology particular to the class of functions).

There is the *holomorphic functional calculus*, which says that if f is analytic in an open neighborhood of $\sigma(T)$ then one can define $f(T)$ by the operator-valued Cauchy integral formula

$$f(T) = \frac{1}{2\pi i} \int_\Gamma f(z)(zI - T)^{-1} dz,$$

where Γ is some appropriate system of curves surrounding $\sigma(T)$ and the integral is interpreted as an operator-valued Riemann integral. If $T \in \mathcal{B}(\mathcal{H})$ is self-adjoint, then there is the *Borel functional calculus*, where $f(T)$ can be meaningfully defined for any bounded Borel function on the real line. Another functional calculus comes from the Sz.-Nagy–Foiaş theory which defines, for a contraction T, the operator $f(T)$ where f belongs to the disk algebra $\mathcal{A}$ (see Definition 5.23). For completely non-unitary contractions, one can meaningfully define $f(T)$ for $f \in H^\infty$. We now proceed to define the H^∞-functional calculus for S_u.

For $\varphi \in H^\infty$, define the operator

$$\varphi(S_u) := P_u T_\varphi|_{\mathcal{K}_u}, \tag{9.9}$$

where T_φ is a Toeplitz operator on H^2 (Definition 4.11). More explicitly, $\varphi(S_u)$ is given by the formula

$$\varphi(S_u)(f) = P_u(\varphi f), \qquad f \in \mathcal{K}_u. \tag{9.10}$$

Another widely used notation for $\varphi(S_u)$ is in terms of the truncated Toeplitz operator A_φ^u on $\mathcal{K}_u$ given by the formula

$$A_\varphi^u : \mathcal{K}_u \to \mathcal{K}_u, \quad A_\varphi^u f = P_u(\varphi f). \tag{9.11}$$

Clearly,

$$\|\varphi(S_u)\| \leqslant \|T_\varphi\| = \|\varphi\|_\infty.$$

The mapping

$$\Lambda : H^\infty \to \mathcal{B}(\mathcal{K}_u), \quad \varphi \mapsto \varphi(S_u)$$

is called the H^∞-*functional calculus* for the operator S_u. It was developed in [184, 186].

Theorem 9.19 *Let u be an inner function. Then the mapping Λ is linear, multiplicative, and contractive. Furthermore, $\Lambda z = A_z^u = S_u$.*

Proof   In the course of defining the functional calculus, we have already seen that Λ is linear and contractive. To prove that Λ is multiplicative, we proceed as follows. Let $\varphi, \psi \in H^\infty$. For $g \in H^2$ we use the formula

$$P_u g = g - uP(\overline{u}g)$$

from Proposition 5.14 to get, for $f \in \mathcal{K}_u$,

$$
\begin{aligned}
A_\varphi^u A_\psi^u f &= P_u \varphi(P_u(\psi f)) \\
&= P_u(\varphi(\psi f - uP(\overline{u}\psi f))) \\
&= P_u(\varphi\psi f) - P_u(u\varphi P(\overline{u}\psi f)) \\
&= A_{\varphi\psi}^u f
\end{aligned}
$$

since $u\varphi P(\overline{u}\psi f) \in uH^2$ and so $P_u(uH^2) = 0$. Thus we have

$$A_\varphi^u A_\psi^u - A_{\varphi\psi}^u, \qquad \varphi, \psi \in H^\infty, \tag{9.12}$$

which completes the proof. □

Here are some other properties of the functional calculus.

Theorem 9.20 *If u is inner and $\varphi \in H^\infty$, then the following assertions hold:*

(i) $\varphi(S_u)^* = T_{\overline{\varphi}}|_{\mathcal{K}_u}$;
(ii) *If φ is in the Wiener algebra, that is, $\sum_{n \geqslant 0} |\widehat{\varphi}(n)| < \infty$, then*

$$\varphi(S_u) = \sum_{n \geqslant 0} \widehat{\varphi}(n) S_u^n,$$

where this series converges in the operator norm;
(iii) $\varphi(S_u) = 0$ *if and only if $\varphi \in uH^\infty$.*

Proof To prove (i), use (9.9) to see that

$$\varphi(S_u)^* = P_u T_\varphi^*|_{\mathcal{K}_u} = P_u T_{\overline{\varphi}}|_{\mathcal{K}_u}.$$

However, by Proposition 5.5, $\mathcal{K}_u$ is invariant under $T_{\overline{\varphi}}$ and so $\varphi(S_u)^* = T_{\overline{\varphi}}|_{\mathcal{K}_u}$.
To prove (ii), set

$$\varphi_N(z) = \sum_{0 \leqslant n \leqslant N} \widehat{\varphi}(n) z^n, \qquad N \geqslant 1.$$

Then we have $\varphi_N(S_u) = \sum_{0 \leqslant n \leqslant N} \widehat{\varphi}(n) S_u^n$ and, using Theorem 9.19, we get

$$
\begin{aligned}
\left\| \sum_{0 \leqslant n \leqslant N} \widehat{\varphi}(n) S_u^n - \varphi(S_u) \right\| &= \|\varphi_N(S_u) - \varphi(S_u)\| \\
&= \|\Lambda(\varphi_N - \varphi)\| \\
&\leqslant \|\varphi_N - \varphi\|_\infty \\
&\leqslant \sum_{n > N} |\widehat{\varphi}(n)| \to 0
\end{aligned}
$$

as $N \to \infty$. Notice in the above how we used the fact that for a function ψ in the Wiener algebra, the Fourier series of ψ converges uniformly on $\mathbb{T}$.

For the proof of (iii), clearly $\varphi(S_u) = 0$ if and only if $\varphi(S_u)' = 0$. By (1), the latter is equivalent to $T_{\bar\varphi}|_{\mathcal{K}_u} = 0$. But, from Proposition 5.8, we know that $\ker T_{\bar\varphi} = \mathcal{K}_v$, where v is the inner factor of φ. Thus, $\varphi(S_u) = 0$ if and only if $\mathcal{K}_u \subset \mathcal{K}_v$, and this inclusion happens precisely when $u|v$ (Corollary 5.9). Since v is the inner factor of φ, we have $u|v$ if and only if $\varphi \in uH^\infty$. $\square$

Theorem 9.21 *Let u be an inner function and $\varphi \in H^\infty$. If $\{\varphi_n\}_{n \geqslant 1} \subset H^\infty$ such that*

$$M = \sup_{n \geqslant 1} \|\varphi_n\|_\infty < \infty,$$

then the following assertions hold:

(i) If

$$\lim_{n \to \infty} \varphi_n(\zeta) = \varphi(\zeta), \qquad a.e.\ on\ \mathbb{T},$$

then $\varphi_n(S_u) \to \varphi(S_u)$ in the strong operator topology;

(ii) If

$$\lim_{n \to \infty} \varphi_n(\lambda) = \varphi(\lambda), \qquad \lambda \in \mathbb{D},$$

then $\varphi_n(S_u) \to \varphi(S_u)$ in the weak operator topology.

Proof   If (i) holds, then, for each $f \in \mathcal{K}_u$,

$$\|\varphi_n(S_u)f - \varphi(S_u)f\| = \|P_u(\varphi_n f) - P_u(\varphi f)\|$$
$$\leqslant \|\varphi_n f - \varphi f\|$$

and

$$\|\varphi_n f - \varphi f\|^2 = \int_{\mathbb{T}} |\varphi_n - \varphi|^2 |f|^2 dm.$$

Since, by assumption, $\varphi_n \to \varphi$, a.e. on $\mathbb{T}$ and

$$|\varphi_n(\zeta) - \varphi(\zeta)|\,|f(\zeta)| \leqslant (\|\varphi_n\|_\infty + \|\varphi\|_\infty)|f(\zeta)| \leqslant 2M|f(\zeta)|,$$

we can apply the Dominated Convergence Theorem to see that

$$\lim_{n \to \infty} \|\varphi_n f - \varphi f\| = 0.$$

Thus $\varphi_n(S_u)f \to \varphi(S_u)f$ in norm for all $f \in H^2$, or equivalently, $\varphi_n(S_u) \to \varphi(S_u)$ in the strong operator topology.

To prove (ii), fix $f, g \in \mathcal{K}_u$ and observe that

$$\langle \varphi_n(S_u)f - \varphi(S_u)f, g \rangle = \langle P_u(\varphi_n f) - P_u(\varphi f), g \rangle$$
$$= \langle \varphi_n f - \varphi f, P_u g \rangle = \langle \varphi_n f - \varphi f, g \rangle.$$
$$= \int_{\mathbb{T}} (\varphi_n - \varphi) f \overline{g} \, dm.$$

To finish the proof, we need a fact from [35, Prop. 2] which says that if $\varphi_n \to \varphi$ pointwise on $\mathbb{D}$ and φ_n is uniformly bounded in H^∞-norm, then $\varphi_n \to \varphi$ in the weak-$*$ topology of H^∞, that is to say,

$$\int_{\mathbb{T}} \varphi_n h \, dm \to \int_{\mathbb{T}} \varphi h \, dm$$

for every $h \in L^1$. Apply this result to $h = f\overline{g} \in L^1$ above to see that $\varphi_n(S_u) \to \varphi(S_u)$ in the weak operator topology. $\qquad\square$

9.6 The spectrum of S_u

In Chapter 7 we defined the spectrum of the inner function $u = B s_\mu$ as the set

$$\sigma(u) = \left\{ \lambda \in \mathbb{D}^- : \lim_{z \to \lambda} |u(z)| = 0 \right\} = (B^{-1}(\{0\}))^- \cup \operatorname{supp} \mu.$$

The main theorem of this section says that $\sigma(u) = \sigma(S_u)$ [125, 140].

Theorem 9.22 (Livšic–Möller) $\quad \sigma(S_u) = \sigma(u)$.

The proof of this theorem needs some preliminary discussion. Recall from Proposition 1.30 that S_u is a contraction and so $\sigma(S_u) \subset \mathbb{D}^-$. Thus $(I - \overline{\lambda}S_u)^{-1}$ is a bounded operator for each $\lambda \in \mathbb{D}$.

Lemma 9.23 $\quad$ *If u is an inner function and $\lambda \in \mathbb{D}$, then*

$$k_\lambda = (I - \overline{\lambda}S_u)^{-1} k_0.$$

Proof $\quad$ From (4.15) we see that $c_\lambda = (I - \overline{\lambda}S)^{-1} c_0$. Hence, for each $f \in \mathcal{K}_u$, we can use (4.15) to get

$$f(\lambda) = \langle f, c_\lambda \rangle = \langle f, (I - \overline{\lambda}S)^{-1} c_0 \rangle = \langle (I - \lambda S^*)^{-1} f, c_0 \rangle.$$

We can now use the identity $P_u c_0 = k_0$ from Proposition 5.13 to get

$$
\begin{aligned}
f(\lambda) - \langle (I - \lambda S_u^*)^{-1} f, c_0 \rangle &= \langle P_u (I - \lambda S_u^*)^{-1} f, c_0 \rangle \\
&= \langle (I - \lambda S_u^*)^{-1} f, P_u c_0 \rangle = \langle (I - \lambda S_u^*)^{-1} f, k_0 \rangle \\
&= \langle f, (I - \overline{\lambda} S_u)^{-1} k_0 \rangle.
\end{aligned}
$$

This proves that $k_\lambda = (I - \overline{\lambda} S_u)^{-1} k_0$.   □

Proof of Theorem 9.22   We follow parts of [166]. As we have seen before, S_u is a contraction and so $\sigma(S_u) \subset \mathbb{D}^-$ (Proposition 1.30). If $\lambda \in \mathbb{D}^- \setminus \sigma(u)$, then from Theorem 7.18 it follows that

$$
u(\lambda) \neq 0 \quad \text{and} \quad Q_\lambda u = \frac{u - u(\lambda)}{z - \lambda} \in H^\infty.
$$

Furthermore, for any $f \in \mathcal{K}_u$,

$$
\begin{aligned}
-\frac{1}{u(\lambda)} A_{Q_\lambda u}^u (S_u - \lambda I) f &= -\frac{1}{u(\lambda)} A_{Q_\lambda u}^u A_{z-\lambda}^u f \\
&= -\frac{1}{u(\lambda)} A_{Q_\lambda u(z-\lambda)}^u f \qquad \text{(Lemma 9.12)} \\
&= -\frac{1}{u(\lambda)} A_{u-u(\lambda)}^u f \\
&= -\frac{1}{u(\lambda)} A_u^u f + A_1^u f \\
&= 0 + f,
\end{aligned}
$$

since $A_u^u = 0$ and $A_1^u = I$ (Theorem 9.20). This means that $(S_u - \lambda I)$ is invertible and hence $\lambda \notin \sigma(S_u)$. Thus $\sigma(S_u) \subset \sigma(u)$.

To show $\sigma(u) \subset \sigma(S_u)$, assume that $S_u - \lambda I$ is invertible. We now argue that λ cannot be a point in $\mathbb{D}$ where $u(\lambda) = 0$. Indeed, by direct computation, we see that if $\lambda \in \mathbb{D}$ with $u(\lambda) = 0$, then $k_\lambda = c_\lambda \in \mathcal{K}_u$ and

$$
S_u^* c_\lambda = S^* c_\lambda = \overline{\lambda} c_\lambda.
$$

This means that $\overline{\lambda}$ is an eigenvalue of S_u^* and thus $\overline{\lambda} \in \sigma(S_u^*)$. Using the fact that $\sigma(S_u^*) = \{\overline{\lambda} : \lambda \in \sigma(S_u)\}$, we see that $\lambda \in \sigma(S_u)$, contradicting our assumption that $S_u - \lambda I$ is invertible. Thus $\lambda \notin \sigma(u) \cap \mathbb{D}$.

To show that $\lambda \notin \mathbb{T} \cap \sigma(u)$ we proceed as follows. For $\eta \in \mathbb{D}$, recall from Lemma 9.23 that

$$
k_\eta = (I - \overline{\eta} S_u)^{-1} k_0.
$$

Observe that the right-hand side of the above equation is a conjugate-analytic (vector-valued) function in a neighborhood of λ (remember that the complement of the spectrum of a bounded operator is an open set). Thus the function

$\eta \mapsto k_\eta$ extends to a (vector-valued) conjugate-analytic function on an open neighborhood of λ. Finally,

$$1 - \overline{u(\eta)}u(0) = \langle k_\eta, k_0 \rangle$$

is conjugate-analytic near λ and thus u must be analytic near λ. By Corollary 7.20, we conclude that $\lambda \notin \sigma(u) \cap \mathbb{T}$. □

We can also determine the point spectrum $\sigma_p(S_u)$ of S_u.

Corollary 9.24 *For an inner function u,*

$$\sigma_p(S_u) = \sigma(u) \cap \mathbb{D} = \{\lambda \in \mathbb{D} : u(\lambda) = 0\}.$$

Proof By Lemma 9.12 we have

$$\lim_{n \to \infty} \|S_u^n f\| \to 0, \qquad f \in \mathcal{K}_u.$$

If $\lambda \in \sigma_p(S_u)$ and $S_u f = \lambda f$ with $\|f\| = 1$, then

$$|\lambda|^n = \|\lambda^n f\| = \|S_u^n f\| \to 0, \quad n \to \infty.$$

Thus $\sigma_p(S_u) \subset \mathbb{D}$. Since $\sigma_p(S_u) \subset \sigma(S_u)$ and $\sigma(S_u) = \sigma(u)$ (Theorem 9.22), we deduce that

$$\sigma_p(S_u) \subset \sigma(u) \cap \mathbb{D} = \{\lambda \in \mathbb{D} : u(\lambda) = 0\}.$$

Now fix $\lambda \in \mathbb{D}$ such that $u(\lambda) = 0$ and set $f = Q_\lambda u$ (see (4.16)). Then by Theorem 9.19 and Theorem 9.20 (iii), along with the identity

$$(z - \lambda)f(z) = u(z),$$

we get

$$(\lambda I - S_u)f(S_u) = -u(S_u) = -A_u^u = 0. \tag{9.13}$$

Note that $f = Q_\lambda u = Ck_\lambda^u \in \mathcal{K}_u$ (Example 8.12). Furthermore, since $f \not\equiv 0$ and $\mathcal{K}_u \cap uH^\infty = \{0\}$, we conclude that $f \notin uH^\infty$ and thus, by Theorem 9.20 (iii), $f(S_u) \neq 0$. Therefore, there is a $g \in \mathcal{K}_u$ such that $h = f(S_u)g \in \mathcal{K}_u$ and $h \neq 0$. The identity (9.13) applied to g reveals that $\lambda \in \sigma_p(S_u)$. □

By Corollary 9.24, we know that

$$\sigma_p(S_u) = \sigma(u) \cap \mathbb{D} = \{\lambda \in \mathbb{D} : u(\lambda) = 0\} \tag{9.14}$$

and hence

$$\sigma_p(S_u^*) = \sigma_p(S_{u^\#}) = \{\overline{\lambda} \in \mathbb{D} : u(\lambda) = 0\} \tag{9.15}$$

(Proposition 9.11). We can now identify the eigenspaces of S_u^* and S_u.

Corollary 9.25 *Let u be an inner function and let $\lambda \in \mathbb{D}$ be such that* $u(\lambda) = 0$. *Then* $\ker(S_u^* - \bar{\lambda}I) = \mathbb{C}k_\lambda$ *and* $\ker(S_u - \lambda I) = \mathbb{C}Ck_\lambda$.

Proof   Since $u(\lambda) = 0$ we have $k_\lambda = c_\lambda \in \mathcal{K}_u$. Moreover, since $S_u^* = S^*|_{\mathcal{K}_u}$,

$$\ker(S_u^* - \bar{\lambda}I) = \ker(S^* - \bar{\lambda}I) \cap \mathcal{K}_u.$$

But since $\ker(S^* - \bar{\lambda}I) = \mathbb{C}c_\lambda$, we see that $\ker(S_u^* - \bar{\lambda}I) = \mathbb{C}c_\lambda$.

For the second equality, we recall the unitary operator $U : \mathcal{K}_u \to \mathcal{K}_{u^\#}$ from (9.6) defined by

$$(Uf)(\zeta) = \bar{\zeta}\overline{f(\bar{\zeta})u(\bar{\zeta})}, \qquad \zeta \in \mathbb{T}.$$

Observe that

$$(U^*g)(\zeta) = \bar{\zeta}\overline{g(\bar{\zeta})}u(\zeta). \tag{9.16}$$

We also recall Proposition 9.11, which says that $US_uU^* = S_{u^\#}^*$. Since

$$S_u - \lambda I = U^*(S_{u^\#}^* - \lambda I)U,$$

we see that

$$\ker(S_u - \lambda I) = U^*\ker(S_{u^\#}^* - \lambda I) = \mathbb{C}U^*c_{\bar{\lambda}}.$$

Using (9.16) and Example 8.12 we get

$$(U^*c_{\bar{\lambda}})(\zeta) = \bar{\zeta}\overline{c_{\bar{\lambda}}(\bar{\zeta})}u(\zeta) = \bar{\zeta}\frac{1}{1 - \bar{\zeta}\lambda}u(\zeta) = \frac{u(\zeta)}{\zeta - \lambda} = Q_\lambda u = Ck_\lambda. \qquad \square$$

Finally, we compute the essential spectrum (see Chapter 1) $\sigma_e(S_u)$ of S_u.

Proposition 9.26 $\sigma_e(S_u) = \sigma(u) \cap \mathbb{T}$.

Proof   Our proof follows [166]. We already know that

$$\sigma_e(S_u) \subset \sigma(S_u) \qquad \text{(Proposition 1.41)}$$
$$= \sigma(u). \qquad \text{(Theorem 9.22)}$$

If $\lambda \in \mathbb{D}$, recall the difference quotient operator

$$Q_\lambda := S^*(I - \lambda S^*)^{-1}$$

on H^2 from (4.16). Also recall that

$$Q_\lambda f = \frac{f - f(\lambda)}{z - \lambda}, \qquad f \in H^2.$$

As we have seen before, $S^*\mathcal{K}_u \subset \mathcal{K}_u$ implies $Q_\lambda\mathcal{K}_u \subset \mathcal{K}_u$. Thus

$$R_\lambda := Q_\lambda|_{\mathcal{K}_u} = S_u^*(I - \lambda S_u^*)^{-1}.$$

defines a bounded operator on $\mathcal{K}_u$. We also see that for $f \in \mathcal{K}_u$,

$$(S_u - \lambda I)R_\lambda f = P_u\left((z - \lambda)\frac{f - f(\lambda)}{z - \lambda}\right)$$
$$= f - f(\lambda)P_u 1$$
$$= f - f(\lambda)k_0 \qquad \text{(by (9.3))}$$
$$= (I - k_0 \otimes k_\lambda)f.$$

Thus

$$(S_u - \lambda I)R_\lambda = I - k_0 \otimes k_\lambda, \qquad (9.17)$$

which means that $S_u - \lambda I$ is right invertible modulo a compact operator (recall that a finite-rank operator is compact).

To show that $S_u - \lambda I$ is left invertible (modulo a compact operator), we first note that

$$CR_\lambda C = CS_u^*(I - \lambda S_u^*)^{-1}C = CS_u^*CC(I - \lambda S_u^*)^{-1}C$$
$$= S_u(I - \overline{\lambda}S_u)^{-1} = R_\lambda^*,$$

where C is the conjugation on $\mathcal{K}_u$. A computation similar to the previous one shows that

$$C(S_u - \lambda I)R_\lambda C = (S_u^* - \overline{\lambda}I)R_\lambda^*. \qquad (9.18)$$

From (9.17) we observe that

$$C(S_u - \lambda I)R_\lambda C = C(I - k_0 \otimes k_\lambda)C \qquad \text{(by (9.17))}$$
$$= I - C(k_0 \otimes k_\lambda)C$$
$$= I - (Ck_0) \otimes (Ck_\lambda) \qquad \text{(by (9.5))}$$
$$= I - S^*u \otimes Q_\lambda u. \qquad \text{(by Example 8.12)}$$

Combine this with (9.18) to conclude

$$(S_u^* - \overline{\lambda}I)R_\lambda^* = I - (Ck_0) \otimes (Ck_\lambda) = I - S^*u \otimes Q_\lambda u.$$

Take adjoints of the previous line to get

$$R_\lambda(S_u - \lambda I) = I - (Ck_\lambda) \otimes (Ck_0) = I - Q_\lambda u \otimes S^*u.$$

Thus $\lambda \notin \sigma_e(S_u)$ and so $\sigma_e(S_u) \cap \mathbb{D} = \varnothing$. We conclude that $\sigma_e(S_u) \subset \mathbb{T}$.

Since every $\lambda \in \mathbb{T} \setminus \sigma(u)$ is in the resolvent set for S_u, and hence $\lambda \notin \sigma_e(S_u)$, we will finish the proof by showing that every point of $\sigma(u) \cap \mathbb{T}$ belongs to $\sigma_e(S_u)$. Indeed, if $\lambda \in \sigma(u)$, then $S_u - \lambda I$ is not invertible (Theorem 9.22). However, since $\lambda \in \mathbb{T}$, $z - \lambda$ is an outer function (Corollary 3.24) and so $S - \lambda I$ has dense range as an operator on H^2 (Corollary 4.5). This implies that

$S_u - \lambda I$ also has dense range as an operator on $\mathcal{K}_u$. By Corollary 9.24, $S_u - \lambda I$ is injective. But since we are assuming that $S_u - \lambda I$ is not invertible, it must be the case that $S_u - \lambda I$ does not have closed range and thus can not be Fredholm. Hence $\lambda \in \sigma_e(S_u)$ (Corollary 1.40). □

We leave it to the reader to use the final part of the proof of the previous proposition, along with the proof of Proposition 4.1, to verify the following.

Corollary 9.27 $\sigma_e(S) = \mathbb{T}$.

9.7 The C^*-algebra generated by S_u

For a family $\mathcal{X}$ of operators in $\mathcal{B}(\mathcal{H})$, let $C^*(\mathcal{X})$ denote the unital C^*-algebra generated by $\mathcal{X}$. In other words, $C^*(\mathcal{X})$ is the closure, in the norm of $\mathcal{B}(\mathcal{H})$, of the unital algebra generated by the operators in $\mathcal{X}$ and their adjoints. Since we are frequently interested in the case where $\mathcal{X} = \{A\}$ is a singleton, we often write $C^*(A)$ in place of $C^*(\{A\})$.

The *commutator ideal* $\mathscr{C}(C^*(\mathcal{X}))$ of $C^*(\mathcal{X})$ is the smallest norm-closed two-sided ideal of $\mathcal{B}(\mathcal{H})$ that contains all of the *commutators*

$$[A, B] := AB - BA,$$

where A and B range over all elements of $C^*(\mathcal{X})$. Since the quotient algebra $C^*(\mathcal{X})/\mathscr{C}(C^*(\mathcal{X}))$ is an abelian C^*-algebra, it is isometrically $*$-isomorphic to $C(Y)$, the set of all continuous functions on some compact Hausdorff space Y [52, Thm. 1.2.1]. We denote this relationship by

$$\frac{C^*(\mathcal{X})}{\mathscr{C}(C^*(\mathcal{X}))} \cong C(Y). \tag{9.19}$$

The *Toeplitz algebra* $C^*(S)$, where S is the unilateral shift on H^2, has been extensively studied since the seminal work of Coburn in the late 1960s [47, 48]. Indeed, the Toeplitz algebra is now one of the standard examples discussed in many well-known texts (for example, [20, Sect. 4.3; 56, Ch. V.1; 59 Ch. 7]). In this setting, we have $\mathscr{C}(C^*(S)) = \mathscr{K}$, the ideal of compact operators on H^2, and $Y = \mathbb{T}$, that is,

$$C^*(S)/\mathscr{K} \cong C(\mathbb{T}).$$

It also follows that

$$C^*(S) = \{T_\varphi + K : \varphi \in C(\mathbb{T}), K \in \mathscr{K}\},$$

and, moreover, that each element of $C^*(S)$ enjoys a unique decomposition of the form $T_\varphi + K$ [20, Thm. 4.3.2].

We now prove the following analogue of Coburn's work where $C^*(S)$ is replaced with $C^*(S_u)$, the C^*-algebra generated by the compressed shift S_u. For $\varphi \in C(\mathbb{T})$ we extend our definition of a truncated Toeplitz operator (defined earlier for $\varphi \in H^\infty$ when discussing the functional calculus for S_u) and set

$$A_\varphi^u : \mathcal{K}_u \to \mathcal{K}_u, \qquad A_\varphi^u f = P_u(\varphi f).$$

We will see A_φ^u when $\varphi \in L^\infty$ (and even sometimes $\varphi \in L^2$) again when we discuss truncated Toeplitz operators more thoroughly in Chapter 13.

Theorem 9.28 *If u is an inner function, then we have the following:*

(i) *The commutator ideal $\mathscr{C}(C^*(S_u))$ of $C^*(S_u)$ is equal to $\mathscr{K}^u$, the algebra of compact operators on $\mathcal{K}_u$;*

(ii) *$C^*(S_u)/\mathscr{K}^u$ is isometrically $*$-isomorphic to $C(\sigma(u) \cap \mathbb{T})$;*

(iii) *If $\varphi \in C(\mathbb{T})$, then A_φ^u is compact if and only if $\varphi|_{\sigma(u) \cap \mathbb{T}} = 0$;*

(iv) *$C^*(S_u) = \{A_\varphi^u + K : \varphi \in C(\mathbb{T}), K \in \mathscr{K}^u\}$;*

(v) *If $\varphi \in C(\mathbb{T})$, then $\sigma_e(A_\varphi^u) = \varphi(\sigma_e(A_z^u)) = \varphi(\sigma(u) \cap \mathbb{T})$;*

(vi) *For $\varphi \in C(\mathbb{T})$, $\|A_\varphi^u\|_e = \sup\{|\varphi(\zeta)| : \zeta \in \sigma(u) \cap \mathbb{T}\}$.*

Remark 9.29

(i) If u is a finite Blaschke product then $\sigma(u) \cap \mathbb{T} = \varnothing$ (Proposition 7.19) and, since $\mathcal{K}_u$ is finite dimensional (Proposition 5.16), A_φ^u is compact for every $\varphi \in C(\mathbb{T})$. Hence in this case $C^*(S_u) = \mathscr{K}^u = \mathcal{B}(\mathcal{K}_u)$. Thus the only interesting cases occur when u is not a finite Blaschke product.

(ii) It should also be noted that many of the statements in Theorem 9.28 can be obtained using the explicit triangularization theory developed by Ahern and Clark in [3] (also see the exposition in [141, Lec. V]).

(iii) When we discuss truncated Toeplitz operators more thoroughly in Chapter 13, we will see that the symbol φ which defines A_φ^u is not unique. In fact $A_\varphi^u = A_\psi^u$ if and only if $\psi - \psi \in uH^2 + \overline{uH^2}$. So if one wanted to be more precise in statement (iv) of Theorem 9.28, one should write $C^*(S_u)$ as the set of all $A \in \mathcal{B}(\mathcal{K}_u)$ such that $A = A_\varphi^u + K$ for some $\varphi \in C(\mathbb{T})$ and some $K \in \mathscr{K}^u$.

Towards a proof of Theorem 9.28, we start with two lemmas.

Lemma 9.30 *If $\varphi \in C(\mathbb{T})$, then A_φ^u is compact if and only if $\varphi|_{\sigma(u) \cap \mathbb{T}} \equiv 0$.*

Proof ($\Leftarrow$) Suppose that $\varphi|_{\sigma(u) \cap \mathbb{T}} \equiv 0$. Let $\epsilon > 0$ and pick ψ in $C(\mathbb{T})$ such that ψ vanishes on an open set containing $\sigma(u) \cap \mathbb{T}$ and $\|\varphi - \psi\|_\infty < \epsilon$. Since $\|A^u_\varphi - A^u_\psi\| \leqslant \|\varphi - \psi\|_\infty < \epsilon$, and the compact operators are norm closed in $\mathcal{B}(\mathcal{K}_u)$ (Proposition 1.38), it suffices to show that A^u_ψ is compact. We will do this by proving that if $\{f_n\}_{n \geqslant 1}$ is a sequence in $\mathcal{K}_u$ that tends weakly to zero, then $A^u_\psi f_n \to 0$ in norm (Proposition 1.37).

To this end, let K denote the closure of $\psi^{-1}(\mathbb{C} \setminus \{0\})$ and observe that $K \subset \mathbb{T} \setminus \sigma(u)$. By Proposition 7.21, each f_n has an analytic continuation across K and so $f_n(\zeta) = \langle f_n, k_\zeta \rangle \to 0$ for each $\zeta \in K$. Since u is analytic on a neighborhood of K we obtain

$$|f_n(\zeta)| = |\langle f_n, k_\zeta \rangle| \leqslant \|f_n\| \, \|k_\zeta\|$$

$$\leqslant \|f_n\| \, |u'(\zeta)|^{\frac{1}{2}} \qquad \text{(by Theorem 7.24)}$$

$$\leqslant \sup_{n \geqslant 1} \|f_n\| \sup_{\xi \in K} |u'(\xi)|^{\frac{1}{2}} < \infty$$

for each ζ in K. By the Dominated Convergence Theorem, it follows that

$$\|A^u_\psi f_n\|^2 = \|P_u(\psi f_n)\|^2 \leqslant \|\psi f_n\|^2 = \int_K |\psi|^2 |f_n|^2 \to 0$$

as $n \to \infty$, whence $A^u_\psi f_n$ tends to zero in norm, as desired.

($\Rightarrow$) Suppose that $\varphi \in C(\mathbb{T})$ and A^u_φ is compact. Let

$$\kappa_\lambda = \frac{k_\lambda}{\|k_\lambda\|}$$

be the normalized reproducing kernels for $\mathcal{K}_u$ and define

$$F_\lambda(z) = |\kappa_\lambda(z)|^2 = \frac{1 - |\lambda|^2}{1 - |u(\lambda)|^2} \left| \frac{1 - \overline{u(\lambda)} u(z)}{1 - \overline{\lambda} z} \right|^2 .$$

Observe that $F_\lambda(z) > 0$ and

$$\frac{1}{2\pi} \int_{-\pi}^{\pi} F_\lambda(e^{it}) \, dt = \|\kappa_\lambda\| = 1.$$

Suppose $\xi = e^{i\alpha} \in \sigma(u) \cap \mathbb{T}$. By Proposition 7.19, there is a sequence $\{\lambda_n\}_{n \geqslant 1} \subset \mathbb{D}$ such that $\lambda_n \to \xi$ and $|u(\lambda_n)| \to 0$. If $|t - \alpha| \geqslant \delta > 0$, then

$$F_{\lambda_n}(e^{it}) \leqslant C_\delta \frac{1 - |\lambda_n|^2}{1 - |u(\lambda_n)|^2} \to 0. \qquad (9.20)$$

First we show that

$$\lim_{n \to \infty} \left| \varphi(\xi) - \frac{1}{2\pi} \int_{-\pi}^{\pi} \varphi(e^{it}) F_{\lambda_n}(e^{it}) \, dt \right| = 0. \qquad (9.21)$$

To do this, note that

$$\left| \varphi(\xi) - \frac{1}{2\pi} \int_{-\pi}^{\pi} \varphi(e^{it}) F_{\lambda_n}(e^{it}) \, dt \right| \leq \frac{1}{2\pi} \int_{|t-\alpha| \leq \delta} |\varphi(\xi) - \varphi(e^{it})| F_{\lambda_n}(e^{it}) \, dt$$
$$+ \frac{1}{2\pi} \int_{\delta \leq |t-\alpha| \leq \pi} |\varphi(\xi) - \varphi(e^{it})| F_{\lambda_n}(e^{it}) \, dt.$$

The first integral can be made small by the continuity of φ (choosing an appropriate δ) and the fact that F_{λ_n} always integrates to one. Once $\delta > 0$ is fixed, the second integral goes to zero by (9.20) and the Dominated Convergence Theorem. This verifies (9.21).

Next we show that

$$\lim_{n \to \infty} \int_{\mathbb{T}} \varphi F_{\lambda_n} \, dm = 0. \tag{9.22}$$

Here is where the compactness of A_φ^u becomes important.

We need the fact that $\kappa_{\lambda_n} \to 0$ weakly in $\mathcal{K}_u$. To prove this, note that if $f \in \mathcal{K}_u \cap H^\infty$ then

$$|\langle f, \kappa_{\lambda_n} \rangle| = \frac{|f(\lambda_n)|}{\|k_{\lambda_n}\|} = |f(\lambda_n)| \sqrt{\frac{1 - |\lambda_n|^2}{1 - |u(\lambda_n)|^2}} \leq \|f\|_\infty \sqrt{\frac{1 - |\lambda_n|^2}{1 - |u(\lambda_n)|^2}},$$

which goes to zero since $|\lambda_n| \to 1$ and $|u(\lambda_n)| \to 0$. To see that $\langle f, \kappa_{\lambda_n} \rangle \to 0$ for a general $f \in \mathcal{K}_u$, we let $\epsilon > 0$ be given and use the density of $\mathcal{K}_u \cap H^\infty$ in $\mathcal{K}_u$ (Proposition 5.21) to produce a $g \in \mathcal{K}_u \cap H^\infty$ with $\|f - g\| < \epsilon$. Then

$$|\langle f, \kappa_{\lambda_n} \rangle| = |\langle f - g, \kappa_\lambda \rangle| + |\langle g, \kappa_{\lambda_n} \rangle|$$
$$\leq \|f - g\| \|\kappa_{\lambda_n}\| + |\langle g, \kappa_{\lambda_n} \rangle|$$
$$\leq \epsilon + |\langle g, \kappa_{\lambda_n} \rangle| \to \epsilon$$

as $n \to \infty$. It follows that $\kappa_{\lambda_n} \to 0$ weakly in $\mathcal{K}_u$.

To verify (9.22), observe that

$$\left| \int_{\mathbb{T}} \varphi F_{\lambda_n} \, dm \right| = |\langle \varphi \kappa_{\lambda_n}, \kappa_{\lambda_n} \rangle| = |\langle \varphi \kappa_{\lambda_n}, P_u \kappa_{\lambda_n} \rangle|$$
$$= |\langle P_u(\varphi \kappa_{\lambda_n}), \kappa_{\lambda_n} \rangle| = |\langle A_\varphi^u \kappa_{\lambda_n}, \kappa_{\lambda_n} \rangle|$$
$$\leq \|A_\varphi^u \kappa_{\lambda_n}\| \|\kappa_{\lambda_n}\| = \|A_\varphi^u \kappa_{\lambda_n}\|.$$

Now use the facts that A_φ^u is compact and $\kappa_{\lambda_n} \to 0$ weakly as $n \to \infty$ to conclude that $\|A_\varphi^u \kappa_{\lambda_n}\| \to 0$. This proves (9.22).

Combining (9.21) with (9.22) shows that $\varphi(\xi) = 0$ and completes the proof of the lemma. $\qquad \square$

Lemma 9.31 *For each $\varphi, \psi \in C(\mathbb{T})$, the semicommutator $A_\varphi^u A_\psi^u - A_{\varphi\psi}^u$ is compact. In particular, the commutator $[A_\varphi^u, A_\psi^u]$ is compact.*

Proof Let $p(z) = \sum_i p_i z^i$ and $q(z) = \sum_j q_j z^j$ be trigonometric polynomials on $\mathbb{T}$ and note that

$$A_p^u A_q^u - A_{pq}^u = \sum_{i,j} p_i q_j (A_{z^i}^u A_{z^j}^u - A_{z^{i+j}}^u).$$

We claim that the preceding operator is compact. Since the sums involved are finite, it suffices to prove that $A_{z^i}^u A_{z^j}^u - A_{z^{i+j}}^u$ is compact for each pair of $i, j \in \mathbb{Z}$.

If $i, j \in \mathbb{Z}$ are of the same sign, then $A_{z^i}^u A_{z^j}^u - A_{z^{i+j}}^u = 0$ (9.12) is trivially compact. If i and j are of different signs, then upon relabeling and taking adjoints (if necessary), it suffices to show that if $n \geqslant m \geqslant 0$, then the operator $A_{z^n}^u A_{\bar{z}^m}^u - A_{z^{n-m}}^u$ is compact (the case $n \leqslant m \leqslant 0$ is done by taking adjoints). In light of the fact that

$$A_{z^n}^u A_{\bar{z}^m}^u - A_{z^{n-m}}^u = A_{z^{n-m}}^u (A_{z^m}^u A_{\bar{z}^m}^u - I),$$

we need only show that $A_{z^m}^u A_{\bar{z}^m}^u - I$ is compact for each $m \geqslant 1$. However, since $A_z^u A_{\bar{z}}^u - I$ has rank one (Lemma 9.9), this follows from the identity

$$A_{z^m}^u A_{\bar{z}^m}^u - I = \sum_{0 \leqslant \ell \leqslant m-1} A_{z^\ell}^u (A_z^u A_{\bar{z}}^u - I) A_{\bar{z}^\ell}^u.$$

Having shown that $A_p^u A_q^u - A_{pq}^u$ is compact for every pair of trigonometric polynomials p and q, the desired result follows since we may uniformly approximate any given φ, ψ in $C(\mathbb{T})$ by trigonometric polynomials (Stone–Weierstrass Theorem) and use the estimate $\|A_\varphi^u\| \leqslant \|\varphi\|_\infty$ for any $\varphi \in C(\mathbb{T})$. $\square$

Lemma 9.32 *If u is an inner function that is not a single Blaschke factor, then k_0 and Ck_0 are linearly independent.*

Proof Suppose $Ck_0 = ak_0$ for some $a \in \mathbb{C}$. A little algebra will show that

$$u = \frac{u(0) + az}{1 + a\overline{z u(0)}}$$

which implies that u is a single Blaschke factor. $\square$

Proof of Theorem 9.28 Before proceeding further, let us remark that statement (iii) has already been shown (see Lemma 9.30). We first claim that

$$C^*(S_u) = C^*(\{A_\varphi^u : \varphi \in C(\mathbb{T})\}), \tag{9.23}$$

noting that the containment $\subset$ in the preceding holds automatically. Since $(A_z^u)^* = A_{\bar{z}}^u$, it follows that $A_p^u \in C^*(A_z^u)$ for any trigonometric polynomial p. We then uniformly approximate any given $\varphi \in C(\mathbb{T})$ by trigonometric polynomials to see that $A_\varphi^u \in C^*(A_z^u)$. This establishes the containment $\supset$ in (9.23).

We next prove statement (i) of Theorem 9.28, that is,

$$\mathscr{C}(C^*(S_u)) = \mathscr{K}^u. \tag{9.24}$$

The containment $\mathscr{C}(C^*(S_u)) \subset \mathscr{K}^u$ follows from (9.23) and Lemma 9.31. On the other hand, Corollary 9.16 says that S_u is irreducible, whence the algebra $C^*(S_u)$ itself is irreducible (that is, contains no non-trivial projections). Lemma 9.9 tells us that

$$[S_u, S_u^*] = Ck_0 \otimes Ck_0 - k_0 \otimes k_0$$

is compact. Furthermore, $[S_u, S_u^*]$ is not the zero operator (Lemma 9.32). It follows that $C^*(S_u) \cap \mathscr{K}^u \neq \{0\}$. We now use the general fact that any irreducible C^*-subalgebra of $\mathscr{B}(\mathcal{H})$ that contains a non-zero compact operator contains all of the compact operators [52, Cor. 3.16.8] to obtain $\mathscr{K}^u \subset \mathscr{C}(C^*(S_u))$, which establishes (9.24).

We claim that

$$C^*(S_u) = \{A_\varphi^u + K : \varphi \in C(\mathbb{T}), K \in \mathscr{K}^u\}, \tag{9.25}$$

which is statement (iv) of Theorem 9.28 (see also Remark 9.29 (iii)). The containment $\supset$ in the preceding follows because $C^*(S_u)$ contains $\mathscr{K}^u$ by (9.24) along with every element of the form A_φ^u with φ in $C(\mathbb{T})$ by (9.23). The containment $\subset$ in (9.25) holds by another application of (9.23).

The map $\gamma : C(\mathbb{T}) \to C^*(S_u)/\mathscr{K}^u$ defined by

$$\gamma(\varphi) = A_\varphi^u + \mathscr{K}^u$$

is a homomorphism (Lemma 9.31) and hence $\gamma(C(\mathbb{T}))$ is a dense subalgebra of $C^*(S_u)/\mathscr{K}^u$ by (9.23). In light of Lemma 9.30, we see that

$$\ker \gamma = \{\varphi \in C(\mathbb{T}) : \varphi|_{\sigma(u) \cap \mathbb{T}} \equiv 0\}, \tag{9.26}$$

whence the map

$$\widetilde{\gamma} : C(\mathbb{T})/\ker \gamma \to C^*(S_u)/\mathscr{K}^u \tag{9.27}$$

defined by

$$\widetilde{\gamma}(\varphi + \ker \gamma) = A_\varphi^u + \mathscr{K}^u$$

is an injective $*$-homomorphism. By [56, Thm. I.5.5], it follows that $\widetilde{\gamma}$ is an isometric $*$-isomorphism.

Since

$$C(\mathbb{T})/\ker\gamma \cong C(\sigma(u) \cap \mathbb{T}) \tag{9.28}$$

by (9.26), we get

$$\sigma_e(A^u_\varphi) = \sigma_{C(\sigma(u)\cap\mathbb{T})}(\varphi) = \varphi(\sigma(u) \cap \mathbb{T}) = \varphi(\sigma_e(A^u_z)),$$

where $\sigma_{C(\sigma(u)\cap\mathbb{T})}(\varphi)$ denotes the spectrum of φ as an element of the Banach algebra $C(\sigma(u) \cap \mathbb{T})$. This yields statement (v). Putting (9.27) and (9.28) together shows that $C^*(A^u_z)/\mathscr{K}^u$ is isometrically $*$-isomorphic to $C(\sigma(u) \cap \mathbb{T})$, which proves statement (ii). The fact that $\widetilde{\gamma}$ is isometric also proves statement (vi).

It remains to justify statement (iv). To this end, we will use a result of Clark to be shown in Chapter 11 (see Theorem 11.4) which asserts that for each $\alpha \in \mathbb{T}$, the operator

$$U_\alpha := S_u + \frac{\alpha}{1 - \overline{u(0)}\alpha}k_0 \otimes Ck_0 \tag{9.29}$$

on $\mathcal{K}_u$ is a cyclic unitary operator. Since

$$U_\alpha \equiv S_u \quad (\mathrm{mod}\ \mathscr{K}^u),$$

we obtain

$$\varphi(U_\alpha) \equiv A^u_\varphi \quad (\mathrm{mod}\ \mathscr{K}^u) \tag{9.30}$$

for every φ in $C(\mathbb{T})$. This last fact follows because the norm on $\mathcal{B}(\mathcal{K}_u)$ dominates the quotient norm on $\mathcal{B}(\mathcal{K}_u)/\mathscr{K}^u$ and since any $\varphi \in C(\mathbb{T})$ can be uniformly approximated by trigonometric polynomials. Since $\mathscr{K}^u \subset C^*(S_u)$, we now have

$$C^*(U_\alpha) + \mathscr{K}^u = C^*(S_u),$$

which yields the desired result. $\qquad\qquad\square$

9.8 Notes

9.8.1 Further references

Some of the proofs in this chapter come from [95, 141, 166, 186]. More advanced ideas can be found in those references.

9.8.2 Vector-valued model spaces

For $n \in \mathbb{N}$, let $H^2_{\mathbb{C}^n}$ denote the vector-valued Hardy space, which consists of functions $f : \mathbb{D} \to \mathbb{C}^n$ that are analytic and for which

$$\sup_{0<r<1} \int_{\mathbb{T}} \|f(r\zeta)\|^2_{\mathbb{C}^n} dm(\zeta) < \infty.$$

Functions in $H^2_{\mathbb{C}^n}$ have many of the same properties as the functions in the classical Hardy space $H^2 = H^2_{\mathbb{C}}$ (boundary values, non-tangential limits, etc.). For a matrix-valued inner function Θ (recall Definition 8.31), define

$$\mathbf{K}_\Theta = (\Theta H^2_{\mathbb{C}^n})^\perp$$

to be the model space corresponding to Θ.

These vector-valued model spaces appear in a generalization of Theorem 9.17 when $\mathrm{rank}(I - T^*T) = \mathrm{rank}(I - TT^*) = n > 1$. Roughly speaking, there is a similar representation in which the compressed shift S_u is replaced by the compression of the shift to $\mathbf{K}_\Theta$. However, in making such a move, one sacrifices a large variety of tools and techniques inherited from classical function theory. For instance, the multiplication of operator-valued inner functions is no longer commutative and the corresponding factorization theory is more complicated.

Still further, the shift compressed to a de Branges–Rovnyak space serves as a model for certain contractions. A good source for all of this material, along with various other generalizations, is [186].

9.9 For further exploration

9.9.1 Matrix representations

When u is a finite Blaschke product, we know that $\mathcal{K}_u$ is finite dimensional and, from Proposition 5.25, we have the TMW basis for $\mathcal{K}_u$. Compute the matrix representation $[S_u]$ of the compressed shift S_u with respect to this basis and verify that (i) $[S_u]$ is lower triangular; (ii) the eigenvalues of S_u are the zeros of u. Further results on matrix representations can be found in [43, 122].

9.9.2 Partial isometry

Show that if $u(0) = 0$, then S_u is a partial isometry. Compute the defect spaces $\ker S_u$ and $(\mathrm{ran}\, S_u)^\perp$.

9.9.3 Unitary equivalence

If u and v are inner and $u(0) = v(0) = 0$, then the defect spaces of the compressed shifts S_u and S_v are one dimensional (see the previous comment). When are S_u and S_v unitarily equivalent? The answer (if and only if $u = e^{i\theta}v$) falls under the general heading of *characteristic functions* which were explored by Livšic [126] (for partial isometries) and more generally (for contractions) by Sz.-Nagy and Foiaş [186].

9.9.4 Compressed shifts with piecewise continuous symbols

In Theorem 9.28 we explored the C^*-algebra generated by $\{A_\varphi^u : \varphi \in C(\mathbb{T})\}$ and connected it to the C^*-algebra generated by $\{T_\varphi : \varphi \in C(\mathbb{T})\}$ explored by Coburn. There is a result of Gohberg and Krupnik [99] which characterizes the C^*-algebra generated by $\{T_\varphi : \varphi \in PC\}$ where PC is the algebra of piecewise continuous functions on $\mathbb{T}$. What is the C^*-algebra generated by $\{A_\varphi^u : \varphi \in PC\}$? The answer, as was the case with $C(\mathbb{T})$, should depend on the boundary properties of u.

10

The commutant lifting theorem

In the previous chapter, we studied the compressed shift S_u. In this chapter, we continue this study with a discussion of its commutant

$$\{S_u\}' = \{A \in \mathcal{B}(\mathcal{K}_u) : AS_u = S_uA\}.$$

The description of $\{S_u\}'$ is a deep result known as the *commutant lifting theorem* for S_u. The term "commutant lifting" stems from the following phenomenon. To find solutions to the operator equation

$$AS_u = S_uA \qquad (10.1)$$

for an $A \in \mathcal{B}(\mathcal{K}_u)$, we "lift it" to the operator equation

$$BS = SB,$$

where S is the unilateral shift on the larger space H^2 and $B \in \mathcal{B}(H^2)$. The Brown–Halmos Theorem (Theorem 4.21) says that B must be a Toeplitz operator T_φ with analytic symbol φ. We then return to (10.1) and prove that

$$A = \varphi(S_u) = A_\varphi^u,$$

where A_φ^u is the analytic truncated Toeplitz operator (9.11).

To accomplish this, we need to further develop certain tools from dilation theory. Even though our goal is to identify the commutant of one particular operator, we broaden our discussion to a general commutant lifting theorem, which is a major result in operator theory. Moreover, expanding our discussion in this way requires only a little extra work.

The description of the commutant $\{S_u\}'$ was obtained by Sarason [162]. His work was motivated by the study of interpolation problems for bounded analytic functions on the open unit disk. The abstract version presented below (Theorem 10.8) was developed by Sz.-Nagy and Foiaş [185]. We follow parts of their presentation.

10.1 Minimal isometric dilations

Let us recall some notation from the previous chapter on compressions and dilations of operators. Assume that $\mathcal{H}$ is a Hilbert space, $\mathcal{M}$ is a subspace of $\mathcal{H}$, $T \in \mathcal{B}(\mathcal{H})$, and $R \in \mathcal{B}(\mathcal{M})$. The operator T is called a *dilation* of R if

$$R^n = PT^n|_{\mathcal{M}}, \qquad n \geqslant 0,$$

where P is the orthogonal projection of $\mathcal{H}$ onto $\mathcal{M}$. If T is an isometry, then it is called an *isometric dilation* of R. An isometric dilation is *minimal* if no restriction of T to a smaller invariant subspace is an isometric dilation of R.

Theorem 10.1 *Let $T \in \mathcal{B}(\mathcal{H})$ be an isometric dilation of $R \in \mathcal{B}(\mathcal{M})$. Then T is a minimal isometric dilation of R if and only if*

$$\bigvee \{T^n \mathcal{M} : n \geqslant 0\} = \mathcal{H}. \tag{10.2}$$

Proof Suppose that T is a minimal isometric dilation of R and let

$$\mathcal{H}' = \bigvee \{T^n \mathcal{M} : n \geqslant 0\}.$$

Observe that $T\mathcal{H}' \subset \mathcal{H}'$. This allows us to define the isometry $T' : \mathcal{H}' \to \mathcal{H}'$ by $T' = T|_{\mathcal{H}'}$. Hence, for each $n \geqslant 0$ and $\mathbf{x} \in \mathcal{M}$, we have

$$PT'^n\mathbf{x} = PT^n\mathbf{x} = R^n\mathbf{x}$$

and so T' is a dilation of R. Since T is a minimal isometric dilation of R, we must have $\mathcal{H}' = \mathcal{H}$.

Conversely, assume that T is an isometric dilation of R satisfying (10.2). Let $\mathcal{M} \subset \mathcal{H}' \subset \mathcal{H}$ be such that $T\mathcal{H}' \subset \mathcal{H}'$ and $T' : \mathcal{H}' \to \mathcal{H}'$, the restriction of T to $\mathcal{H}'$, is an isometric dilation of R. Since $\mathcal{M} \subset \mathcal{H}'$, it follows that

$$T^n \mathcal{M} \subset T^n \mathcal{H}' \subset \mathcal{H}', \qquad n \geqslant 0.$$

However, by (10.2) we have

$$\mathcal{H} = \bigvee \{T^n \mathcal{M} : n \geqslant 0\} \subset \mathcal{H}',$$

which implies that $\mathcal{H} = \mathcal{H}'$. Thus T is a minimal isometric dilation of R. $\square$

Corollary 10.2 *The unilateral shift S on H^2 is a minimal isometric dilation of S_u on $\mathcal{K}_u$.*

Proof We already know from Theorem 9.6 that S is an isometric dilation of S_u. We just need to verify that S is minimal. The space $\mathcal{K}_u$ contains the outer function $k_0 = 1 - \overline{u(0)}u$ (see Corollary 3.24) and so

$$\bigvee \{S^n \mathcal{K}_u : n \geqslant 0\} \supset \bigvee \{S^n k_0 : n \geqslant 0\} = H^2$$

by Corollary 4.5. Now invoke Theorem 10.1 to see that S is a minimal isometric dilation of S_u. □

10.2 Existence and uniqueness

The unilateral shift S on H^2 is an explicit minimal isometric dilation of S_u and it is unique (up to unitary equivalence). For future applications, we need to show that every contraction on a Hilbert space has a unique, up to unitary equivalence, minimal isometric dilation. The reader should review the discussion in Chapter 1 of contractions, partial isometries, the defect space $\mathscr{D}_T$, and the defect operator D_T.

Since

$$
\begin{aligned}
\|D_T \mathbf{x}\|^2 = \langle D_T \mathbf{x}, D_T \mathbf{x} \rangle &= \langle D_T^2 \mathbf{x}, \mathbf{x} \rangle \\
&= \langle (I - T^*T)\mathbf{x}, \mathbf{x} \rangle \\
&= \langle \mathbf{x}, \mathbf{x} \rangle - \langle T^*T\mathbf{x}, \mathbf{x} \rangle \\
&= \|\mathbf{x}\|^2 - \|T\mathbf{x}\|^2
\end{aligned}
$$

for each $\mathbf{x} \in \mathcal{H}$, we obtain the identity

$$\|\mathbf{x}\|^2 = \|T\mathbf{x}\|^2 + \|D_T \mathbf{x}\|^2, \qquad \mathbf{x} \in \mathcal{H}, \tag{10.3}$$

which will be exploited several times below.

Theorem 10.3 *Every contraction has a minimal isometric dilation.*

Proof Let $R \in \mathcal{B}(\mathcal{M})$ be a contraction. We now explicitly construct a minimal isometric dilation T for R. Define a new Hilbert space $\mathcal{H}$ by

$$\mathcal{H} = \mathcal{M} \oplus \mathscr{D}_R \oplus \mathscr{D}_R \oplus \cdots , \tag{10.4}$$

where we write the elements of $\mathcal{H}$ as sequences of vectors

$$\mathbf{x} = (\mathbf{x}_0, \mathbf{x}_1, \mathbf{x}_2, \dots), \quad \mathbf{x}_0 \in \mathcal{M}, \ \mathbf{x}_n \in \mathscr{D}_R \subset \mathcal{M} \, (n \geqslant 1),$$

and assume that $\sum_{n \geqslant 0} \|\mathbf{x}_n\|^2 < \infty$. We define a norm on $\mathcal{H}$ by setting

$$\|\mathbf{x}\| = \sqrt{\sum_{n \geqslant 0} \|\mathbf{x}_n\|^2}. \tag{10.5}$$

We can regard $\mathcal{M}$ as a subspace of $\mathcal{H}$ by identifying each vector $\mathbf{x}_0 \in \mathcal{M}$ with the vector $(\mathbf{x}_0, \mathbf{0}, \mathbf{0}, \ldots) \in \mathcal{H}$. Abusing notation a bit, we regard $\mathcal{M}$ to be this copy, that is to say,

$$\mathcal{M} = \mathcal{M} \oplus \{\mathbf{0}\} \oplus \{\mathbf{0}\} \oplus \cdots.$$

With this identification, we can also equate the norm $\|\mathbf{x}\|$ of $\mathbf{x} \in \mathcal{M}$ with

$$\|(\mathbf{x}, \mathbf{0}, \mathbf{0}, \ldots)\|,$$

the norm in $\mathcal{H}$. We also identify $\mathcal{N} = \mathcal{M}^\perp$ as

$$\mathcal{N} = \{\mathbf{0}\} \oplus \mathscr{D}_R \oplus \mathscr{D}_R \oplus \cdots$$

and the projection P of $\mathcal{H}$ onto $\mathcal{M}$ by

$$P(\mathbf{x}_0, \mathbf{x}_1, \mathbf{x}_2, \ldots) = (\mathbf{x}_0, \mathbf{0}, \mathbf{0}, \ldots).$$

Define

$$T : \mathcal{H} \to \mathcal{H}, \quad T(\mathbf{x}_0, \mathbf{x}_1, \mathbf{x}_2, \ldots) = (R\mathbf{x}_0, D_R\mathbf{x}_0, \mathbf{x}_1, \mathbf{x}_2, \ldots) \tag{10.6}$$

and observe that

$$\begin{aligned}
\|T\mathbf{x}\|^2 &= \|R\mathbf{x}_0\|^2 + \|D_R\mathbf{x}_0\|^2 + \sum_{n \geqslant 1} \|\mathbf{x}_n\|^2 \\
&= \|\mathbf{x}_0\|^2 + \sum_{n \geqslant 1} \|\mathbf{x}_n\|^2 \qquad \text{(by (10.3))} \\
&= \sum_{n \geqslant 0} \|\mathbf{x}_n\|^2 \\
&= \|\mathbf{x}\|^2.
\end{aligned}$$

This shows that T is an isometry. Moreover, $R = PT|_{\mathcal{M}}$, $T\mathcal{N} \subset \mathcal{N}$, and $\mathcal{H} = \mathcal{M} \oplus \mathcal{N}$. Therefore T is a dilation of R (Theorem 9.3).

To finish the proof, we need to show that T is minimal. For each $\mathbf{x}_0 \in \mathcal{M}$, we have $T\mathbf{x}_0 = (R\mathbf{x}_0, D_R\mathbf{x}_0, \mathbf{0}, \mathbf{0}, \ldots)$, and thus

$$\mathcal{M} \vee T\mathcal{M} \subset \mathcal{M} \oplus \mathscr{D}_R \oplus \{\mathbf{0}\} \oplus \{\mathbf{0}\} \oplus \cdots. \tag{10.7}$$

On the other hand, for each

$$\mathbf{x} = (\mathbf{x}_0, D_R\mathbf{x}_1, \mathbf{0}, \mathbf{0}, \ldots) \in \mathcal{M} \oplus \mathscr{D}_R \oplus \{\mathbf{0}\} \oplus \{\mathbf{0}\} \oplus \cdots,$$

where $\mathbf{x}_0, \mathbf{x}_1 \in \mathcal{M}$, we can write

$$\begin{aligned}
\mathbf{x} &= (\mathbf{x}_0 - R\mathbf{x}_1, \mathbf{0}, \mathbf{0}, \ldots) + (R\mathbf{x}_1, D_R\mathbf{x}_1, \mathbf{0}, \mathbf{0}, \ldots) \\
&= (\mathbf{x}_0 - R\mathbf{x}_1, \mathbf{0}, \mathbf{0}, \ldots) + T\mathbf{x}_1.
\end{aligned}$$

This representation shows that $\mathbf{x} \in \mathcal{M} \vee T\mathcal{M}$ and so the containment in (10.7) is an equality, that is to say,

$$T\mathcal{M} = \mathcal{M} \oplus \mathscr{D}_R \oplus \{\mathbf{0}\} \oplus \{\mathbf{0}\} \oplus \cdots .$$

By induction, we obtain the identity

$$\bigvee_{0 \leqslant j \leqslant n} T^j \mathcal{M} = \mathcal{M} \oplus \Big(\bigoplus_{1 \leqslant k \leqslant n} \mathscr{D}_R \Big) \oplus \{\mathbf{0}\} \oplus \{\mathbf{0}\} \oplus \cdots . \tag{10.8}$$

From here we have

$$\bigvee \{T^j \mathcal{M} : j \geqslant 0\} = \mathcal{H},$$

which shows that T is minimal (Theorem 10.1). $\square$

The following uniqueness result complements Theorem 10.3.

Theorem 10.4   *Let $R \in \mathcal{B}(\mathcal{M})$ be a contraction and $T \in \mathcal{B}(\mathcal{H})$, $T' \in \mathcal{B}(\mathcal{H}')$ be two minimal isometric dilations of R. Then there is a unitary operator $U : \mathcal{H} \to \mathcal{H}'$ such that $U|_\mathcal{M} = I_\mathcal{M}$, the identity operator on $\mathcal{M}$, and $T'U = UT$.*

The theorem above is described by the following commutative diagram:

$$
\begin{array}{ccc}
\mathcal{H} & \xrightarrow{\ \ T\ \ } & \mathcal{H} \\
\big\downarrow{\scriptstyle U} & & \big\downarrow{\scriptstyle U} \\
\mathcal{H}' & \xrightarrow{\ \ T'\ \ } & \mathcal{H}'
\end{array}
$$

Proof   Since T and T' are minimal isometric dilations of R, we have, via Theorem 10.1,

$$\mathcal{H} = \bigvee \{T^n \mathcal{M} : n \geqslant 0\} \quad \text{and} \quad \mathcal{H}' = \bigvee \{T'^n \mathcal{M} : n \geqslant 0\}.$$

If $\{\mathbf{x}_n\}_{n \geqslant 0}$ is a finitely supported sequence (all but a finite number of terms are zero) with components in $\mathcal{M}$, then

$$\Big\| \sum_{n \geqslant 0} T^n \mathbf{x}_n \Big\|^2 = \sum_{m \geqslant 0} \sum_{n \geqslant 0} \langle T^m \mathbf{x}_m, T^n \mathbf{x}_n \rangle.$$

Because T is an isometry, it follows that

$$\langle T^m \mathbf{x}_m, T^n \mathbf{x}_n \rangle = \begin{cases} \langle T^{m-n} \mathbf{x}_m, \mathbf{x}_n \rangle & \text{if } m \geqslant n, \\ \langle \mathbf{x}_m, T^{n-m} \mathbf{x}_n \rangle & \text{if } m < n. \end{cases}$$

Since T is a dilation of R, we get

$$\left\| \sum_{n \geqslant 0} T^n \mathbf{x}_n \right\|^2 = \sum_{m \geqslant 0} \sum_{n \geqslant 0} \langle T^m \mathbf{x}_m, T^n \mathbf{x}_n \rangle$$

$$= \sum_{0 \leqslant n \leqslant m} \langle T^{m-n} \mathbf{x}_m, \mathbf{x}_n \rangle + \sum_{0 \leqslant m < n} \langle \mathbf{x}_m, T^{n-m} \mathbf{x}_n \rangle$$

$$= \sum_{0 \leqslant n \leqslant m} \langle P_{\mathcal{M}} T^{m-n} \mathbf{x}_m, \mathbf{x}_n \rangle + \sum_{0 \leqslant m < n} \langle \mathbf{x}_m, P_{\mathcal{M}} T^{n-m} \mathbf{x}_n \rangle$$

$$= \sum_{0 \leqslant n \leqslant m} \langle R^{m-n} \mathbf{x}_m, \mathbf{x}_n \rangle + \sum_{0 \leqslant m < n} \langle \mathbf{x}_m, R^{n-m} \mathbf{x}_n \rangle.$$

We can do the analogous computation with T' in place of T and obtain the same expression in terms of R at the end. Therefore,

$$\left\| \sum_{n \geqslant 0} T^n \mathbf{x}_n \right\|^2 = \left\| \sum_{n \geqslant 0} T'^n \mathbf{x}_n \right\|^2 \tag{10.9}$$

for any finitely supported sequence $\{\mathbf{x}_n\}_{n \geqslant 0}$. Therefore, if we initially define a linear transformation U on finitely supported sequences $\{\mathbf{x}_n\}_{n \geqslant 0}$ whose components are in $\mathcal{M}$ by

$$U\left(\sum_{n \geqslant 0} T^n \mathbf{x}_n \right) = \sum_{n \geqslant 0} T'^n \mathbf{x}_n,$$

then (10.9) implies that U is well-defined and, moreover, that it extends to an isometry from $\mathcal{H}$ onto $\mathcal{H}'$.

On the set of all finitely supported sequences $\{\mathbf{x}_n\}_{n \geqslant 0}$, we have

$$UT\left(\sum_{n \geqslant 0} T^n \mathbf{x}_n \right) = U\left(\sum_{n \geqslant 0} T^{n+1} \mathbf{x}_n \right)$$

$$= \sum_{n \geqslant 0} T'^{n+1} \mathbf{x}_n$$

$$= T' \sum_{n \geqslant 0} T'^n \mathbf{x}_n$$

$$= T' U\left(\sum_{n \geqslant 0} T^n \mathbf{x}_n \right).$$

Thus $T'U = UT$. Finally, by considering the sequence $(\mathbf{x}, 0, 0, \dots)$, where $\mathbf{x} \in \mathcal{M}$, we see that U satisfies $U\mathbf{x} = \mathbf{x}$ for all $\mathbf{x} \in \mathcal{M}$ and so $U|_{\mathcal{M}} = I_{\mathcal{M}}$. $\quad\square$

To better understand the construction presented in the previous two theorems, we now provide a concrete example.

Example 10.5 Let $R = S_u$ on $M = \mathcal{K}_u$, where $u = z^N$, and note that $\mathcal{K}_u = \bigvee\{1, z, z^2, \ldots, z^{N-1}\}$ (Proposition 5.16). The defect operator D_{S_u} can be computed as

$$
\begin{aligned}
D_{S_u} &= (I - S_u^* S_u)^{1/2} \\
&= (S^* u \otimes S^* u)^{1/2} \qquad \text{(Lemma 9.9)} \\
&= (z^{N-1} \otimes z^{N-1})^{1/2} \\
&= z^{N-1} \otimes z^{N-1}.
\end{aligned}
$$

The last equality follows since $z^{N-1} \otimes z^{N-1}$ is an orthogonal projection. The defect space $\mathscr{D}_{S_u}$ is $\mathbb{C}z^{N-1}$. The Hilbert space $\mathcal{H}$ from (10.4) becomes

$$
\mathcal{H} = \mathcal{K}_u \oplus \mathbb{C}z^{N-1} \oplus \mathbb{C}z^{N-1} \oplus \cdots
$$

and the dilation T from (10.6) is

$$
\begin{aligned}
&T\Big(\sum_{0 \leqslant j \leqslant N-1} a_j z^j, c_1 z^{N-1}, c_2 z^{N-1}, c_3 z^{N-1}, \ldots \Big) \\
&= \Big(P_u\Big(z \sum_{0 \leqslant j \leqslant N-1} a_j z^j\Big), (z^{N-1} \otimes z^{N-1})\Big(\sum_{0 \leqslant j \leqslant N-1} a_j z^j\Big), c_1 z^{N-1}, c_2 z^{N-1}, \ldots \Big) \\
&= \Big(a_0 z + a_1 z^2 + \cdots + a_{N-2} z^{N-1}, a_{N-1} z^{N-1}, c_1 z^{N-1}, c_2 z^{N-1}, \ldots \Big).
\end{aligned}
$$

By the definition of the norm from (10.5) and the orthogonality of z^j in H^2 for $j \geqslant 0$, we see that T is isometric. Furthermore, $T|_M = S_u = R$ (just look at the first entry in the vector above).

By Corollary 10.2, the unilateral shift S on H^2 is a minimal isometric dilation of S_u. By Theorem 10.4, the operator T on $\mathcal{H}$ should be unitarily equivalent to S with the additional condition that the intertwining unitary restricts to the identity operator on M. To see this, define $U : \mathcal{H} \to H^2$ by

$$
U\Big(\sum_{0 \leqslant j \leqslant N-1} a_j z^j, c_1 z^{N-1}, c_2 z^{N-1}, c_3 z^{N-1}, \ldots \Big) = \sum_{0 \leqslant j \leqslant N-1} a_j z^j + \sum_{N \leqslant j \leqslant \infty} c_{j-N+1} z^j
$$

and observe that U is unitary and $U|_{\mathcal{K}_u} = I_{\mathcal{K}_u}$. Finally,

$$
\begin{aligned}
&UT\Big(\sum_{0 \leqslant j \leqslant N-1} a_j z^j, c_1 z^{N-1}, c_2 z^{N-1}, c_3 z^{N-1}, \cdots \Big) \\
&= (a_0 z + a_1 z^2 + a_2 z^3 + \cdots + a_{N-2} z^{N-1}) + (a_{N-1} z^N + c_1 z^{N+1} + c_2 z^{N+2} + \cdots) \\
&= z\big((a_0 + a_1 z + a_2 z^2 + \cdots + a_{N-2} z^{N-2}) + (a_{N-1} z^{N-1} + c_1 z^N + c_2 z^{N+1} + \cdots)\big) \\
&= z U\Big(\sum_{0 \leqslant j \leqslant N-1} a_j z^j, c_1 z^{N-1}, c_2 z^{N-1}, c_3 z^{N-1}, \ldots \Big) \\
&= S U\Big(\sum_{0 \leqslant j \leqslant N-1} a_j z^j, c_1 z^{N-1}, c_2 z^{N-1}, c_3 z^{N-1}, \ldots \Big).
\end{aligned}
$$

10.3 Strong convergence

Let $T \in \mathcal{B}(\mathcal{H})$ denote the minimal isometric dilation of the contraction $R \in \mathcal{B}(\mathcal{M})$ constructed in Theorem 10.3. Let $\mathcal{M}_0 = \mathcal{M}$ and, for $n \geqslant 1$, define

$$\mathcal{M}_n = \mathcal{M} \oplus \left(\bigoplus_{1 \leqslant k \leqslant n} \mathcal{D}_R \right) \oplus \{0\} \oplus \{0\} \oplus \cdots, \qquad (10.10)$$

$$T_n = P_{\mathcal{M}_n} T | \mathcal{M}_n, \qquad n \geqslant 0. \qquad (10.11)$$

The orthogonal complement of $\mathcal{M}_n$ in $\mathcal{M}_{n+1}$ is invariant under T_{n+1} and $T_n = P_{\mathcal{M}_n} T_{n+1} |_{\mathcal{M}_n}$ and thus we can appeal to Theorem 9.3 to see that T_{n+1} is a non-isometric dilation of T_n.

Proposition 10.6 *Let $R \in \mathcal{B}(\mathcal{M})$ be a contraction and let $T \in \mathcal{B}(\mathcal{H})$ be the minimal isometric dilation of R. For each $\mathbf{x} \in \mathcal{H}$ we have the following:*

(i) *For each $n \geqslant 0$, $\mathcal{M}_n$ is invariant under T^*;*
(ii) $\lim_{n \to \infty} P_{\mathcal{M}_n} \mathbf{x} = \mathbf{x}$;
(iii) $T_n P_{\mathcal{M}_n} = P_{\mathcal{M}_n} T$ *and thus* $T\mathbf{x} = \lim_{n \to \infty} T_n P_{\mathcal{M}_n} \mathbf{x}$;
(iv) $T^* \mathbf{x} = \lim_{n \to \infty} T_n^* \mathbf{x}$.

Proof (i): Since $\mathcal{M}_0^\perp = \mathcal{N}$ and $T\mathcal{N} \subset \mathcal{N}$, we see that $T^* \mathcal{M}_0 \subset \mathcal{M}_0$. Thus the result holds when $n = 0$. The verification for $n \geqslant 1$ is based on two facts. First, since T is an isometry, we have $T^* T = I$. Second, by (10.8), the subspace $\mathcal{M}_n$ can be written as

$$\mathcal{M}_n = \bigvee \{ T^j \mathcal{M} : 0 \leqslant j \leqslant n \}$$

and therefore

$$\begin{aligned} T^* \mathcal{M}_n &\subset \bigvee \{ T^* T^j \mathcal{M} : 0 \leqslant j \leqslant n \} \\ &\subset \bigvee \{ T^j \mathcal{M} : 0 \leqslant j \leqslant n - 1 \} \\ &= \mathcal{M}_{n-1} \subset \mathcal{M}_n. \end{aligned}$$

To prove (ii), write $\mathbf{x} = (\mathbf{x}_0, \mathbf{x}_1, \dots) \in \mathcal{H}$. Then

$$P_{\mathcal{M}_n} \mathbf{x} = (\mathbf{x}_0, \mathbf{x}_1, \dots, \mathbf{x}_n, \mathbf{0}, \mathbf{0}, \dots),$$

and

$$\mathbf{x} - P_{\mathcal{M}_n} \mathbf{x} = (\mathbf{0}, \mathbf{0}, \dots, \mathbf{0}, \mathbf{x}_{n+1}, \mathbf{x}_{n+2}, \dots).$$

Hence

$$\| \mathbf{x} - P_{\mathcal{M}_n} \mathbf{x} \|^2 = \sum_{k \geqslant n+1} \| \mathbf{x}_k \|^2 \to 0, \qquad n \to \infty$$

by the definition of the norm from (10.5).

To prove (iii), note that $TM_n^\perp \subset M_n^\perp$ (by (i)) and so

$$T_n P_{M_n} \mathbf{x} = P_{M_n} T P_{M_n} \mathbf{x}$$
$$= P_{M_n} T(\mathbf{x} - P_{M_n^\perp} \mathbf{x})$$
$$= P_{M_n} T \mathbf{x} - P_{M_n} T P_{M_n^\perp} \mathbf{x}$$
$$= P_{M_n} T \mathbf{x}.$$

To finish the proof, apply (ii).

For the proof of (iv), use (i) and (10.11) to get

$$T_n^* = P_{M_n} T^*|_{M_n} = T^*|_{M_n}.$$

Fix $m \geqslant 0$. Use the containment $M_m \subset M_n$ when $n \geqslant m$ to see that for each $\mathbf{x} \in M_m$,

$$T_n^* \mathbf{x} = T^* \mathbf{x}.$$

In particular, this yields

$$T^* \mathbf{x} = \lim_{n \to \infty} T_n^* \mathbf{x}.$$

Now use the two facts $\mathcal{H} = \bigvee \{M_m : m \geqslant 0\}$ and $\|T_n^*\| \leqslant \|T^*\|$ for all $n \geqslant 1$ to prove the result. □

10.4 An associated partial isometry

Given a Hilbert space $\mathcal{H}$ and a contraction $T \in \mathcal{B}(\mathcal{H})$, we define the operator T° on $\mathcal{H} \oplus \mathcal{D}_T$ by

$$T^\circ(\mathbf{x}_0, \mathbf{x}_1) = (T\mathbf{x}_0, D_T \mathbf{x}_0). \tag{10.12}$$

By (10.3) we have

$$\|T^\circ(\mathbf{x}_0, 0)\|^2 = \|T\mathbf{x}_0\|^2 + \|D_T \mathbf{x}_0\|^2 = \|\mathbf{x}_0\|^2 = \|(\mathbf{x}_0, 0)\|^2$$

for every $\mathbf{x}_0 \in \mathcal{H}$. Hence

$$\ker T^\circ = \{0\} \oplus \mathcal{D}_T$$

and so

$$(\ker T^\circ)^\perp = \mathcal{H} \oplus \{0\}.$$

We conclude that T° is a partial isometry (isometric on $(\ker T^\circ)^\perp$) and, by (1.27),

$$\mathcal{D}_{T^\circ} = \ker T^\circ = \{0\} \oplus \mathcal{D}_T$$

and

$$D_{T^\circ}(\mathbf{x}_0, \mathbf{x}_1) = P_{\ker T^\circ}(\mathbf{x}_0, \mathbf{x}_1) = (\mathbf{0}, \mathbf{x}_1),$$

where $P_{\ker T^\circ}$ is the orthogonal projection onto $\ker T^\circ$. The definition in (10.12) says that $T^{2\circ} = (T^\circ)^\circ$ is defined on

$$(\mathcal{H} \oplus \mathcal{D}_T) \oplus \mathcal{D}_{T^\circ}$$

by

$$T^{2\circ}((\mathbf{x}_0, \mathbf{x}_1), (\mathbf{0}, \mathbf{x}_2)) = (T^\circ(\mathbf{x}_0, \mathbf{x}_1), D_{T^\circ}(\mathbf{0}, \mathbf{x}_2)) = ((T\mathbf{x}_0, D_T\mathbf{x}_0), (\mathbf{0}, \mathbf{x}_2)).$$

After concatenating and shifting the indices, we can say that $T^{2\circ}$ is defined on

$$\mathcal{H} \oplus \mathcal{D}_T \oplus \mathcal{D}_T$$

by

$$T^{2\circ}(\mathbf{x}_0, \mathbf{x}_1, \mathbf{x}_2) = (T\mathbf{x}_0, D_T\mathbf{x}_0, \mathbf{x}_1).$$

By induction, we can define $T^{n\circ}$ on

$$\mathcal{H} \oplus \mathcal{D}_T \oplus \cdots \oplus \mathcal{D}_T$$

by

$$T^{n\circ}(\mathbf{x}_0, \mathbf{x}_1, \ldots, \mathbf{x}_n) = (T\mathbf{x}_0, D_T\mathbf{x}_0, \mathbf{x}_1, \ldots, \mathbf{x}_{n-1}).$$

Recall the subspace $\mathcal{M}_n$ defined in (10.10). Abusing notation, we may also write

$$\mathcal{M}_n = \mathcal{M} \oplus \left(\bigoplus_{1 \leqslant k \leqslant n} \mathcal{D}_R \right).$$

With this convention, the operator T_n acts as $T_0 = R$ and, for $n \geqslant 1$,

$$T_n(\mathbf{x}_0, \mathbf{x}_1, \ldots, \mathbf{x}_n) = (R\mathbf{x}_0, D_R\mathbf{x}_0, \mathbf{x}_1, \ldots, \mathbf{x}_{n-1}) = R^{n\circ}(\mathbf{x}_0, \mathbf{x}_1, \ldots, \mathbf{x}_n).$$

10.5 The commutant lifting theorem

We now apply the results of the preceding sections to establish a general commutant lifting theorem. The first step in this process is the following.

Proposition 10.7 *Let $R \in \mathcal{B}(\mathcal{M})$ and $R' \in \mathcal{B}(\mathcal{M}')$ be two contractions. Assume there is an $X \in \mathcal{B}(\mathcal{M}, \mathcal{M}')$ such that $R'X = XR$. Then there is a bounded operator $Y : \mathcal{M} \oplus \mathcal{D}_R \rightarrow \mathcal{M}' \oplus \mathcal{D}_{R'}$ that satisfies the following properties:*

(i) $Y(\{\mathbf{0}\} \oplus \mathscr{D}_R) \subset \{\mathbf{0}\} \oplus \mathscr{D}_{R'}$;

(ii) $P_{\mathcal{M}'} Y|_{\mathcal{M}} = X$;

(iii) $\|Y\| = \|X\|$;

(iv) $R'^{\circ} Y = Y R^{\circ}$, where R° and R'° are the dilations of R and R' defined by (10.12).

This is illustrated by the following commutative diagrams:

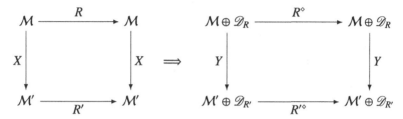

In other words, if the diagram on the left commutes, then so does the one on the right.

Proof   Let us first construct the operator Y. Without loss of generality, we can rescale X and assume that $\|X\| = 1$. By (10.3) and the operator identity $XR = R'X$, we have

$$\|(D_X R\mathbf{x}, D_R \mathbf{x})\|^2 - \|D_{R'} X\mathbf{x}\|^2$$
$$= \|D_X R\mathbf{x}\|^2 + \|D_R \mathbf{x}\|^2 - \|D_{R'} X\mathbf{x}\|^2$$
$$= \|R\mathbf{x}\|^2 - \|XR\mathbf{x}\|^2 + \|\mathbf{x}\|^2 - \|R\mathbf{x}\|^2 - \|X\mathbf{x}\|^2 + \|R'X\mathbf{x}\|^2$$
$$= \|\mathbf{x}\|^2 - \|X\mathbf{x}\|^2 \geqslant 0$$

for each $\mathbf{x} \in \mathcal{M}$. Thus,

$$\|D_{R'} X\mathbf{x}\| \leqslant \|(D_X R\mathbf{x}, D_R \mathbf{x})\|, \qquad \mathbf{x} \in \mathcal{M}.$$

This allows us to define a linear transformation V on the linear manifold

$$\mathscr{E} = \{(D_X R\mathbf{x}, D_R \mathbf{x}) : \mathbf{x} \in \mathcal{M}\} \subset \mathcal{M} \oplus \mathscr{D}_R$$

by

$$V(D_X R\mathbf{x}, D_R \mathbf{x}) = D_{R'} X\mathbf{x}, \qquad \mathbf{x} \in \mathcal{M}. \tag{10.13}$$

Since V is contractive on $\mathscr{E}$, we can extend it to the closure of $\mathscr{E}$. We extend V to $\mathcal{M} \oplus \mathscr{D}_R$ by defining V to be zero on $\mathscr{E}^{\perp}$. Thus V is a contraction from $\mathcal{M} \oplus \mathscr{D}_R$ to $\mathscr{D}_{R'}$.

Now define $Y : \mathcal{M} \oplus \mathscr{D}_R \to \mathcal{M}' \oplus \mathscr{D}_{R'}$ by

$$Y(\mathbf{x}_0, \mathbf{x}_1) = (X\mathbf{x}_0, V(D_X \mathbf{x}_0, \mathbf{x}_1)), \qquad (\mathbf{x}_0, \mathbf{x}_1) \in \mathcal{M} \oplus \mathscr{D}_R. \tag{10.14}$$

The required properties (i)–(iv) in the statement can now be verified. Indeed:

(i): Taking $\mathbf{x}_0 = \mathbf{0}$ in (10.14), we see that

$$Y(\mathbf{0}, \mathbf{x}_1) = (\mathbf{0}, V(\mathbf{0}, \mathbf{x}_1)) \in \{\mathbf{0}\} \oplus \mathscr{D}_{R'}.$$

(ii): Taking $\mathbf{x}_1 = 0$ in (10.14), we get

$$Y(\mathbf{x}_0, \mathbf{0}) = (X\mathbf{x}_0, V(D_X\mathbf{x}_0, \mathbf{0})),$$

and thus

$$P_{\mathcal{M}'} Y(\mathbf{x}_0, \mathbf{0}) = X\mathbf{x}_0, \qquad \mathbf{x}_0 \in \mathcal{M}.$$

(iii): Since V is a contraction, we have

$$\begin{aligned}
\|Y(\mathbf{x}_0, \mathbf{x}_1)\|^2 &= \|X\mathbf{x}_0\|^2 + \|V(D_X\mathbf{x}_0, \mathbf{x}_1)\|^2 \\
&\leqslant \|X\mathbf{x}_0\|^2 + \|(D_X\mathbf{x}_0, \mathbf{x}_1)\|^2 \\
&= \|X\mathbf{x}_0\|^2 + \|D_X\mathbf{x}_0\|^2 + \|\mathbf{x}_1\|^2 \\
&= \|\mathbf{x}_0\|^2 + \|\mathbf{x}_1\|^2 \qquad \text{(by (10.3))} \\
&= \|(\mathbf{x}_0, \mathbf{x}_1)\|^2
\end{aligned}$$

for each $(\mathbf{x}_0, \mathbf{x}_1) \in \mathcal{M} \oplus \mathscr{D}_R$. Thus $\|Y\| \leqslant 1$. Moreover,

$$\begin{aligned}
\|Y(\mathbf{x}_0, \mathbf{0})\|^2 &= \|(X\mathbf{x}_0, V(D_X\mathbf{x}_0, \mathbf{0}))\|^2 \\
&= \|X\mathbf{x}_0\|^2 + \|V(D_X\mathbf{x}_0, \mathbf{0})\|^2 \\
&\geqslant \|X\mathbf{x}_0\|^2
\end{aligned}$$

for each $\mathbf{x}_0 \in \mathcal{M}$. Taking a supremum with respect to all unit vectors $\mathbf{x}_0 \in \mathcal{M}$ on both sides of the inequality above yields $\|Y\| \geqslant \|X\| = 1$ and hence $\|Y\| = \|X\| = 1$.

(iv): For $(\mathbf{x}_0, \mathbf{x}_1) \in \mathcal{M} \oplus \mathscr{D}_R$, we have

$$R'^{\circ} Y(\mathbf{x}_0, \mathbf{x}_1) = R'^{\circ}(X\mathbf{x}_0, V(D_X\mathbf{x}_0, \mathbf{x}_1)) = (R'X\mathbf{x}_0, D_{R'}X\mathbf{x}_0)$$

and, by (10.13) and (10.14), it follows that

$$\begin{aligned}
YR^{\circ}(\mathbf{x}_0, \mathbf{x}_1) &= Y(R\mathbf{x}_0, D_R\mathbf{x}_0) \\
&= (XR\mathbf{x}_0, V(D_X R\mathbf{x}_0, D_R\mathbf{x}_0)) \\
&= (XR\mathbf{x}_0, D_{R'}X\mathbf{x}_0).
\end{aligned}$$

Since $R'X = XR$, we deduce $R'^{\circ}Y = YR^{\circ}$. $\qquad\qquad \square$

The dilation produced in Proposition 10.7 is not an isometry. However, if we repeat the process infinitely many times, in a certain controlled way, we obtain an isometric dilation. This is the content of our next result.

Theorem 10.8   *Let $R \in \mathcal{B}(M)$ and $R' \in \mathcal{B}(M')$ be two contractions and let $T \in \mathcal{B}(\mathcal{H})$ and $T' \in \mathcal{B}(\mathcal{H}')$ be their respective minimal isometric dilations. Assume that $X \in \mathcal{B}(M, M')$ satisfies $R'X = XR$. Then there is a $Y \in \mathcal{B}(\mathcal{H}, \mathcal{H}')$ which satisfies the following properties:*

(i) $Y(\mathcal{H} \ominus M) \subset \mathcal{H}' \ominus M'$;
(ii) $X = P_{M'} Y|_M$;
(iii) $\|Y\| = \|X\|$;
(iv) $T'Y = YT$.

Pictorially this is explained with the following commutative diagrams:

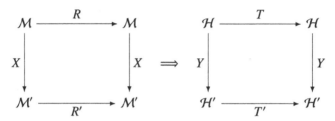

In other words, if the diagram on the left commutes, so does the one on the right.

Proof   Set $Y_0 = X$. Then for each $n \geqslant 1$ we can apply Proposition 10.7 n times to produce operators $Y_n \in \mathcal{B}(M_n, M'_n)$ satisfying the conditions

$$Y_n(M_n \ominus M_{n-1}) \subset M'_n \ominus M'_{n-1},$$
$$P_{M'_{n-1}} Y_n|_{M_{n-1}} = Y_{n-1},$$
$$\|Y_n\| = \|X\|,$$
$$R'^{n \diamond} Y_n = Y_n R^{n \diamond}.$$

The equations above imply that

$$Y_{n-1}^* = Y_n^*|_{M'_{n-1}},$$
$$Y_n^* M'_{n-1} \subset M_{n-1},$$
$$\|Y_n^*\| = \|X\|,$$
$$Y_n^* (R'^{n \diamond})^* = (R^{n \diamond})^* Y_n^*.$$

These conditions will be exploited below.
     Fix $m \geqslant 0$. For each $\mathbf{x} \in M'_m$ we use the facts

$$M'_m \subset M'_n, \qquad n \geqslant m,$$

and $Y_{n-1}^* = Y_n^*|_{\mathcal{M}_{n-1}'}$ to see that

$$Y_n^* \mathbf{x} = Y_m^* \mathbf{x}, \qquad n \geqslant m.$$

Hence we can define the mapping

$$Z : \bigcup_{m \geqslant 0} \mathcal{M}_m' \to \mathcal{H}$$

by

$$Z\mathbf{x} = \lim_{n \to \infty} Y_n^* \mathbf{x}, \qquad \mathbf{x} \in \bigcup_{m \geqslant 0} \mathcal{M}_m'.$$

Since

$$\|Y_n^* \mathbf{x}\| \leqslant \|Y_n^*\| \, \|\mathbf{x}\| = \|X\| \, \|\mathbf{x}\|,$$

Z satisfies the growth restriction

$$\|Z\mathbf{x}\| \leqslant \|X\| \, \|\mathbf{x}\|, \qquad \mathbf{x} \in \bigcup_{m \geqslant 0} \mathcal{M}_m'.$$

Because $\bigcup_{m \geqslant 0} \mathcal{M}_m'$ is dense in $\mathcal{H}'$, the map Z extends to a bounded operator from $\mathcal{H}'$ into $\mathcal{H}$, with $\|Z\| \leqslant \|X\|$. Moreover, for each $\mathbf{x} \in \mathcal{M}' = \mathcal{M}_0'$ and $n \geqslant 0$, we have

$$Y_n^* \mathbf{x} = Y_0^* \mathbf{x} = X^* \mathbf{x}.$$

Let $n \to \infty$ in the previous equation to get

$$Z\mathbf{x} = X^* \mathbf{x}. \tag{10.15}$$

The last identity can be rewritten as

$$P_{\mathcal{M}} Z|_{\mathcal{M}'} = X^*. \tag{10.16}$$

Hence $\|X\| = \|X^*\| \leqslant \|Z\|$, which yields $\|Z\| = \|X\|$.

After taking adjoints, the equality (10.16) becomes

$$X = P_{\mathcal{M}'} Z^*|_{\mathcal{M}}.$$

Moreover, (10.15) implies that $Z\mathcal{M}' \subset \mathcal{M}$, which is equivalent to

$$Z^*(\mathcal{H} \ominus \mathcal{M}) \subset \mathcal{H}' \ominus \mathcal{M}'.$$

Finally, consider $Y_n^*(R'^{n\circ})^* = (R^{n\circ})^* Y_n^*$ and let $n \to \infty$. By Proposition 10.6 we obtain

$$ZT'^* = T^* Z,$$

which can be written as

$$T'Z^* = Z^* T.$$

In other words, the operator Y that we are looking for is precisely Z^*. $\qquad \square$

10.6 The characterization of $\{S_u\}'$

We are now all set up to apply the commutant lifting theorem to describe the commutant $\{S_u\}'$ of the compressed shift S_u. So far, we know from (9.12) and Theorem 9.19 that

$$\{\varphi(S_u) = A_\varphi^u : \varphi \in H^\infty\} \subset \{S_u\}'.$$

It remains to show that equality holds. Note that the symbol φ of $\varphi(S_u) = A_\varphi^u$ is not uniquely determined. In fact (Theorem 9.20),

$$\varphi_1(S_u) = \varphi_2(S_u) \iff \varphi_1 - \varphi_2 \in uH^\infty.$$

Theorem 10.9 *Let u be an inner function and let $A \in \{S_u\}'$. Then there is a $\varphi \in H^\infty$ such that $A = \varphi(S_u)$. Moreover, for any representation $A = \varphi(S_u)$*

$$\|A\| = \mathrm{dist}(\varphi, uH^\infty) \leqslant \|\varphi\|_\infty \qquad (10.17)$$

and there is a symbol $\varphi \in H^\infty$ for which $\|A\| = \|\varphi\|_\infty$.

Proof By Corollary 10.2, the unilateral shift S is the minimal isometric dilation of S_u. Hence Theorem 10.8 yields a bounded operator $B : H^2 \to H^2$ such that $BS = SB$, $\|B\| = \|A\|$, and $A = P_u B|_{\mathcal{K}_u}$. According to Theorem 4.21 and Proposition 4.12, there is a $\varphi \in H^\infty$ such that $B = T_\varphi$ and $\|B\| = \|\varphi\|_\infty$. Thus, $A = P_u T_\varphi|_{\mathcal{K}_u} = \varphi(S_u)$ and $\|A\| = \|\varphi\|_\infty$.

The inequality in (10.17) is established with the following observation. By Theorem 9.20, if $\varphi_1(S_u) = \varphi_2(S_u)$ there is an $f \in H^\infty$ such that $\varphi_1 - \varphi_2 = uf$. Thus the quantity $\mathrm{dist}(\varphi_1, uH^\infty)$ remains the same as φ_1 ranges over all symbols φ_1 with $A = \varphi_1(S_u)$. Since the mapping $\varphi_1 \mapsto \varphi_1(S_u)$ is contractive (Theorem 9.19), we see that $\|A\| \leqslant \|\varphi_1\|_\infty$ holds for all such symbols. Replacing φ_1 with $\varphi_1 - uf$, we get

$$\|A\| \leqslant \mathrm{dist}(\varphi_1, uH^\infty) \leqslant \|\varphi_1\|_\infty, \qquad \varphi_1 \in H^\infty.$$

However, for the particular choice of symbol which was obtained in the first paragraph, we have $\|A\| = \mathrm{dist}(\varphi, uH^\infty) = \|\varphi\|_\infty$. $\qquad\square$

Let us provide a concrete example of $\{S_u\}'$.

Example 10.10 If $u = z^4$ then, with respect to the orthonormal basis $\{1, z, z^2, z^3\}$ for $\mathcal{K}_u$, S_u has the matrix representation

$$[S_u] = \begin{bmatrix} 0 & 0 & 0 & 0 \\ 1 & 0 & 0 & 0 \\ 0 & 1 & 0 & 0 \\ 0 & 0 & 1 & 0 \end{bmatrix}.$$

The commutant of S_u is the set $\varphi(S_u)$, where $\varphi \in H^\infty$, which, as we will see in (13.3), all have the matrix form

$$\begin{bmatrix} a_0 & 0 & 0 & 0 \\ a_1 & a_0 & 0 & 0 \\ a_2 & a_1 & a_0 & 0 \\ a_3 & a_2 & a_1 & a_0 \end{bmatrix}.$$

10.7 Notes

10.7.1 References

The commutant of S_u was first described by Sarason in [162]. See [186] for more on the commutant lifting theorem.

11

Clark measures

In this chapter we develop a family of unitary operators $\{U_\alpha : \alpha \in \mathbb{T}\}$ (Clark unitary operators) on $\mathcal{K}_u$ along with their corresponding family of spectral measures $\{\sigma_\alpha : \alpha \in \mathbb{T}\}$ (Clark measures). The Clark operators are the rank-one unitary perturbations of the compressed shift S_u and, as we will see in this chapter, the corresponding Clark measures yield a lot of information about the inner function u.

11.1 The family of Clark measures

To simplify the formulas in this chapter, let us assume that u is an inner function satisfying

$$u(0) = 0.$$

In the endnotes for this chapter, we will discuss what changes when $u(0) \neq 0$.

For each fixed α in $\mathbb{T}$, the function

$$z \mapsto \frac{1 + \overline{\alpha}u(z)}{1 - \overline{\alpha}u(z)}$$

is analytic on $\mathbb{D}$ and thus

$$z \mapsto \operatorname{Re}\left(\frac{1 + \overline{\alpha}u(z)}{1 - \overline{\alpha}u(z)}\right) = \frac{1 - |u(z)|^2}{|\alpha - u(z)|^2}$$

is a positive harmonic function on $\mathbb{D}$. By Herglotz's Theorem (Theorem 1.18), there is a unique probability measure σ_α on $\mathbb{T}$ (that is, $\sigma_\alpha \in M_+(\mathbb{T})$ and $\sigma_\alpha(\mathbb{T}) = 1$) such that

$$\frac{1 - |u(z)|^2}{|\alpha - u(z)|^2} = \int_{\mathbb{T}} P_z(\zeta)\, d\sigma_\alpha(\zeta), \qquad z \in \mathbb{D}. \tag{11.1}$$

Moreover, Corollary 1.20 yields the formula

$$\frac{1 + \overline{\alpha}u(z)}{1 - \overline{\alpha}u(z)} = \int_{\mathbb{T}} \frac{\zeta + z}{\zeta - z} d\sigma_\alpha(\zeta), \qquad z \in \mathbb{D}. \qquad (11.2)$$

Note the two uses of our assumption that $u(0) = 0$ in the computations above. The first is to ensure that σ_α is a probability measure while the second is to eliminate the imaginary constant at the end of the integral formula in Corollary 1.20.

Definition 11.1 The measures $\{\sigma_\alpha : \alpha \in \mathbb{T}\}$ are called the *Clark measures corresponding to u.*

The following proposition summarizes some of the basic properties of Clark measures. The reader might want to review the concepts of a carrier of a measure from (1.1) and the angular derivative of a function from Definition 2.20.

Proposition 11.2 *For an inner function u with $u(0) = 0$, the corresponding family of Clark measures $\{\sigma_\alpha : \alpha \in \mathbb{T}\}$ satisfies the following properties:*

(i) $\sigma_\alpha \perp m$ for all α;

(ii) $\sigma_\alpha \perp \sigma_\beta$ for $\alpha \neq \beta$;

(iii) σ_α has a point mass at $\zeta \in \mathbb{T}$ if and only if

$$u(\zeta) := \lim_{r \to 1^-} u(r\zeta) = \alpha$$

and u has a finite angular derivative $u'(\zeta)$ in the sense of Carathéodory at ζ. Furthermore,

$$\sigma_\alpha(\{\zeta\}) = \frac{1}{|u'(\zeta)|} \quad \text{and} \quad u'(\zeta) = \frac{\alpha\overline{\zeta}}{\sigma_\alpha(\{\zeta\})};$$

(iv) *A carrier for σ_α is the set*

$$\left\{ \zeta \in \mathbb{T} : \lim_{r \to 1^-} u(r\zeta) = \alpha \right\}.$$

Proof For the proof of (i), note that for m-almost every $w \in \mathbb{T}$ we have

$$(D\sigma_\alpha)(w) = \lim_{r \to 1^-} \int_{\mathbb{T}} P_{rw}(\xi) d\sigma_\alpha(\xi) \qquad \text{(by Theorem 1.10)}$$

$$= \lim_{r \to 1^-} \frac{1 - |u(rw)|^2}{|\alpha - u(rw)|^2}. \qquad \text{(by (11.1))}$$

Since the set

$$\left\{w \in \mathbb{T} : \lim_{r \to 1^-} u(rw) = \alpha\right\}$$

has Lebesgue measure zero (Corollary 3.21) and

$$\lim_{r \to 1^-} u(rw) \in \mathbb{T}$$

for m-almost every $w \in \mathbb{T}$, we see that

$$\lim_{r \to 1^-} \frac{1 - |u(rw)|^2}{|\alpha - u(rw)|^2} = 0$$

m-almost everywhere. Thus $\sigma_\alpha \perp m$ (Theorem 1.2).

For the proofs of (ii) and (iv), note that

$$\frac{1 - |u(rw)|^2}{|\alpha - u(rw)|^2} = \mathscr{P}(\sigma_\alpha)(rw), \qquad 0 < r < 1, \ w \in \mathbb{T}.$$

Using Remark 1.11 we get

$$\left\{w : (\underline{D}\sigma_\alpha)(w) = \infty\right\} \subset \left\{w : \lim_{r \to 1^-} \mathscr{P}(\sigma_\alpha)(rw) = \infty\right\}$$

$$\subset \left\{w : \lim_{r \to 1^-} u(rw) = \alpha\right\}.$$

From (i) it follows that $\sigma_\alpha \perp m$. Applying this to Theorem 1.2 we see that $\{u = \alpha\}$ is a carrier for σ_α. This proves (iv) as well as (ii).

The proof of (iii) needs a bit of setup. By (11.2) we have the formula

$$\frac{\alpha + u(z)}{\alpha - u(z)} = \int_\mathbb{T} \frac{\xi + z}{\xi - z} d\sigma_\alpha(\xi).$$

For fixed $\zeta \in \mathbb{T}$, multiply both sides of the formula above by $(\zeta - z)$ to obtain

$$(\alpha + u(z))\frac{\zeta - z}{\alpha - u(z)} = \int_\mathbb{T} (\xi + z)\frac{\zeta - z}{\xi - z} d\sigma_\alpha(\xi). \qquad (11.3)$$

Notice how for any $\xi \in \mathbb{T}$ and for any z belonging to the Stolz domain $\Gamma_c(\zeta)$ we have

$$\left|\frac{\zeta - z}{\xi - z}\right| \leqslant \frac{|\zeta - z|}{1 - |z|} \leqslant c.$$

Since

$$\angle\lim_{z \to \zeta} (\xi + z)\frac{\zeta - z}{\xi - z} = \begin{cases} 2\zeta & \text{if } \xi = \zeta, \\ 0 & \text{otherwise,} \end{cases}$$

we can apply the Dominated Convergence Theorem to obtain

$$\angle\lim_{z \to \zeta} \int_\mathbb{T} (\xi + z)\frac{\zeta - z}{\xi - z} d\sigma_\alpha(\xi) = \int_\mathbb{T} 2\xi\chi_\zeta(\xi) d\sigma_\alpha(\xi) = 2\zeta\sigma_\alpha(\{\zeta\}),$$

where χ_ζ is the characteristic function for the set $\{\zeta\}$. Now apply (11.3) to see that

$$\angle \lim_{z \to \zeta}(\alpha + u(z))\frac{\zeta - z}{\alpha - u(z)} = 2\zeta\sigma_\alpha(\{\zeta\}). \qquad (11.4)$$

We are now ready for the proof of (iii). If $\sigma_\alpha(\{\zeta\}) > 0$, we use (11.4) to see that

$$\angle \lim_{z \to \zeta} u(z) = \alpha$$

and

$$u'(\zeta) = \angle \lim_{z \to \zeta} \frac{u(z) - \alpha}{z - \zeta} = \frac{\alpha\overline{\zeta}}{\sigma_\alpha(\{\zeta\})}.$$

Hence u has a finite angular derivative in the sense of Carathéodory at ζ.

Conversely, if u has an angular derivative in the sense of Carathéodory at ζ, we see from Theorem 7.24 that $u'(\zeta) \neq 0$. If we also assume that

$$\angle \lim_{z \to \zeta} u(z) = \alpha,$$

we can use (11.4) again to conclude that

$$0 \neq \frac{1}{u'(\zeta)} = \overline{\alpha}\zeta\sigma_\alpha(\{\zeta\}). \qquad \square$$

At this point, one might think that Clark measures are very special and belong to an exclusive class of singular probability measures on $\mathbb{T}$. They are not. Indeed, start with *any* singular probability measure μ on $\mathbb{T}$ (for example, $\mu = \delta_1$) and form the *Herglotz integral*

$$H(z) = \int_\mathbb{T} \frac{\zeta + z}{\zeta - z} d\mu(\zeta), \qquad z \in \mathbb{D}.$$

This function is analytic on $\mathbb{D}$ and has positive real part since

$$\mathrm{Re}(H(z)) = \int_\mathbb{T} \mathrm{Re}\left(\frac{\zeta + z}{\zeta - z}\right) d\mu(\zeta) = \int_\mathbb{T} P_z(\zeta) d\mu(\zeta) \geqslant 0.$$

Furthermore, $H(0) = \mu(\mathbb{T}) = 1$. This means that the function

$$u(z) = \frac{H(z) - 1}{H(z) + 1}, \qquad z \in \mathbb{D},$$

satisfies $u(0) = 0$ as well as the identity

$$\frac{1 - |u(z)|^2}{|1 - u(z)|^2} = \mathrm{Re}\left(\frac{1 + u(z)}{1 - u(z)}\right) = \mathrm{Re}(H(z)) = \int_\mathbb{T} P_z(\zeta) d\mu(\zeta).$$

Using the fact that

$$\lim_{r \to 1^-} \int_{\mathbb{T}} P_{rw}(\zeta) d\mu(\zeta) = 0$$

for m-almost every $w \in \mathbb{T}$, we see that u is inner and μ is the Clark measure with $\alpha = 1$ corresponding to u. Thus every singular probability measure on $\mathbb{T}$ arises as a Clark measure for some inner function which vanishes at the origin.

Example 11.3

(i) For the inner function $u = z^n$ and $\alpha = 1$, let ξ be a primitive nth root of unity. The solutions to $u = 1$ are just the points $1, \xi, \xi^2, \dots, \xi^{n-1}$. Moreover, $|u'(\xi^k)| = n$ for every k. Thus

$$\sigma_1 = \frac{1}{n}\delta_1 + \frac{1}{n}\delta_\xi + \frac{1}{n}\delta_{\xi^2} + \cdots + \frac{1}{n}\delta_{\xi^{n-1}}.$$

(ii) For the inner function $u = \exp\left(\frac{z+1}{z-1}\right)$ and $\alpha = 1$, the solutions to $u = 1$ are

$$\zeta_n := \frac{2n\pi - i}{2n\pi + i}, \qquad n \in \mathbb{Z}.$$

Notice how these points lie on $\mathbb{T}$ and cluster only at 1. Another computation shows that

$$\frac{1}{|u'(\zeta_n)|} = \frac{8}{1 + 4\pi^2 n^2}$$

and thus

$$\sigma_1 = \sum_{n \in \mathbb{Z}} \frac{8}{1 + 4\pi^2 n^2} \delta_{\zeta_n}.$$

Though this example does not satisfy the property that $u(0) = 0$, it will be used several times in later chapters.

11.2 The Clark unitary operators

Maintaining our standing assumption that $u(0) = 0$, recall the compressed shift

$$S_u : \mathcal{K}_u \to \mathcal{K}_u, \quad S_u f = P_u(zf)$$

from Chapter 9. Also recall the kernel function $k_0 = 1$ and the conjugate kernel function $Ck_0 = u(z)/z$. From Lemma 9.9 we know that

$$S_u S_u^* = I - k_0 \otimes k_0, \qquad S_u^* S_u = I - Ck_0 \otimes Ck_0 \tag{11.5}$$

and so the compressed shift S_u is "close" to being a unitary operator. One might suspect, correctly, that a suitable rank-one perturbation of S_u should yield a unitary operator. This is the content of the following theorem [46]. For each $\alpha \in \mathbb{T}$, define

$$U_\alpha : \mathcal{K}_u \to \mathcal{K}_u, \quad U_\alpha := S_u + \alpha(k_0 \otimes Ck_0). \tag{11.6}$$

Theorem 11.4 (Clark)   *For each $\alpha \in \mathbb{T}$, the operator U_α is a cyclic unitary operator. The eigenvalues of U_α are the points $\zeta \in \mathbb{T}$ such that $u(\zeta) = \alpha$ and u has an angular derivative in the sense of Carathéodory at ζ. The corresponding eigenvectors are the boundary kernels*

$$k_\zeta(z) = \frac{1 - \overline{\alpha}u(z)}{1 - \overline{\zeta}z}.$$

Note that the boundary kernels belong to $\mathcal{K}_u$ (Theorem 7.24). The unitary operators $\{U_\alpha : \alpha \in \mathbb{T}\}$ on $\mathcal{K}_u$ are called the *Clark unitary operators* associated with the inner function u.

Proof of Theorem 11.4   There are two ways to prove that each U_α is unitary. The first is to note that since $u(0) = 0$, the compressed shift S_u is a partial isometry with $\ker S_u = \mathbb{C}Ck_0$ and $\ker S_u^* = \mathbb{C}k_0$ (Corollary 9.25). Furthermore, since $u(0) = 0$, we have $\|k_0\| = \|Ck_0\| = 1$. Now use Theorem 1.36 to prove that U_α is unitary. As an additional byproduct of Theorem 1.36, we see that any unitary rank-one perturbation of S_u must be equal to U_α for some $\alpha \in \mathbb{T}$.

The second way to prove that U_α is unitary uses the conjugation operator C from Chapter 8. Note from Proposition 9.8 that S_u is C-symmetric (that is, $CS_uC = S_u^*$) and a computation using (9.5) will show that $k_0 \otimes Ck_0$ is C-symmetric as well. This means that U_α is also C-symmetric. Thus to show that U_α is unitary, it suffices to show that U_α is an isometry (note the use of the fact that C is an isometric mapping). Observe that

$$U_\alpha(Ck_0) = S_u(Ck_0) + \alpha\langle Ck_0, Ck_0 \rangle k_0 = \alpha k_0.$$

This follows from the fact that $Ck_0 = u/z$ and thus

$$S_u(Ck_0) = S_u(zu/z) = S_u u = 0.$$

Hence U_α is an isometry on $\mathbb{C}Ck_0$.

If $\langle f, Ck_0 \rangle = 0$, then $(Cf)(0) = 0$ so that

$$f = \overline{zCfu} = \overline{z^2 hu}$$

for some $h \in H^2$. Thus $zf = \overline{zhu}$ and hence $zf \in K_u$ (Proposition 5.4). It follows that

$$U_\alpha f = S_u f + \alpha\langle f, Ck_0 \rangle k_0 = P_u(zf) + 0 = zf,$$

and so U_α is an isometry on $\mathcal{K}_u \ominus \mathbb{C}Ck_0$. Since $U_\alpha(\mathbb{C}Ck_0) = \mathbb{C}k_0$ is orthogonal to

$$U_\alpha(\mathcal{K}_u \ominus \mathbb{C}Ck_0) = \{zf : \langle f, Ck_0 \rangle = 0\},$$

we see that U_α is an isometry.

To show that U_α is cyclic, we will show that $k_0 = 1$ is a cyclic vector for U_α. For notational convenience, let

$$X := (k_0 \otimes Ck_0)$$

and note that

$$
\begin{aligned}
XS_u^n k_0 = (k_0 \otimes Ck_0)S_u^n k_0 &= \langle S_u^n k_0, Ck_0 \rangle k_0 \\
&= \langle k_0, S_u^{*n}Ck_0 \rangle k_0 = \langle k_0, S^{*(n+1)}u \rangle k_0 \\
&= \overline{(S^{*(n+1)}u)(0)}k_0, \quad n \geqslant 0.
\end{aligned}
$$

Use the previous identity to prove

$$
\begin{aligned}
U_\alpha^0 k_0 &= k_0, \\
U_\alpha^1 k_0 &= S_u k_0 + c_{1,1}k_0, \\
U_\alpha^2 k_0 &= (S_u + X)^2 k_0 \\
&= (S_u^2 + S_u X + X S_u + X^2)k_0 \\
&= S_u^2 k_0 + c_{2,1} S_u k_0 + c_{2,2}k_0, \\
U_\alpha^3 k_0 &= (S_u + X)^3 k_0 \\
&= (S_u^3 + S_u^2 X + S_u X^2 + X^3 + X S_u^2 + X^2 S_u + S_u X S_u + X S_u X)k_0 \\
&= S_u^3 k_0 + c_{3,1} S_u^2 k_0 + c_{3,2} S_u k_0 + c_{3,3}k_0,
\end{aligned}
$$

and so forth. It follows that

$$\bigvee\{U_\alpha^n k_0 : n \geqslant 0\} \supset \bigvee\{S_u^n k_0 : n \geqslant 0\} = \mathcal{K}_u$$

(Proposition 9.13), which proves that k_0 is a cyclic vector for U_α.

We now discuss the eigenvalues and eigenvectors of U_α. As noted earlier, for fixed $\zeta \in \mathbb{T}$, we have

$$\lim_{r \to 1^-} u(r\zeta) = u(\zeta) = \alpha \quad \text{and} \quad |u'(\zeta)| < \infty$$

if and only if

$$k_\zeta = \frac{1 - \overline{\alpha}u}{1 - \overline{\zeta}z} \in \mathcal{K}_u.$$

Use the fact that U_α is unitary to prove that $f \in \mathcal{K}_u$ is a solution to the equation $U_\alpha f = \zeta f$ if and only if f is a solution to

$$U_\alpha^* f = \overline{\zeta}f.$$

Since

$$U_\alpha^* f = S^* f + \overline{\alpha}(Ck_0 \otimes k_0)f = \frac{f - f(0)}{z} + \overline{\alpha}\frac{u}{z}f(0),$$

we see that $U_\alpha^* f = \overline{\zeta} f$ if and only if

$$\overline{\zeta} f = \frac{f - f(0)}{z} + \overline{\alpha}\frac{u}{z}f(0) \iff f = f(0)\frac{1 - \overline{\alpha}u}{1 - \overline{\zeta}z}. \qquad \square$$

Here is a concrete example of Clark's result.

Example 11.5   Let $u = z^3$ and consider the corresponding model space $\mathcal{K}_u = \bigvee\{1, z, z^2\}$. A computation shows that

$$S_u 1 = z, \quad S_u z = z^2, \quad S_u z^2 = 0,$$

and so the matrix representation of S_u with respect to the orthonormal basis $\{1, z, z^2\}$ is

$$\begin{bmatrix} 0 & 0 & 0 \\ 1 & 0 & 0 \\ 0 & 1 & 0 \end{bmatrix}.$$

In this case,

$$U_\alpha = S_u + \alpha(1 \otimes z^2).$$

Similarly, since

$$U_\alpha 1 = z, \quad U_\alpha z = z^2, \quad U_\alpha z^2 = \alpha,$$

the matrix representation of U_α is

$$\begin{bmatrix} 0 & 0 & \alpha \\ 1 & 0 & 0 \\ 0 & 1 & 0 \end{bmatrix}.$$

This matrix is unitary since its columns form an orthonormal basis for $\mathbb{C}^3$. The eigenvalues of U_α are precisely the solutions to $u = \alpha$ (the three cube roots of α) and corresponding eigenvectors are the boundary kernels

$$k_\beta(z) = \frac{1 - \overline{\beta}^3 z^3}{1 - \overline{\beta}z} = 1 + \overline{\beta}z + \overline{\beta}^2 z^2,$$

where $\beta^3 = \alpha$. Each eigenvector k_β has norm $\sqrt{3}$, which is precisely the square root of the modulus of the angular derivative (namely $3z^2$) of z^3 at the point β in $\mathbb{T}$. Moreover, the normalized eigenvectors of U_α form an orthonormal basis for $\mathcal{K}_u$ as expected since U_α is a unitary operator on a finite-dimensional space. This example can be easily generalized to $\mathcal{K}_u$, where $u = z^n$.

11.3 Spectral representation of the Clark operator

Since U_α is a cyclic unitary operator, we know, from the Spectral Theorem for cyclic unitary operators (Theorem 1.31), that there exists a $\mu_\alpha \in M_+(\mathbb{T})$ such that U_α is unitarily equivalent to the multiplication operator $f \mapsto \zeta f$ on $L^2(\mu_\alpha)$. The following theorem of Clark tells us that $\mu_\alpha = \sigma_\alpha$ and also gives us a concrete spectral representation for U_α [46].

Theorem 11.6 (Clark) *For an inner function u with $u(0) = 0$ and $\alpha \in \mathbb{T}$, the operator defined by*

$$(V_\alpha f)(z) = (1 - \overline{\alpha} u(z)) \int_\mathbb{T} \frac{f(\zeta)}{1 - \overline{\zeta} z}\, d\sigma_\alpha(\zeta), \qquad z \in \mathbb{D},$$

is a unitary operator from $L^2(\sigma_\alpha)$ onto $\mathcal{K}_u$. Furthermore, if

$$Z_\alpha : L^2(\sigma_\alpha) \to L^2(\sigma_\alpha), \quad (Z_\alpha f)(\zeta) = \zeta f(\zeta),$$

then $V_\alpha Z_\alpha = U_\alpha V_\alpha$. In other words, U_α is unitarily equivalent to Z_α.

The theorem above can be expressed with the commutative diagram:

$$
\begin{array}{ccc}
L^2(\sigma_\alpha) & \xrightarrow{\ Z_\alpha\ } & L^2(\sigma_\alpha) \\[2pt]
\Big\downarrow{\scriptstyle V_\alpha} & & \Big\downarrow{\scriptstyle V_\alpha} \\[2pt]
\mathcal{K}_u & \xrightarrow{\ U_\alpha\ } & \mathcal{K}_u
\end{array}
$$

The proof of this theorem is involved and needs some preliminary results. We start with a few technical identities involving c_λ, the Cauchy kernels, and k_λ, the reproducing kernels for $\mathcal{K}_u$. Notice that $C(\mathbb{D}^-) \subset L^2(\sigma_\alpha)$. In particular, the Cauchy kernels c_λ belong to $L^2(\sigma_\alpha)$.

Lemma 11.7 *If $u(0) = 0$ and $\lambda, \mu \in \mathbb{D}$, $\alpha \in \mathbb{T}$, then*

$$\langle c_\lambda, c_\mu \rangle_{L^2(\sigma_\alpha)} = \left\langle \frac{k_\lambda}{1 - \alpha \overline{u(\lambda)}}, \frac{k_\mu}{1 - \alpha \overline{u(\mu)}} \right\rangle_{H^2}.$$

Proof For $\beta_1, \beta_2 \in \mathbb{C}$, note that

$$\frac{1 + \overline{\beta_1}}{1 - \overline{\beta_1}} + \frac{1 + \beta_2}{1 - \beta_2} = 2 \frac{1 - \overline{\beta_1}\beta_2}{(1 - \overline{\beta_1})(1 - \beta_2)}.$$

Thus

$$
\begin{aligned}
\langle c_\lambda, c_\mu \rangle_{L^2(\sigma_\alpha)} &= \int_{\mathbb{T}} \frac{1}{1 - \overline{\lambda}\zeta} \frac{1}{1 - \mu\overline{\zeta}} d\sigma_\alpha(\zeta) \\
&= \frac{1}{2(1 - \overline{\lambda}\mu)} \int_{\mathbb{T}} \left(\frac{1 + \overline{\lambda}\zeta}{1 - \overline{\lambda}\zeta} + \frac{1 + \overline{\zeta}\mu}{1 - \overline{\zeta}\mu} \right) d\sigma_\alpha(\zeta) \\
&= \frac{1}{2(1 - \overline{\lambda}\mu)} \left(\frac{1 + \alpha\overline{u(\lambda)}}{1 - \alpha\overline{u(\lambda)}} + \frac{1 + \overline{\alpha}u(\mu)}{1 - \overline{\alpha}u(\mu)} \right) \quad \text{(by (11.2))} \\
&= \frac{1}{(1 - \alpha\overline{u(\lambda)})(1 - \overline{\alpha}u(\mu))} \frac{1 - \overline{u(\lambda)}u(\mu)}{1 - \overline{\lambda}\mu} \\
&= \frac{1}{(1 - \alpha\overline{u(\lambda)})(1 - \overline{\alpha}u(\mu))} \langle k_\lambda, k_\mu \rangle_{H^2} \\
&= \left\langle \frac{k_\lambda}{1 - \alpha\overline{u(\lambda)}}, \frac{k_\mu}{1 - \alpha\overline{u(\mu)}} \right\rangle_{H^2}.
\end{aligned}
$$

$\square$

Up to now we know that V_α maps $L^2(\sigma_\alpha)$ to a linear manifold contained in the space of analytic functions on $\mathbb{D}$. The next result is the first step in identifying the range of V_α.

Corollary 11.8 *If u is inner with $u(0) = 0$, $\lambda \in \mathbb{D}$, and $\alpha \in \mathbb{T}$, then*

$$
V_\alpha c_\lambda = \frac{1}{1 - \alpha\overline{u(\lambda)}} k_\lambda.
$$

Proof   Indeed,

$$
\begin{aligned}
(V_\alpha c_\lambda)(z) &= (1 - \overline{\alpha}u(z)) \int_{\mathbb{T}} \frac{1}{1 - \zeta\overline{\lambda}} \frac{1}{1 - \overline{\zeta}z} d\sigma_\alpha(\zeta) \\
&= (1 - \overline{\alpha}u(z))\langle c_\lambda, c_z \rangle_{L^2(\sigma_\alpha)} \\
&= (1 - \overline{\alpha}u(z)) \left\langle \frac{k_\lambda}{1 - \alpha\overline{u(\lambda)}}, \frac{k_z}{1 - \alpha\overline{u(z)}} \right\rangle_{H^2} \quad \text{(Lemma 11.7)} \\
&= \frac{1}{1 - \alpha\overline{u(\lambda)}} \langle k_\lambda, k_z \rangle_{H^2} \\
&= \frac{1}{1 - \alpha\overline{u(\lambda)}} k_\lambda(z).
\end{aligned}
$$

$\square$

Remark 11.9   When $\lambda = 0$ in the computation above, we have $c_0 = 1$ and $u(0) = 0$. Thus $V_\alpha 1 = 1$. This will play an important role later on.

We know that V_α maps c_λ to a constant multiple of k_λ (Corollary 11.8). Moreover, if

$$g = \sum_{1 \leqslant j \leqslant n} a_j c_{\lambda_j}$$

is a finite linear combination of Cauchy kernels, then

$$
\begin{aligned}
\|g\|^2_{L^2(\sigma_\alpha)} &= \langle g, g \rangle_{L^2(\sigma_\alpha)} \\
&= \sum_{1 \leqslant j, \ell \leqslant n} a_j \overline{a_\ell} \langle c_{\lambda_j}, c_{\lambda_\ell} \rangle_{L^2(\sigma_\alpha)} \\
&= \sum_{1 \leqslant j, \ell \leqslant n} a_j \overline{a_\ell} \left\langle \frac{k_{\lambda_j}}{1 - \alpha \overline{u(\lambda_j)}}, \frac{k_{\lambda_\ell}}{1 - \alpha \overline{u(\lambda_\ell)}} \right\rangle_{H^2} \quad \text{(Lemma 11.7)} \\
&= \sum_{1 \leqslant j, \ell \leqslant n} a_j \overline{a_\ell} \langle V_\alpha c_{\lambda_j}, V_\alpha c_{\lambda_\ell} \rangle_{H^2} \quad \text{(Corollary 11.8)} \\
&= \langle V_\alpha g, V_\alpha g \rangle_{H^2} \\
&= \|V_\alpha g\|^2_{H^2}.
\end{aligned}
$$

Thus V_α is an isometry from

$$\bigvee \{c_\lambda : \lambda \in \mathbb{D}\} \subset L^2(\sigma_\alpha) \quad \text{onto} \quad \bigvee \{k_\lambda : \lambda \in \mathbb{D}\} = \mathcal{K}_u$$

(Proposition 5.20).

Lemma 11.10

$$\bigvee \{c_\lambda : \lambda \in \mathbb{D}\} = L^2(\sigma_\alpha).$$

Proof   Suppose that $g \in L^2(\sigma_\alpha)$ and

$$\langle g, c_\lambda \rangle_{L^2(\sigma_\alpha)} = 0 \qquad \forall \lambda \in \mathbb{D}. \tag{11.7}$$

Expanding $c_\lambda(z) = (1 - \overline{\lambda}z)^{-1}$ as

$$c_\lambda(z) = \sum_{n \geqslant 0} \overline{\lambda}^n z^n,$$

we see that (11.7) can be restated as

$$\sum_{n \geqslant 0} \overline{\lambda}^n \int_{\mathbb{T}} \zeta^n g(\zeta) \, d\sigma_\alpha = 0 \quad \forall \lambda \in \mathbb{D}$$

and thus, by the uniqueness of Taylor series coefficients,

$$\int_{\mathbb{T}} \zeta^n g(\zeta) \, d\sigma_\alpha(\zeta) = 0 \qquad \forall n \geqslant 0.$$

The F. and M. Riesz Theorem (Theorem 1.21) now says that $g\,d\sigma_\alpha \ll m$ which, since $\sigma_\alpha \perp m$ (Proposition 11.2), can only happen when $g \equiv 0$. To finish, use Theorem 1.27.   □

The previous lemma and the two corollaries above imply the following.

Proposition 11.11 *If u is inner with $u(0) = 0$, the operator V_α is a unitary operator from $L^2(\sigma_\alpha)$ onto $\mathcal{K}_u$.*

We are now ready to prove Clark's Theorem (Theorem 11.6). Recall the bounded multiplication operator

$$Z_\alpha : L^2(\sigma_\alpha) \to L^2(\sigma_\alpha), \qquad (Z_\alpha g)(\zeta) = \zeta g(\zeta)$$

and notice that the adjoint Z_α^* is given by

$$(Z_\alpha^* g)(\zeta) = \overline{\zeta} g(\zeta).$$

Proof of Theorem 11.6   It suffices to prove that

$$V_\alpha Z_\alpha^* V_\alpha^* = U_\alpha^*.$$

If $g \in L^1(\sigma_\alpha)$, let us use the shorthand

$$(\mathscr{C}_{\sigma_\alpha} g)(z) := \int_{\mathbb{T}} \frac{g(\zeta)}{1 - \overline{\zeta}z} d\sigma_\alpha(\zeta) \tag{11.8}$$

for the *Cauchy transform*. We will make use of the "product rule" identity

$$S^*(f_1 f_2) = f_1 S^* f_2 + f_2(0) S^* f_1 \tag{11.9}$$

from (4.19) as well as the formal operator identity

$$\mathscr{C}_{\sigma_\alpha} Z_\alpha^* = S^* \mathscr{C}_{\sigma_\alpha}, \tag{11.10}$$

that is to say,

$$\int_{\mathbb{T}} \frac{\overline{\zeta} g(\zeta)}{1 - \overline{\zeta}z} d\sigma_\alpha(\zeta) = \frac{1}{z} \Big((\mathscr{C}_{\sigma_\alpha} g)(z) - (\mathscr{C}_{\sigma_\alpha} g)(0) \Big), \quad g \in L^2(\sigma).$$

Also remember that $u(0) = 0$ and so $S^* u = u/z$. Thus for $g \in L^2(\sigma_\alpha)$,

$$\begin{aligned}
V_\alpha Z_\alpha^* g &= (1 - \overline{\alpha}u)\mathscr{C}_{\sigma_\alpha} Z_\alpha^* g \\
&= (1 - \overline{\alpha}u)S^* \mathscr{C}_{\sigma_\alpha} g && \text{(by (11.10))} \\
&= S^*(1 - \overline{\alpha}u)\mathscr{C}_{\sigma_\alpha} g + \overline{\alpha} S^* u \langle g, 1 \rangle_{L^2(\sigma_\alpha)} && \text{(by (11.9))} \\
&= S^* V_\alpha g + \overline{\alpha}\langle g, 1 \rangle_{L^2(\sigma_\alpha)} \frac{u}{z}
\end{aligned}$$

$$= S^* V_\alpha g + \overline{\alpha} \langle V_\alpha g, V_\alpha 1 \rangle_{H^2} \frac{u}{z} \qquad \text{(Proposition 11.11)}$$

$$= S^* V_\alpha g + \overline{\alpha} \langle V_\alpha g, 1 \rangle_{H^2} \frac{u}{z} \qquad (V_\alpha 1 = 1)$$

$$= S^* V_\alpha g + \overline{\alpha}(S^* u \otimes 1) V_\alpha g.$$

This proves the identity

$$V_\alpha Z_\alpha^* g = S^* V_\alpha g + \overline{\alpha}(S^* u \otimes 1) V_\alpha g.$$

Now plug $g = V_\alpha^* f$ ($f \in \mathcal{K}_u$) into the identity above to see that

$$V_\alpha Z_\alpha^* V_\alpha^* f = S^* f + \overline{\alpha}(S^* u \otimes 1) f = U_\alpha^* f. \qquad \square$$

Remark 11.12

(i) One can also recognize the eigenvalues and eigenvectors of U_α using V_α. Indeed, by Theorem 11.6, U_α is unitarily equivalent to Z_α. Thus if ζ is an eigenvalue of U_α it is also one for Z_α and so

$$z f(z) = (Z_\alpha f)(z) = \zeta f(z), \qquad z \in \mathbb{T}.$$

The only way this can be true for an $f \in L^2(\sigma_\alpha) \setminus \{0\}$ is when $f = \chi_\zeta$ (the characteristic function for the singleton $\{\zeta\}$) and ζ is a point mass of σ_α. By Proposition 11.2, this holds precisely when $u(\zeta) = \alpha$ and $|u'(\zeta)| < \infty$. To get a formula for the corresponding eigenvector for U_α we just need to compute $V_\alpha \chi_\zeta$. For this, we use the formula for V_α from Theorem 11.6 to see that

$$V_\alpha \chi_\zeta = (1 - \overline{\alpha} u) \int_{\mathbb{T}} \frac{\chi_\zeta(w)}{1 - \overline{w} z} d\sigma_\alpha(w)$$

$$= \frac{1 - \overline{\alpha} u}{1 - \overline{\zeta} z} \sigma_\alpha(\{\zeta\})$$

$$= \frac{1}{|u'(\zeta)|} k_\zeta.$$

(ii) We can also use the multiplication operator Z_α to give another proof that U_α is cyclic with cyclic vector 1. From Remark 11.9 and Theorem 11.6, it suffices to show that 1 is a cyclic vector for Z_α on $L^2(\sigma_\alpha)$, or equivalently,

$$\bigvee \{\zeta^n : n \geqslant 0\} = L^2(\sigma_\alpha).$$

Indeed, if $f \in L^2(\sigma_\alpha)$ satisfies $f \perp \{\zeta^n : n \geqslant 0\}$, then

$$\int_{\mathbb{T}} f(\zeta) \overline{\zeta}^n d\sigma_\alpha(\zeta) = 0, \qquad n \geqslant 0.$$

The F. and M. Riesz Theorem (Theorem 1.21) now says that $f d\sigma_\alpha \ll m$ which, since $\sigma_\alpha \perp m$ (Proposition 11.2), can only happen when $f = 0$. To finish, use Theorem 1.27.

Example 11.13 Returning to Example 11.5, let $\zeta_1, \zeta_2, \zeta_3$ be the three solutions to $u = \alpha$ (that is, to the equation $z^3 = \alpha$). Then the discrete measure

$$d\sigma_\alpha = \tfrac{1}{3}\delta_{\zeta_1} + \tfrac{1}{3}\delta_{\zeta_2} + \tfrac{1}{3}\delta_{\zeta_3}$$

satisfies

$$\frac{1 + \overline{\alpha}z^3}{1 - \overline{\alpha}z^3} = \int_{\mathbb{T}} \frac{\zeta + z}{\zeta - z} \, d\sigma_\alpha(\zeta).$$

From the theorem above, the operator V_α is given by

$$V_\alpha : L^2(\sigma_\alpha) \to \mathcal{K}_{z^3}, \quad (V_\alpha f)(z) = (1 - \overline{\alpha}z^3) \int_{\mathbb{T}} \frac{f(\zeta)}{1 - \overline{\zeta}z} \, d\sigma_\alpha(\zeta).$$

To see this worked out, note that if

$$f = c_1 \chi_{\zeta_1} + c_2 \chi_{\zeta_2} + c_3 \chi_{\zeta_3}, \quad c_1, c_2, c_3 \in \mathbb{C},$$

is a generic element of $L^2(\sigma_\alpha)$, then

$$(V_\alpha f)(z) = (1 - \overline{\alpha}z^3)\left(c_1 \frac{1/3}{1 - \overline{\zeta_1}z} + c_2 \frac{1/3}{1 - \overline{\zeta_2}z} + c_3 \frac{1/3}{1 - \overline{\zeta_3}z}\right). \tag{11.11}$$

Since $\zeta_j^3 = \alpha$, we have

$$(1 - \overline{\alpha}z^3) = (1 - \overline{\zeta_1}z)(1 - \overline{\zeta_2}z)(1 - \overline{\zeta_3}z)$$

and thus the rational function above is actually a polynomial of degree at most 2, which are precisely the elements of $\mathcal{K}_{z^3}$.

Furthermore, $\{\sqrt{3}\chi_{\zeta_1}, \sqrt{3}\chi_{\zeta_2}, \sqrt{3}\chi_{\zeta_3}\}$ is an orthonormal basis for $L^2(\sigma_\alpha)$ and

$$V_\alpha(\sqrt{3}\chi_{\zeta_j}) = \frac{\sqrt{3}}{3}k_{\zeta_j}.$$

But since

$$\left\{\frac{\sqrt{3}}{3}k_{\zeta_j} : j = 1, 2, 3\right\}$$

is an orthonormal basis for $\mathcal{K}_{z^3}$ we see that V_α is unitary. From here one can see via linear algebra that V_α intertwines U_α with Z_α.

The radial boundary values of $V_\alpha f$ are of particular interest. For example, we know that $V_\alpha f$ belongs to $\mathcal{K}_u \subset H^2$ for every $f \in L^2(\sigma_\alpha)$ and, as such, $V_\alpha f$

has a radial limit almost everywhere with respect to Lebesgue measure m. The following theorem of Poltoratski [148] says much more.

Theorem 11.14 (Poltoratski)   *For $f \in L^2(\sigma_\alpha)$,*

$$\lim_{r \to 1^-} (V_\alpha f)(r\zeta) = f(\zeta)$$

for σ_α-almost every $\zeta \in \mathbb{T}$.

Remark 11.15   One can see how this theorem works with $\zeta = \zeta_1, \zeta_2, \zeta_3$ in (11.11).

Theorem 11.14 is significant for it says that functions in model spaces (the functions $V_\alpha f$ for $f \in L^2(\sigma_\alpha)$ in the previous theorem) have non-tangential limits on finer sets than generic functions from H^2. We have seen a result along these lines already in Theorem 7.24 in which, under certain circumstances (existence of angular derivatives for u), all functions in $\mathcal{K}_u$ have non-tangential limits at a fixed boundary point. The previous theorem says a bit more in that all functions from $\mathcal{K}_u$ have non-tangential limits σ_α-almost everywhere. For example, if σ_α has an isolated point mass at $\zeta \in \mathbb{T}$, then every function in $\mathcal{K}_u$ has a non-tangential limit at ζ. In this particular case, the result should not be surprising since, by Proposition 11.2, u will have a finite angular derivative at ζ and so by Theorem 7.24 every function in $\mathcal{K}_u$ has a non-tangential limit at ζ.

Theorem 11.14 is especially notable since $\sigma_\alpha \perp m$ and a classical result [127, 129] says that if E is any closed subset of $\mathbb{T}$ with Lebesgue measure zero, then there exists an $f \in H^2$ (which can be taken to be inner) whose radial limits do not exist on E.

11.4 The Aleksandrov disintegration theorem

Here is one more fascinating and useful gem about Clark measures [7] (see also [42]).

Theorem 11.16 (Aleksandrov)   *Suppose u is an inner function with $u(0) = 0$ and with its associated family of Clark measures $\{\sigma_\alpha : \alpha \in \mathbb{T}\}$. Then*

$$\int_\mathbb{T} \left(\int_\mathbb{T} f(\zeta) d\sigma_\alpha(\zeta) \right) dm(\alpha) = \int_\mathbb{T} f(\xi) \, dm(\xi) \qquad (11.12)$$

for all $f \in C(\mathbb{T})$.

Proof   For a fixed $z \in \mathbb{D}$ we can integrate both sides of (11.1) with respect to Lebesgue measure m to get

$$\int_{\mathbb{T}}\left(\int_{\mathbb{T}} P_z(\zeta)d\sigma_\alpha(\zeta)\right)dm(\alpha) = \int_{\mathbb{T}}\frac{1 - |u(z)|^2}{|\alpha - u(z)|^2}\,dm(\alpha)$$

$$= \int_{\mathbb{T}} P_{u(z)}(\alpha)\,dm(\alpha)$$

$$= 1 \qquad\qquad \text{(by (1.15))}$$

$$= \int_{\mathbb{T}} P_z(\xi)\,dm(\xi).$$

By linearity, (11.12) holds for finite linear combinations of Poisson kernels.

To obtain the result for any $f \in C(\mathbb{T})$, let $\{f_n\}_{n\geqslant 1}$ be a sequence of finite linear combinations of Poisson kernels that converge uniformly to f (Proposition 1.17). Let us first note that for each n, the function

$$g_n(\alpha) = \int_{\mathbb{T}} f_n\,d\sigma_\alpha$$

is continuous on $\mathbb{T}$. To see this write

$$f_n = \sum_{1\leqslant j\leqslant N} c_j P_{z_j}$$

and, by the definition of a Clark measure from (11.1), observe that

$$\int_{\mathbb{T}} f_n\,d\sigma_\alpha = \sum_{1\leqslant j\leqslant N} c_j \frac{1 - |u(z_j)|^2}{|\alpha - u(z_j)|^2},$$

which is continuous in the variable α. We can extend this to show that the function g defined by

$$g(\alpha) = \int_{\mathbb{T}} f\,d\sigma_\alpha$$

is also continuous. Indeed, still assuming $u(0) = 0$, we know that each σ_α is a probability measure and so

$$|g(\alpha) - g_n(\alpha)| = \left|\int_{\mathbb{T}} f\,d\sigma_\alpha - \int_{\mathbb{T}} f_n\,d\sigma_\alpha\right| \leqslant \|f - f_n\|_\infty.$$

Since $f_n \to f$ uniformly on $\mathbb{T}$, we see that $g_n \to g$ uniformly on $\mathbb{T}$. Hence g is continuous. This ensures that the left-hand side of (11.12) is well-defined.

Finally,

$$\int_{\mathbb{T}} f\,dm = \lim_{n\to\infty} \int_{\mathbb{T}} f_n\,dm \qquad\qquad \text{(uniform convergence)}$$

$$= \lim_{n\to\infty} \int_{\mathbb{T}}\left(\int_{\mathbb{T}} f_n(\zeta)d\sigma_\alpha(\zeta)\right)dm(\alpha) \qquad \text{(disintegration formula)}$$

$$= \int_{\mathbb{T}}\left(\int_{\mathbb{T}} f(\zeta)d\sigma_\alpha(\zeta)\right)dm(\alpha). \qquad \text{(uniform convergence)}$$

This proves the result. $\qquad\qquad\qquad\qquad\qquad\qquad\qquad\qquad\qquad\qquad$ □

Remark 11.17   Aleksandrov [7] proved the stronger result

$$\int_{\mathbb{T}}\left(\int_{\mathbb{T}} f(\zeta)d\sigma_\alpha(\zeta)\right)dm(\alpha) = \int_{\mathbb{T}} f(\xi)\,dm(\xi), \quad f \in L^1.$$

There are a few non-trivial technical issues to work out here. For example, the inner integrals

$$\int_{\mathbb{T}} f(\zeta)\,d\sigma_\alpha(\zeta)$$

do not seem to be well-defined for L^1 functions since the measure σ_α may contain point masses on $\mathbb{T}$ while L^1 functions are defined merely m-almost everywhere. Amazingly though, for a fixed $f \in L^1$, the function

$$\alpha \mapsto \int_{\mathbb{T}} f(\zeta)\,d\sigma_\alpha(\zeta)$$

is defined for m-almost every α and is (Lebesgue) integrable. An argument with the monotone class theorem is used to prove this more general result. See [42] for the details.

Example 11.18   Let $u = z^n$ and note that if $\alpha = e^{i\theta}$, then

$$\sigma_\alpha = \frac{1}{n} \sum_{0 \leqslant k \leqslant n-1} \delta_{e^{i\frac{\theta+2\pi k}{n}}}.$$

Moreover, for any $f \in C(\mathbb{T})$ we have

$$\int_0^{2\pi}\left(\int_0^{2\pi} f(e^{it})d\sigma_\alpha(e^{it})\right)\frac{d\theta}{2\pi} = \int_0^{2\pi} \frac{1}{n} \sum_{0 \leqslant k \leqslant n-1} f(e^{i\frac{\theta+2\pi k}{n}})\frac{d\theta}{2\pi}$$

$$= \frac{1}{n} \sum_{0 \leqslant k \leqslant n-1} \int_0^{2\pi} f(e^{i\frac{\theta+2\pi k}{n}})\frac{d\theta}{2\pi}$$

$$= \sum_{0 \leqslant k \leqslant n-1} \int_{\frac{2\pi k}{n}}^{\frac{2\pi k+2\pi}{n}} f(e^{it})\frac{dt}{2\pi}$$

$$= \int_0^{2\pi} f(e^{it})\frac{dt}{2\pi}.$$

This is precisely the formula (11.12).

11.5 A connection to composition operators

When $\varphi \in H^\infty$, $\varphi(0) = 0$, $\|\varphi\|_\infty \leqslant 1$, and $\alpha \in \mathbb{T}$, one can still use Herglotz's Theorem, as was used to prove (11.2), to obtain the identity

$$\frac{1 + \overline{\alpha}\varphi(z)}{1 - \overline{\alpha}\varphi(z)} = \int_{\mathbb{T}} \frac{\zeta + z}{\zeta - z} d\mu_\alpha(\zeta), \quad z \in \mathbb{D},$$

for some $\mu_\alpha \in M_+(\mathbb{T})$. The measures μ_α might not be singular with respect to m. However, they do form an interesting class of measures explored by Aleksandrov and others (see [42] for further references). They are often called *Aleksandrov–Clark* measures. One can identify all the parts of these measures (absolutely continuous, singular continuous, point masses, etc.) and develop a disintegration theorem as before. Versions of these measures also appear in mathematical physics through some papers of B. Simon and T. Wolff [178, 180].

Recall our earlier discussion of composition operators

$$C_\varphi : H^2 \to H^2$$

from Chapter 6. Let us mention two connections that Aleksandrov–Clark measures make to these operators.

For the first connection, let φ be an analytic self-map of $\mathbb{D}$ with its associated family of Aleksandrov–Clark measures $\{\mu_\alpha : \alpha \in \mathbb{T}\}$. For $f \in C(\mathbb{T})$, define

$$(A_\varphi f)(\alpha) = \int_{\mathbb{T}} f(\zeta) \, d\mu_\alpha(\zeta), \quad \alpha \in \mathbb{T}.$$

One can see from the proof of Theorem 11.16 that $A_\varphi f \in C(\mathbb{T})$. The operator above is called the *Aleksandrov operator* [7].

Theorem 11.19 (Aleksandrov)   *If $\varphi(0) = 0$, then the operator A_φ extends to a bounded operator on H^2.*

The connection the Aleksandrov operator A_φ makes with composition operators is contained in the following theorem.

Theorem 11.20   *If $\varphi(0) = 0$ then $A_\varphi = C_\varphi^*$.*

Proof   Assuming that $\varphi(0) = 0$, one can manipulate the identity

$$\int_{\mathbb{T}} \frac{\zeta + \lambda}{\zeta - \lambda} d\mu_\alpha(\zeta) = \frac{\alpha + \varphi(\lambda)}{\alpha - \varphi(\lambda)}, \quad \lambda \in \mathbb{D}, \, \alpha \in \mathbb{T},$$

to get the formula

$$\int_{\mathbb{T}} \frac{1}{1 - \overline{\lambda}\zeta} d\mu_\alpha(\zeta) = \frac{1}{1 - \overline{\varphi(\lambda)}\alpha}. \tag{11.13}$$

Indeed,

$$
\begin{aligned}
\int_{\mathbb{T}} \frac{1}{1 - \bar{\zeta}\lambda} d\mu_\alpha(\zeta) &= \frac{1}{2} \int_{\mathbb{T}} \left(\frac{\zeta + \lambda}{\zeta - \lambda} + 1 \right) d\mu_\alpha(\zeta) \\
&= \frac{1}{2} \left(\frac{\alpha + \varphi(\lambda)}{\alpha - \varphi(\lambda)} + 1 \right) \\
&= \frac{\alpha}{\alpha - \varphi(\lambda)} \\
&= \frac{1}{1 - \bar{\alpha}\varphi(\lambda)}.
\end{aligned}
$$

Now take complex conjugates to get the formula in (11.13). This identity can be written as

$$
A_\varphi c_\lambda = c_{\varphi(\lambda)},
$$

which allows us to relate this to the composition operator C_φ by means of the computation

$$
\begin{aligned}
(C_\varphi^* c_\lambda)(z) = \langle C_\varphi^* c_\lambda, c_z \rangle &= \langle c_\lambda, C_\varphi c_z \rangle \\
&= \langle c_\lambda, c_z \circ \varphi \rangle = \overline{\langle c_z \circ \varphi, c_\lambda \rangle} = \overline{c_z(\varphi(\lambda))} \\
&= \frac{1}{1 - \overline{\varphi(\lambda)}z} \\
&= c_{\varphi(\lambda)}(z).
\end{aligned}
$$

Thus,

$$
A_\varphi c_\lambda = C_\varphi^* c_\lambda, \qquad \lambda \in \mathbb{D}.
$$

Using the fact that the Cauchy kernels c_λ have dense linear span in H^2, the desired identity follows. □

The second connection Aleksandrov–Clark measures make with composition operators concerns essential norms. Although the essential norm of C_φ can be computed in terms of the Nevanlinna counting function [53, 174], Cima and Matheson [44] computed it in terms of σ_α, where $\mu_\alpha = h_\alpha dm + \sigma_\alpha$ is the Lebesgue decomposition (see (1.5)) of the Aleksandrov–Clark measure μ_α.

Theorem 11.21 (Cima–Matheson) *For an analytic self-map φ of $\mathbb{D}$,*

$$
\|C_\varphi\|_e = \left(\sup_{\alpha \in \mathbb{T}} \sigma_\alpha(\mathbb{T}) \right)^{\frac{1}{2}}.
$$

See [160] for other connections to composition operators.

11.6 Carleson measures

For $\mu \in M_+(\mathbb{D}^-)$, let $\|f\|_\mu$ denote the $L^2(\mu)$ norm of f and $B(\zeta, a)$ denote the open disk of radius a about $\zeta \in \mathbb{T}$. A well-known result of Carleson is the following.

Theorem 11.22 (Carleson) *For $\mu \in M_+(\mathbb{D}^-)$ the following are equivalent:*

(i) $\|f\|_\mu \lesssim \|f\|$ for all $f \in H^2 \cap C(\mathbb{D}^-)$;
(ii) $\mu(B(\zeta, a)) \lesssim a$ for all $\zeta \in \mathbb{T}$ and $a \in (0, 1)$.

Notice how we are testing the condition $\|f\|_\mu \lesssim \|f\|$ on H^2 functions that are continuous on $\mathbb{D}^-$ in order to make sense of the integrals $\|f\|_\mu$ since μ might place mass on $\mathbb{T}$. However, at the end of the day, the embedding $\|f\|_\mu \lesssim \|f\|$ extends to *every* $f \in H^2$ (defining f on $\mathbb{T}$ m-almost everywhere via radial boundary values) due to the fact that the condition $\mu(B(\zeta, a)) \lesssim a$ for all $\zeta \in \mathbb{T}$ and $a \in (0, 1)$ implies that $\mu|_\mathbb{T} \ll m$ with a bounded Radon–Nikodym derivative.

Recall that $\mathcal{K}_u \cap C(\mathbb{D}^-)$ is dense in $\mathcal{K}_u$ (Theorem 5.24). So one can ask for which $\mu \in M_+(\mathbb{D}^-)$ do we have $\|f\|_\mu \lesssim \|f\|$ for all $f \in \mathcal{K}_u \cap C(\mathbb{D}^-)$? A result of Aleksandrov says that if this condition holds then *every* function in $\mathcal{K}_u$ has a radial limit μ-almost everywhere (making the integrals $\|f\|_\mu$ defined for all $f \in \mathcal{K}_u$, even when μ places mass on $\mathbb{T}$). Furthermore, the embedding can be extended to all of $\mathcal{K}_u$. We refer the reader to the papers [10, 12, 49, 189, 190] for other results on Carleson measures for $\mathcal{K}_u$.

Instead of giving a survey of these results for model spaces, which would take us too far afield, we focus our attention on a connection between Carleson measures for $\mathcal{K}_u$ and Clark theory. Let $\sigma = \sigma_1$ be the Clark measure for u at 1 and recall the unitary operator $V = V_1 : L^2(\sigma) \to \mathcal{K}_u$ defined by

$$Vg = (1 - u) \int_\mathbb{T} \frac{g(\zeta)}{1 - \bar{\zeta}z} \, d\sigma(\zeta) = (1 - u)\mathscr{C}_\sigma g,$$

where $\mathscr{C}_\sigma$ is the Cauchy transform defined in (11.8). Since every function $f \in \mathcal{K}_u$ can be written as $f = Vg$ for some $g \in L^2(\sigma)$ (Theorem 11.6), we see that μ is a Carleson measure for $\mathcal{K}_u$ if and only if

$$\|Vg\|_\mu = \|f\|_\mu \lesssim \|f\| = \|Vg\| = \|g\|_\sigma, \qquad g \in L^2(\sigma).$$

Let us assume that the radial limits of u exist μ-almost everywhere so that the measure $\nu_{u,\mu} := |1 - u|^2 d\mu$ is well-defined. Then we have

$$\|Vg\|_\mu^2 = \int_{\mathbb{D}^-} |1 - u|^2 |\mathscr{C}_\sigma g|^2 \, d\mu = \|\mathscr{C}_\sigma g\|_{\nu_{u,\mu}}^2,$$

which yields the following:

Theorem 11.23 *Let $\mu \in M_+(\mathbb{D}^-)$ and u be inner with radial boundary values that exist μ-almost everywhere. When $\nu_{u,\mu} := |1 - u|^2 d\mu$, the following are equivalent:*

(i) *The measure μ is a Carleson measure for $\mathcal{K}_u$;*
(ii) *The Cauchy transform $\mathcal{C}_\sigma$ is a bounded linear operator from $L^2(\sigma)$ into $L^2(\mathbb{D}^-, \nu_{u,\mu})$.*

Condition (ii) is characterized in [121, Theorem 1.7].

11.7 Isometric embeddings

In the previous section, we made some general comments about Carleson embedding theorems for model spaces. In this section, we focus on one particular case of an embedding that brings in our earlier discussion of Aleksandrov–Clark measures.

Define

$$H_1^\infty = \{b \in H^\infty : \|b\|_\infty \leqslant 1\}$$

to be the closed unit ball in H^∞. For any $b \in H_1^\infty$ and $\alpha \in \mathbb{T}$, there corresponds a unique $\sigma_b^\alpha \in M_+(\mathbb{T})$ for which

$$\frac{1 - |b(z)|^2}{|\alpha - b(z)|^2} = \int_\mathbb{T} P_z(\zeta)\, d\sigma_b^\alpha(\zeta). \tag{11.14}$$

Recall that these are the Aleksandrov–Clark measures $\{\sigma_b^\alpha : \alpha \in \mathbb{T}\}$ associated with b. By the uniqueness of the measure, ultimately stemming from Herglotz's Theorem (Theorem 1.18), observe that

$$\sigma_b^\alpha = \sigma_{\bar\alpha b}^1. \tag{11.15}$$

This formula will help make some of the discussion below more concise.

If $\sigma = \sigma_b^\alpha$ is any Clark measure associated with an inner function b, Clark's Theorem (Theorem 11.6) says that every $g \in \mathcal{K}_b$ is equal to $V_\alpha f$ for some unique $f \in L^2(\sigma)$ and $\|g\| = \|f\|_{L^2(\sigma)}$. Furthermore, Poltoratski's Theorem (Theorem 11.14) says that the radial limits of $g = Vf$ exist and are equal to f σ-almost everywhere. Combining these two theorems, and defining g to be equal to its radial limiting value on the carrier of σ (that is, $g = f$ on a carrier of σ), we see that

$$\|g\|^2 = \int_\mathbb{T} |g|^2 d\sigma.$$

In other words, $\mathcal{K}_u$ embeds isometrically into $L^2(\sigma)$. In fact, if v is inner, then $\mathcal{K}_u \subset \mathcal{K}_{uv}$ (Corollary 5.9). Thus if $\sigma = \sigma_{uv}^\alpha$ is one of the Clark measures for uv, we have the isometric embedding of $\mathcal{K}_u$ into $L^2(\sigma)$.

The full classification of such measures is due to Aleksandrov [10].

Theorem 11.24 (Aleksandrov) *For an inner function u and $\mu \in M_+(\mathbb{T})$, the following are equivalent:*

(i) $\mathcal{K}_u$ embeds isometrically into $L^2(\mu)$;
(ii) There exists a $\varphi \in H_1^\infty$ such that $\mu = \mu_{\varphi u}^1$.

The proof of this theorem (from [31]) requires us to revisit the endnote discussion of de Branges–Rovnyak spaces $\mathcal{H}(b)$ from Chapter 5. Indeed, for $b \in H_1^\infty$, recall that $\mathcal{H}(b)$ is the Hilbert space of analytic functions on $\mathbb{D}$ whose reproducing kernel is

$$k_\lambda^b(z) = \frac{1 - \overline{b(\lambda)}b(z)}{1 - \overline{\lambda}z}.$$

Clark [46] and Ball [22, 23] (see also [165]), proved that for each fixed $\alpha \in \mathbb{T}$, the operator

$$\omega_b^\alpha : L^2(\sigma_b^\alpha) \to \mathcal{H}(b), \quad \omega_b^\alpha h = (1 - \overline{\alpha}b)\,\mathscr{C}_{\sigma_b^\alpha}h, \tag{11.16}$$

is a surjective partial isometry whose kernel is $(H^2(\sigma_b^\alpha))^\perp$, where $H^2(\sigma_b^\alpha)$ is the closure of the analytic polynomials in the $L^2(\sigma_b^\alpha)$-norm.

A generalization of Poltoratski's Theorem (Theorem 11.14) shows that for each $h \in L^2(\sigma_b^\alpha)$,

$$\lim_{r \to 1^-} (\omega_b^\alpha h)(r\xi) = h(\xi), \quad (\sigma_b^\alpha)_s\text{-a.e.},$$

where $(\sigma_b^\alpha)_s$ is the singular part of σ_b^α with respect to Lebesgue measure m. In particular, if b is an inner function, then σ_b^α is singular and so $H^2(\sigma_b^\alpha) = L^2(\sigma_b^\alpha)$ (this follows using annihilators and Theorem 1.21). Putting this all together, we recover the isometric embedding discussed earlier.

We will also make use of the following formula. For $a \in \mathbb{D}$, let

$$\tau_a(z) = \frac{a - z}{1 - \overline{a}z}$$

be one of the disk automorphisms from (2.1). Straightforward computations [8] show that

$$\sigma_{\tau_a \circ b}^{\tau_a(\alpha)} = \frac{1}{|\tau_a'(\alpha)|}\sigma_b^\alpha. \tag{11.17}$$

Proof of Theorem 11.24   First assume that $\mu = \sigma^1_{ub}$ for some $b \in H^\infty_1$. By a slight generalization of Corollary 2.24 from [97, p. 6] there exists a sequence of Blaschke products $\{B_n\}_{n \geqslant 1}$ that converges pointwise to b. In particular, by (11.14), we get that for every $z \in \mathbb{D}$,

$$\int_{\mathbb{T}} P_z(\zeta)\, d\sigma^1_{uB_n}(\zeta) \to \int_{\mathbb{T}} P_z(\zeta)\, d\mu(\zeta), \qquad n \to \infty.$$

Note that $\sigma^1_{uB_n}(\mathbb{T})$ is bounded. Indeed, apply (11.14) to $b = uB_n$ and $z = 0$ and use the fact that $(uB_n)(0) \to u(0)b(0)$. Since the closed linear span of $\{P_z : z \in \mathbb{D}\}$ is dense in $C(\mathbb{T})$ (Proposition 1.17), we get that $\sigma^1_{uB_n} \to \mu$ in the weak−∗ topology. In other words, for any $f \in C(\mathbb{T})$, we have

$$\int_{\mathbb{T}} f\, d\sigma^1_{uB_n} \to \int_{\mathbb{T}} f\, d\mu, \qquad n \to \infty. \tag{11.18}$$

Clark's Theorem, along with our discussion at the beginning of this section, says that the embedding of $\mathcal{K}_{uB_n}$ into $L^2(\sigma^1_{uB_n})$ is isometric. Since $\mathcal{K}_u \subset \mathcal{K}_{uB_n}$, as discussed earlier, we have the isometric embedding of $\mathcal{K}_u$ into $L^2(\sigma^1_{uB_n})$. Thus, according to (11.18), for any $f \in \mathcal{K}_u \cap C(\mathbb{D}^-)$, we have

$$\|f\|^2 = \int_{\mathbb{T}} |f|^2\, d\sigma^1_{uB_n} \to \int_{\mathbb{T}} |f|^2\, d\mu.$$

By the Aleksandrov embedding theorem discussed earlier, each function $f \in \mathcal{K}_u$ has a finite radial limit at μ-almost every point of $\mathbb{T}$ so that $\|f\| = \|f\|_{L^2(\mu)}$ for every $f \in \mathcal{K}_u$.

Conversely, suppose that $\mu \in M_+(\mathbb{T})$ and that the embedding of $\mathcal{K}_u$ into $L^2(\mu)$ is isometric. The same proof which showed that any singular measure is the Clark measure for some inner function at $\alpha = 1$ will show that there exists a $b_0 \in H^\infty_1$ such that

$$\mu = \sigma^1_{b_0}. \tag{11.19}$$

We will now show that u divides b_0 in the sense that $b_0/u = \varphi \in H^\infty_1$.

We first do this when $u = B$ is a Blaschke product with simple zeros $\Lambda = \{\lambda_n\}_{n \geqslant 1}$. By our earlier discussion just before the proof of this theorem, we have the surjective partial isometry

$$\omega_{b_0} : L^2(\mu) \to \mathcal{H}(b_0), \qquad \omega_{b_0} h = (1 - b_0) C_\mu h.$$

Note that for any $n \geqslant 1$,

$$k^B_{\lambda_n}(z) = c_{\lambda_n}(z) = \frac{1}{1 - \overline{\lambda_n} z} \in \mathcal{K}_B \cap C(\mathbb{D}^-),$$

which yields, since $c_{\lambda_n} \in H^2(\mu) = (\ker \omega_{b_0})^{\perp}$,

$$\langle \omega_{b_0} c_{\lambda_n}, \omega_{b_0} c_{\lambda_\ell} \rangle_{\mathcal{H}(b_0)} = \langle c_{\lambda_n}, c_{\lambda_\ell} \rangle_{L^2(\mu)} = \langle c_{\lambda_n}, c_{\lambda_\ell} \rangle, \quad n, \ell \geq 1. \tag{11.20}$$

A standard computation (see [165, III.6]) shows that

$$\omega_{b_0} c_{\lambda_n} = (1 - \overline{b_0(\lambda_n)})^{-1} k_{\lambda_n}^{b_0}, \quad n \geq 1.$$

Apply formula (11.20) to the above identity and use the reproducing property to get

$$\begin{aligned}
\langle c_{\lambda_n}, c_{\lambda_\ell} \rangle &= (1 - \overline{b_0(\lambda_n)})^{-1} (1 - b_0(\lambda_\ell))^{-1} \langle k_{\lambda_n}^{b_0}, k_{\lambda_\ell}^{b_0} \rangle_{\mathcal{H}(b_0)} \\
&= (1 - \overline{b_0(\lambda_n)})^{-1} (1 - b_0(\lambda_\ell))^{-1} k_{\lambda_n}^{b_0}(\lambda_\ell) \\
&= (1 - \overline{b_0(\lambda_n)})^{-1} (1 - b_0(\lambda_\ell))^{-1} (1 - \overline{b_0(\lambda_n)} b_0(\lambda_\ell)) c_{\lambda_n}(\lambda_\ell).
\end{aligned}$$

Since

$$\langle c_{\lambda_n}, c_{\lambda_\ell} \rangle = c_{\lambda_n}(\lambda_\ell) = (1 - \overline{\lambda_n} \lambda_\ell)^{-1} \neq 0,$$

we obtain

$$(1 - \overline{b_0(\lambda_n)})(1 - b_0(\lambda_\ell)) = 1 - \overline{b_0(\lambda_n)} b_0(\lambda_\ell),$$

which, after a little algebra, gives us

$$\overline{b_0(\lambda_n)} - 2\overline{b_0(\lambda_n)} b_0(\lambda_\ell) + b_0(\lambda_\ell) = 0.$$

The preceding identity can be rearranged as

$$\overline{b_0(\lambda_n)}(1 - b_0(\lambda_\ell)) = -b_0(\lambda_\ell)(1 - \overline{b_0(\lambda_n)}), \quad n, \ell \geq 1.$$

Setting $f := b_0/(1 - b_0)$, the last equality implies that

$$f(\lambda_n) = -\overline{f(\lambda_1)} = c, \tag{11.21}$$

and so

$$b_0(\lambda_n) = \frac{c}{1 + c} =: \delta, \quad n \geq 1. \tag{11.22}$$

Note that $\tau_\delta \circ b_0$ vanishes on Λ and so B, since it has *simple* zeros, divides $\tau_\delta \circ b_0$. This implies the existence of a $\vartheta \in H_1^\infty$ such that $B\vartheta = \tau_\delta \circ b_0$. Since $\tau_\delta \circ \tau_\delta$ is the identity we get $b_0 = \tau_\delta \circ (B\vartheta)$, which shows that

$$\mu = \sigma^1_{\tau_\delta \circ (B\vartheta)}.$$

To finish this off, we use (11.17) and (11.15) to get

$$\mu = \frac{1}{|\tau_\delta'(\tau_\delta(1))|} \sigma_{B\vartheta}^{\tau_\delta(1)} = \frac{1 - |\delta|^2}{|1 - \delta|^2} \sigma_{B\vartheta}^{\tau_\delta(1)} = \frac{1 - |\delta|^2}{|1 - \delta|^2} \sigma_{Bb}^1,$$

where $b \in H_1^\infty$ is defined by $b = \overline{\tau_\delta(1)}\vartheta$ (note that $|\tau_\delta(1)| = 1$). Finally, using (11.21), we easily see that c is imaginary. Thus from the definition of δ in (11.22) we have

$$\frac{1 - |\delta|^2}{|1 - \delta|^2} = 1,$$

which concludes the proof in the case when $u = B$ is a Blaschke product with simple zeros.

For a general inner function u, use Remark 2.23 to produce a sequence $\{\lambda_n\}_{n \geqslant 1}$ in $\mathbb{D}$ with $\lambda_n \to 0$ and such that

$$B_n := \tau_{\lambda_n} \circ (-u)$$

is a Blaschke product with simple zeros. Corollary 2.24 shows that

$$\|B_n - u\|_\infty \leqslant \frac{2|\lambda_n|}{1 - |\lambda_n|} \to 0, \qquad n \to \infty. \tag{11.23}$$

Fix $n \geqslant 1$. Again, since $\tau_{\lambda_n} \circ \tau_{\lambda_n}$ is the identity, we see that the Crofoot transform

$$J_{\lambda_n} f = \frac{\sqrt{1 - |\lambda_n|^2}}{1 - \overline{\lambda_n} B_n} f$$

from (6.9) is a unitary operator from $\mathcal{K}_{B_n}$ onto $\mathcal{K}_u$. If we define $\mu_n \in M_+(\mathbb{T})$ by

$$d\mu_n := \frac{1 - |\lambda_n|^2}{|1 - \overline{\lambda_n} B_n|^2} d\mu, \tag{11.24}$$

then $\mathcal{K}_{B_n}$ embeds isometrically into $L^2(\mu_n)$. Indeed, for $g \in \mathcal{K}_{B_n}$, note that $J_{\lambda_n} g \in \mathcal{K}_u$ and $\mathcal{K}_u$ embeds into $L^2(\mu)$ isometrically. Thus,

$$\int_{\mathbb{T}} |g|^2 dm = \int_{\mathbb{T}} |J_{\lambda_n} g|^2 dm = \int_{\mathbb{T}} |J_{\lambda_n} g|^2 d\mu = \int_{\mathbb{T}} |g|^2 d\mu_n.$$

By the first case of the proof (remember that B_n has simple zeros), there is a $b_n \in H_1^\infty$ such that $\mu_n = \sigma_{B_n b_n}^1$, that is to say,

$$\mu = \frac{|1 - \overline{\lambda_n} B_n|^2}{1 - |\lambda_n|^2} \sigma_{B_n b_n}^1, \qquad n \geqslant 1. \tag{11.25}$$

By the Banach–Alaoglu Theorem, there is a subsequence $\{b_{n_\ell}\}_{\ell \geqslant 1}$ and a $b \in H_1^\infty$ such that b_{n_ℓ} converges to b in the weak-$*$ topology of H^∞, which means that

$$\int_{\mathbb{T}} f b_{n_\ell} dm \to \int_{\mathbb{T}} f b\, dm, \qquad \ell \to \infty,$$

for any $f \in L^1(\mathbb{T})$. In particular (letting f be Cauchy kernels), for any $z \in \mathbb{D}$, $b_{n_\ell}(z) \to b(z)$ as $\ell \to \infty$. We now argue that $\mu = \sigma_{ub}^1$. Indeed, according to (11.25), we have

$$\int_{\mathbb{T}} P_z(\zeta) \, d\mu(\zeta) = \int_{\mathbb{T}} P_z(\zeta) \left(\frac{|1 - \overline{\lambda_{n_\ell}} B_{n_\ell}(\zeta)|^2}{1 - |\lambda_{n_\ell}|^2} - 1 \right) d\sigma_{B_{n_\ell} b_{n_\ell}}^1(\zeta)$$
$$+ \int_{\mathbb{T}} P_z(\zeta) \, d\sigma_{B_{n_\ell} b_{n_\ell}}^1(\zeta).$$

Observe that

$$\frac{|1 - \overline{\lambda_n} B_n(\zeta)|^2}{1 - |\lambda_n|^2} - 1 \to 0, \quad n \to \infty,$$

uniformly on $\mathbb{T}$. According to (11.23), B_{n_ℓ} tends to u uniformly on $\mathbb{D}^-$ and $b_{n_\ell} \to b$ pointwise on $\mathbb{D}$ as $\ell \to \infty$. Hence,

$$\int_{\mathbb{T}} P_z(\zeta) \, d\sigma_{B_{n_\ell} b_{n_\ell}}^1(\zeta) = \frac{1 - |B_{n_\ell}(z) b_{n_\ell}(z)|^2}{|1 - B_{n_\ell}(z) b_{n_\ell}(z)|^2}$$

tends to

$$\frac{1 - |u(z) b(z)|^2}{|1 - u(z) b(z)|^2} = \int_{\mathbb{T}} P_z(\zeta) \, d\sigma_{ub}^1(\zeta).$$

Thus for any $z \in \mathbb{D}$ we see that

$$\int_{\mathbb{T}} P_z(\zeta) \, d\mu(\zeta) = \int_{\mathbb{T}} P_z(\zeta) \, d\sigma_{ub}^1(\zeta).$$

But the closed linear span of $\{P_z : z \in \mathbb{D}\}$ is dense in $C(\mathbb{T})$ (Proposition 1.17), which yields that

$$\mu = \sigma_{ub}^1,$$

and thus concludes the proof. $\qquad\qquad\square$

11.8 Notes

11.8.1 References

The study of Aleksandrov–Clark measures dates back to the seminal work of Clark [46] and it continued with deep work of Aleksandrov [7, 8, 12] and Poltoratski [148]. Since then, Aleksandrov–Clark measures have appeared in the study of spectral theory, composition operators, completeness problems, and mathematical physics. Several sources for this important topic, including references to their applications and the connections mentioned above, are [42, 146, 160].

11.8.2 A technical detail

In order to make the formulas in this chapter easier, we made the assumption that u is an inner function with $u(0) = 0$. What happens when $u(0) \neq 0$? For $w \in \mathbb{T}$ define

$$U_w = S_u + \frac{w}{1 - |u(0)|^2} k_0 \otimes C k_0.$$

Clark [46] proved that U_w is unitary and that every rank-one unitary perturbation of S_u is of this form for some $w \in \mathbb{T}$. A point $\zeta \in \mathbb{T}$ is an eigenvalue of U_w if and only if

$$|u'(\zeta)| < \infty \quad \text{and} \quad \lim_{r \to 1^-} u(r\zeta) = \alpha_w := \frac{u(0) + w}{1 + \overline{u(0)}w}.$$

The corresponding eigenvector is

$$k_\zeta(z) = \frac{1 - \overline{\alpha_w} u}{1 - \overline{\zeta} z}.$$

The measure $\sigma_{\alpha_w} \in M_+(\mathbb{T})$ is defined so that

$$\frac{1 - |u(z)|^2}{|\alpha_w - u(z)|^2} = \int_{\mathbb{T}} P_z(\xi)\, d\sigma_{\alpha_w}(\xi), \quad z \in \mathbb{D}$$

Then $\sigma_{\alpha_w} \perp m$ and U_w is unitarily equivalent to the multiplication operator $g \mapsto \xi g$ on $L^2(\sigma_{\alpha_w})$.

11.8.3 The general disintegration theorem

We point out a further generalization in [133] of the disintegration theorem which works for vector-valued model spaces where the Clark measures are matrix-valued measures.

11.8.4 Measure preserving

In the proof of Proposition 6.5, we mentioned a classical result of Riesz which says that if u is inner and $E \subset \mathbb{T}$ with $m(E) = 0$, then $m(u^{-1}(E)) = 0$. One can use the Aleksandrov disintegration theorem to prove a generalization of this, namely, if u is inner and $u(0) = 0$, then $m(E) = m(u^{-1}(E))$ for any Lebesgue measurable set $E \subset \mathbb{T}$ (see [42, Rem. 9.4.6]).

11.8.5 Clark theory again

As to be expected, there is a Clark theory for $\mathscr{K}_\Theta$, the model space of the upper half plane $\mathbb{C}_+$. Indeed, for an inner function Θ on $\mathbb{C}_+$, the function

$$M_\Theta = i\frac{1 + \Theta}{1 - \Theta}$$

is analytic on $\mathbb{C}_+$ and satisfies $\operatorname{Im} M_\Theta \geqslant 0$ (that is, M_Θ is a *Herglotz function*). In this setting, the Herglotz Representation Theorem guarantees the existence of parameters $b \geqslant 0, c \in \mathbb{R}$, and a positive measure μ_Θ on the real line such that

$$M_\Theta(\lambda) = b\lambda + c + \frac{1}{\pi}\int_\mathbb{R}\left(\frac{1}{x - \lambda} - \frac{x}{1 + x^2}\right)d\mu_\Theta(x).$$

Using a version of Poltoratski's result (Theorem 11.14) we can define the operator

$$Q_\Theta : \mathscr{K}_\Theta \to L^2(\mu_\Theta), \quad Q_\Theta f = f|_{C_\Theta}, \tag{11.26}$$

where C_Θ is a carrier for μ_Θ. This operator turns out to be unitary.

This current discussion probably reminds the reader of our earlier treatment of the family of Clark measures. Indeed, if one begins with an inner Θ and looks at the family of inner functions

$$\{\overline{\alpha}\Theta : \alpha \in \mathbb{T}\},$$

there is an associated family of Poisson finite measures

$$\{\mu_{\overline{\alpha}\Theta} : \alpha \in \mathbb{T}\},$$

the Clark measures associated with Θ. Many of the properties of Clark measures that hold for inner function u on $\mathbb{D}$ have direct analogues for inner functions Θ on $\mathbb{C}_+$.

11.9 For further exploration

11.9.1 Eigenvalues of the Clark unitary operator

For a finite Blaschke product u, the eigenvalues of the Clark operator U_α form a finite subset of $\mathbb{T}$. Show that the eigenvalues of two Clark unitaries U_α and U_β interlace as one moves around $\mathbb{T}$. Show that $\{\sigma(U_\alpha) : \alpha \in \mathbb{T}\}$ covers $\mathbb{T}$ exactly once. There are related results of Krein in [100] (see also [131, 132]).

11.9.2 Clark families

For two positive, finite, singular measures μ, ν on $\mathbb{T}$, is there a useful criterion to determine whether or not μ and ν belong to the same Clark family? In other words, when is there an inner function u and $\alpha, \beta \in \mathbb{T}$ such that

$$\frac{1 - |u(z)|^2}{|\alpha - u(z)|^2} = \int_{\mathbb{T}} P_z(\zeta)d\mu(\zeta) \quad \text{and} \quad \frac{1 - |u(z)|^2}{|\beta - u(z)|^2} = \int_{\mathbb{T}} P_z(\zeta)d\nu(\zeta)?$$

11.9.3 The compressed shift again

In this chapter we discussed the Clark unitary operator U_α for inner u with $u(0) = 0$. We showed that U_α was unitarily equivalent to $Z_\alpha f = \zeta f$ on $L^2(\sigma_\alpha)$. What is the relationship between S_u and Z_α? Work through the following details to find out.

Define

$$\mathcal{D}_\alpha = \left\{ f \in L^2(\sigma) : \int_{\mathbb{T}} \zeta f(\zeta)d\sigma_\alpha(\zeta) = 0 \right\}$$

and show that $\mathcal{D}_\alpha^\perp = \mathbb{C}\overline{\zeta}$. Now define the operator

$$Z_\alpha' : L^2(\sigma_\alpha) \to L^2(\sigma_\alpha), \quad Z_\alpha' f = \begin{cases} Z_\alpha f & \text{if } f \in \mathcal{D}_\alpha, \\ 0 & \text{if } f \in \mathcal{D}_\alpha^\perp, \end{cases}$$

and show that $Z_\alpha = Z_\alpha' + (1 \otimes \overline{\zeta})$. Prove the adjoint formula

$$(Z_\alpha')^* f = \overline{\zeta}\left(f(\zeta) - \int_{\mathbb{T}} f d\sigma_\alpha \right).$$

Poltoratski's Theorem (Theorem 11.14) shows that for any $f \in L^2(\sigma_\alpha)$ we have

$$\lim_{r \to 1^-} (V_\alpha f)(r\zeta) = f(\zeta)$$

for σ_α-almost every $\zeta \in \mathbb{T}$. This means that if we abuse notation a bit and equate $g \in \mathcal{K}_u$ with its σ_α-almost every defined boundary values on $\mathbb{T}$ (via V_α) we get

$$V_\alpha^* g = g, \quad g \in \mathcal{K}_u.$$

Use this line of reasoning to show that

$$g(0) = \int_{\mathbb{T}} g \, d\sigma_\alpha, \quad g \in \mathcal{K}_u.$$

Finally, use the fact that $S_u^* = S^*|_{\mathcal{K}_u}$ and the calculations above to prove the formula $V_\alpha^* S_u^* = (Z_\alpha')^* V_\alpha^*$, that is, $S_u \cong Z_\alpha'$. A further discussion of operators related to Z_α' can be found in [133].

12

Riesz bases

In previous chapters, we discussed spanning sets and bases for $\mathcal{K}_u$, often consisting of reproducing kernels. In this chapter, we examine circumstances under which a sequence of (normalized) reproducing kernels form a Riesz basis. These are natural relaxations of orthonormal bases and appear in many settings. Consequently, our discussion will be very general.

12.1 Minimal sequences

Let $\{\mathbf{x}_n\}_{n \geqslant 1}$ be a sequence of vectors in a separable complex Hilbert space $\mathcal{H}$.

Definition 12.1   The sequence $\{\mathbf{x}_n\}_{\geqslant 1}$ is said to be *complete* in $\mathcal{H}$ if

$$\bigvee \{\mathbf{x}_n : n \geqslant 1\} = \mathcal{H},$$

and *minimal* if, for any $n \geqslant 1$,

$$\mathbf{x}_n \notin \bigvee \{\mathbf{x}_k : k \neq n\}.$$

A sequence $\{\widehat{\mathbf{x}}_n\}_{n \geqslant 1} \subset \mathcal{H}$ is called a *biorthogonal sequence* associated with $\{\mathbf{x}_n\}_{n \geqslant 1}$ if

$$\langle \mathbf{x}_m, \widehat{\mathbf{x}}_n \rangle = \delta_{mn}, \qquad m, n \geqslant 1.$$

For example, if $\{\mathbf{x}_n\}_{n \geqslant 1}$ is a sequence of orthonormal vectors, then $\widehat{\mathbf{x}}_n = \mathbf{x}_n$ is a biorthogonal sequence sequence associated with $\{\mathbf{x}_n\}_{n \geqslant 1}$. By Theorem 1.27, we know that if $\mathcal{M}$ is a subspace of $\mathcal{H}$ and $\mathbf{x}_0 \in \mathcal{H}$, then $\mathbf{x}_0 \notin \mathcal{M}$ if and only if there exists a $\mathbf{y}_0 \in \mathcal{H}$ such that $\langle \mathbf{x}_0, \mathbf{y}_0 \rangle = 1$ and $\langle \mathbf{x}, \mathbf{y}_0 \rangle = 0$ for all $\mathbf{x} \in \mathcal{M}$. This theorem, when applied to minimal sequences, yields the following.

Proposition 12.2 *Let* $\{\mathbf{x}_n\}_{n \geqslant 1}$ *be a sequence of vectors in a Hilbert space* $\mathcal{H}$. *Then* $\{\mathbf{x}_n\}_{n \geqslant 1}$ *is minimal if and only if it has a biorthogonal sequence. Moreover, the biorthogonal sequence is unique if and only if* $\{\mathbf{x}_n\}_{n \geqslant 1}$ *is complete in* $\mathcal{H}$.

We now apply the concepts above to Cauchy kernels and Blaschke sequences (see Chapters 2 and 3).

Theorem 12.3 *Let* $\{\lambda_n\}_{n \geqslant 1}$ *be a sequence of distinct points in* $\mathbb{D}$.

(i) *If* $\{\lambda_n\}_{n \geqslant 1}$ *is not a Blaschke sequence, then* $\{c_{\lambda_n}\}_{n \geqslant 1}$ *is complete in* H^2;

(ii) *If* $\{\lambda_n\}_{n \geqslant 1}$ *is a Blaschke sequence, then* $\{c_{\lambda_n}\}_{n \geqslant 1}$ *is complete in* $\mathcal{K}_B$, *where* B *is the Blaschke product with zeros* $\{\lambda_n\}_{n \geqslant 1}$;

(iii) *The sequence* $\{c_{\lambda_n}\}_{n \geqslant 1}$ *is minimal if and only if* $\{\lambda_n\}_{n \geqslant 1}$ *is a Blaschke sequence.*

Proof (i) is a restatement of Corollary 3.18 while (ii) is a restatement of Proposition 5.22.

To prove one direction of (iii), assume that $\{\lambda_n\}_{n \geqslant 1}$ is not a Blaschke sequence. Then for any $m \geqslant 1$, $\{\lambda_n\}_{n \geqslant 1, n \neq m}$ is still not a Blaschke sequence. Hence by (i) we have

$$c_{\lambda_m} \in H^2 = \bigvee \{c_{\lambda_n} : n \neq m\}. \tag{12.1}$$

In other words, the sequence $\{c_{\lambda_n}\}_{n \geqslant 1}$ is not minimal in H^2.

Conversely if $\{\lambda_n\}_{n \geqslant 1}$ is a Blaschke sequence, then $c_{\lambda_n} = k_{\lambda_n}$ and we have already seen that $\{k_{\lambda_n}\}_{n \geqslant 1}$ is complete in $\mathcal{K}_B$. To prove minimality, let $m \geqslant 1$ and let

$$B_m = \frac{B}{b_{\lambda_m}} = \prod_{\substack{n \geqslant 1 \\ n \neq m}} b_{\lambda_n} \tag{12.2}$$

be the Blaschke product formed by the sequence $\{\lambda_n\}_{n \geqslant 1, n \neq m}$. We now exploit these Blaschke products to obtain the corresponding biorthogonal sequence for $\{k_{\lambda_n}\}_{n \geqslant 1}$ in H^2. Proposition 12.2 will then show that the sequence is minimal. Indeed, since $\{\lambda_n\}_{n \geqslant 1}$ consists of distinct points, we know that $B_m(\lambda_m) \neq 0$. Moreover, for each $m, n \geqslant 1$, we have

$$\langle B_m, k_{\lambda_n} \rangle = B_m(\lambda_n) = B_m(\lambda_m) \delta_{mn}. \tag{12.3}$$

Hence $\{B_n/B_n(\lambda_n)\}_{n \geqslant 1}$ is a biorthogonal sequence of $\{k_{\lambda_n}\}_{n \geqslant 1}$ in H^2 and thus $\{k_{\lambda_n}\}_{n \geqslant 1}$ is minimal. $\square$

Since $\{k_{\lambda_n}\}_{n \geqslant 1}$, where $\{\lambda_n\}_{n \geqslant 1}$ is a Blaschke sequence, is complete and minimal in $\mathcal{K}_B$, Proposition 12.2 ensures that $\{k_{\lambda_n}\}_{n \geqslant 1}$ has a unique biorthogonal sequence $\{\widehat{k_{\lambda_n}}\}_{n \geqslant 1}$ in $\mathcal{K}_B$. We now obtain an explicit formula for $\widehat{k_{\lambda_n}}$. Note that the biorthogonal sequence presented via (12.3) belongs to H^2. Hence we need to find its projection onto $\mathcal{K}_B$.

Theorem 12.4 *Let $\{\lambda_n\}_{n\geqslant 1}$ be a Blaschke sequence of distinct points in $\mathbb{D}$ and let B denote the corresponding Blaschke product. Then the unique biorthogonal sequence $\{\widehat{k}_{\lambda_n}\}_{n\geqslant 1}$ associated with $\{k_{\lambda_n}\}_{n\geqslant 1}$ in $\mathcal{K}_B$ is*

$$\widehat{k}_{\lambda_n} = \frac{(1 - |\lambda_n|^2)^{1/2}}{B_n(\lambda_n)} \frac{\lambda_n}{|\lambda_n|} Ck_{\lambda_n} = \frac{(1 - |\lambda_n|^2)}{B_n(\lambda_n)} B_n k_{\lambda_n}, \qquad n \geqslant 1, \qquad (12.4)$$

where B_n is given by (12.2) and C is the conjugation operator on $\mathcal{K}_B$ from (8.6).

Proof   Note that

$$Ck_{\lambda_m} = \frac{B}{z - \lambda_m}$$

(Example 8.13) and so

$$\langle Ck_{\lambda_m}, k_{\lambda_n} \rangle = \frac{B(\lambda_n)}{\lambda_n - \lambda_m} = 0, \qquad n \neq m,$$

and

$$\langle Ck_{\lambda_m}, k_{\lambda_m} \rangle = B'(\lambda_m).$$

Thus

$$\widehat{k}_{\lambda_n} = \frac{1}{B'(\lambda_n)} Ck_{\lambda_n} \qquad (12.5)$$

is the biorthogonal sequence associated with $\{k_{\lambda_n}\}_{n\geqslant 1}$.
    With

$$B(z) = \prod_{n\geqslant 1} b_{\lambda_n}(z)$$

we observe by formal logarithmic differentiation that

$$\frac{B'(z)}{B(z)} = \sum_{n\geqslant 1} \frac{b'_{\lambda_n}(z)}{b_{\lambda_n}(z)}$$

and hence

$$B(z) = \sum_{n\geqslant 1} B_n(z) b'_{\lambda_n}(z) = -\sum_{n\geqslant 1} B_n(z) \frac{|\lambda_n|}{\lambda_n} \frac{1 - |\lambda_n|^2}{(1 - \overline{\lambda}_n z)^2}.$$

This implies that

$$B'(\lambda_n) = -B_n(\lambda_n) \frac{|\lambda_n|}{\lambda_n} \frac{1}{1 - |\lambda_n|^2}.$$

Furthermore,

$$Ck_{\lambda_n} = \frac{B}{z - \lambda_n} = -B_n \frac{|\lambda_n|}{\lambda_n} k_{\lambda_n}.$$

Combine the two previous identities to get

$$\widehat{k}_{\lambda_n} = \frac{Ck_{\lambda_n}}{B'(\lambda_n)} = \frac{1 - |\lambda_n|^2}{B_n(\lambda_n)} B_n k_{\lambda_n}. \qquad \square$$

In general, even when $\{\mathbf{x}_n\}_{n \geqslant 1}$ is complete, the biorthogonal sequence $\{\widehat{\mathbf{x}}_n\}_{n \geqslant 1}$ need not be complete in $\mathcal{H}$. The story is different for sequences of reproducing kernels in $\mathcal{K}_B$.

Corollary 12.5 *Let $\{\lambda_n\}_{n \geqslant 1}$ be a Blaschke sequence of distinct points in $\mathbb{D}$, let B be the corresponding Blaschke product, and let $\{\widehat{k}_{\lambda_n}\}_{n \geqslant 1}$ be the unique biorthogonal sequence associated with $\{k_{\lambda_n}\}_{n \geqslant 1}$ in $\mathcal{K}_B$. Then $\{\widehat{k}_{\lambda_n}\}_{n \geqslant 1}$ is complete in $\mathcal{K}_B$.*

Proof   From (12.5) we have

$$\widehat{k}_{\lambda_n} = \frac{1}{B'(\lambda_n)} Ck_{\lambda_n}.$$

The result now follows, since $\{k_{\lambda_n}\}_{n \geqslant 1}$ is complete in $\mathcal{K}_B$ and the conjugation C is a surjective conjugate-linear isometry. $\qquad \square$

12.2 Uniformly minimal sequences

A sequence $\{\mathbf{x}_n\}_{n \geqslant 1} \subset \mathcal{H}$ is said to be *uniformly minimal* if

$$\delta := \inf_{n \geqslant 1} \text{dist}\left(\frac{\mathbf{x}_n}{\|\mathbf{x}_n\|}, \bigvee \{\mathbf{x}_k : k \neq n\}\right) > 0.$$

The constant δ is called the *constant of uniform minimality* of $\{\mathbf{x}_n\}_{n \geqslant 1}$. A uniformly minimal sequence is trivially minimal (recall Definition 12.1).

Proposition 12.6 *Let $\{\mathbf{x}_n\}_{n \geqslant 1}$ be a minimal sequence in a Hilbert space $\mathcal{H}$. Then for each biorthogonal sequence $\{\widehat{\mathbf{x}}_n\}_{n \geqslant 1}$, we have*

$$\delta \geqslant \frac{1}{\sup_{n \geqslant 1} \|\mathbf{x}_n\| \|\widehat{\mathbf{x}}_n\|}. \qquad (12.6)$$

Moreover, there is a particular biorthogonal sequence $\{\widehat{\mathbf{x}}_n\}_{n \geqslant 1}$ for which

$$\delta = \frac{1}{\sup_{n \geqslant 1} \|\mathbf{x}_n\| \|\widehat{\mathbf{x}}_n\|}. \qquad (12.7)$$

Proof   For each $n \geqslant 1$, apply the Hahn–Banach Theorem [158, p. 104] to get

$$\frac{1}{\text{dist}(\mathbf{x}_n, \bigvee_{k \neq n} \mathbf{x}_k)} = \inf\{\|\mathbf{x}\| : \langle \mathbf{x}_n, \mathbf{x} \rangle = 1 \text{ and } \langle \mathbf{x}_k, \mathbf{x} \rangle = 0, k \neq n\} \qquad (12.8)$$

and the infimum is attained. Hence, for any biorthogonal sequence $\{\widehat{\mathbf{x}}_n\}_{n\geqslant 1}$, we can put $\mathbf{x} = \widehat{\mathbf{x}}_n$ in the identity above to obtain

$$\|\widehat{\mathbf{x}}_n\| \geqslant \frac{1}{\mathrm{dist}(\mathbf{x}_n, \bigvee_{k\neq n} \mathbf{x}_k)}, \quad n \geqslant 1.$$

The preceding inequality can be rewritten as

$$\mathrm{dist}\left(\frac{\mathbf{x}_n}{\|\mathbf{x}_n\|}, \bigvee\{\mathbf{x}_k : k \neq n\}\right) \geqslant \frac{1}{\|\mathbf{x}_n\| \|\widehat{\mathbf{x}}_n\|}, \quad n \geqslant 1.$$

If we take the infimum with respect to n, we deduce (12.6). On the other hand, since the infimum in (12.8) is attained, there is a biorthogonal sequence $\{\widehat{\mathbf{x}}_n\}_{n\geqslant 1}$ such that

$$\|\widehat{\mathbf{x}}_n\| = \frac{1}{\mathrm{dist}(\mathbf{x}_n, \bigvee_{k\neq n} \mathbf{x}_k)}, \quad n \geqslant 1.$$

Rewrite this as

$$\mathrm{dist}\left(\frac{\mathbf{x}_n}{\|\mathbf{x}_n\|}, \bigvee\{\mathbf{x}_k : k \neq n\}\right) = \frac{1}{\|\mathbf{x}_n\| \|\widehat{\mathbf{x}}_n\|}, \quad n \geqslant 1,$$

and then take the infimum with respect to n to obtain (12.7).   □

In practice, one usually applies the following weaker version of Proposition 12.6.

Corollary 12.7 *Let $\{\mathbf{x}_n\}_{n\geqslant 1}$ be a minimal sequence in $\mathcal{H}$. Then $\{\mathbf{x}_n\}_{n\geqslant 1}$ is uniformly minimal if and only if it has a biorthogonal sequence $\{\widehat{\mathbf{x}}_n\}_{n\geqslant 1}$ such that*

$$\sup_{n\geqslant 1} \|\mathbf{x}_n\| \|\widehat{\mathbf{x}}_n\| < \infty.$$

If the sequence $\{\mathbf{x}_n\}_{n\geqslant 1}$ is minimal and complete, Proposition 12.2 says it has a unique biorthogonal sequence. In this situation, Proposition 12.6 can be stated as follows.

Corollary 12.8 *Let $\{\mathbf{x}_n\}_{n\geqslant 1} \subset \mathcal{H}$ be minimal and complete and let $\{\widehat{\mathbf{x}}_n\}_{n\geqslant 1}$ denote its unique biorthogonal sequence. Then we have*

$$\delta = \frac{1}{\sup_{n\geqslant 1} \|\mathbf{x}_n\| \|\widehat{\mathbf{x}}_n\|}.$$

In Theorem 12.3, we saw that $\{k_{\lambda_n}\}_{n\geqslant 1}$ is minimal if and only if $\{\lambda_n\}_{n\geqslant 1}$ is a Blaschke sequence. In Theorem 12.4, for the latter case, we obtained the corresponding unique biorthogonal sequence in the associated model space $\mathcal{K}_B$. By applying Corollary 12.8, we now provide a characterization of the uniformly minimal sequences of kernels. We also compute the constant δ of uniform minimality.

Theorem 12.9 *Let* $\{\lambda_n\}_{n\geqslant 1}$ *be a Blaschke sequence of distinct points in* $\mathbb{D}$ *and* B *be the corresponding Blaschke product. Then*

$$\delta = \inf_{n\geqslant 1} |B_n(\lambda_n)|, \tag{12.9}$$

where $B_n = B/b_{\lambda_n}$. *Thus* $\{k_{\lambda_n}\}_{n\geqslant 1}$ *is uniformly minimal if and only if*

$$\inf_{n\geqslant 1} |B_n(\lambda_n)| > 0.$$

Proof   By Theorem 12.3, $\{k_{\lambda_n}\}_{n\geqslant 1}$ is minimal. Moreover, by (12.4) we have

$$\begin{aligned}
\|\widehat{k_{\lambda_n}}\| &= \frac{1-|\lambda_n|^2}{|B_n(\lambda_n)|}\left\|\frac{B_n}{1-\overline{\lambda}_n z}\right\| = \frac{1-|\lambda_n|^2}{|B_n(\lambda_n)|}\left\|\frac{1}{1-\overline{\lambda}_n z}\right\| \\
&= \frac{1-|\lambda_n|^2}{|B_n(\lambda_n)|}\|k_{\lambda_n}\| = \frac{\sqrt{1-|\lambda_n|^2}}{|B_n(\lambda_n)|} \\
&= \frac{1}{\|k_{\lambda_n}\|}\frac{1}{|B_n(\lambda_n)|}.
\end{aligned}$$

From here we see that

$$\|k_{\lambda_n}\|\,\|\widehat{k_{\lambda_n}}\| = \frac{1}{|B_n(\lambda_n)|}.$$

Now apply Corollary 12.8 to obtain the desired formula for δ.   □

12.3 Uniformly separated sequences

A sequence $\{\lambda_n\}_{n\geqslant 1}$ satisfying (12.9) is called a *uniformly separated sequence*. In this situation, the constant of uniform minimality δ is also called the *separation constant* for $\{\lambda_n\}_{n\geqslant 1}$. The reader could use some specific examples of uniformly separated sequences.

Definition 12.10   A sequence $\{\lambda_n\}_{n\geqslant 1} \subset \mathbb{D}$ is called *exponential* if there is a constant $0 < c < 1$ such that

$$1 - |\lambda_{n+1}| \leqslant c\,(1 - |\lambda_n|), \quad n \geqslant 1.$$

Proposition 12.11   *An exponential sequence is uniformly separated.*

Proof   By induction,

$$1 - |\lambda_n| \leqslant c^{n-k}(1-|\lambda_k|), \quad n > k \geqslant 1.$$

This inequality shows that

$$\sum_{n\geqslant 1}(1 - |\lambda_n|) \leqslant \sum_{n\geqslant 1} c^{n-1}(1 - |\lambda_1|) - \frac{1 - |\lambda_1|}{1 - c} < \infty$$

and so $\{\lambda_n\}_{n\geqslant 1}$ is a Blaschke sequence. Moreover,

$$|\lambda_n| - |\lambda_k| \geqslant (1 - c^{n-k})(1 - |\lambda_k|), \quad n > k \geqslant 1,$$

and similarly

$$1 - |\lambda_k\lambda_n| = 1 - |\lambda_n| + |\lambda_n|(1 - |\lambda_k|) \leqslant (1 + c^{n-k})(1 - |\lambda_k|), \quad n > k \geqslant 1.$$

Since

$$1 - \left|\frac{w - z}{1 - \overline{w}z}\right|^2 = \frac{(1 - |z|^2)(1 - |w|^2)}{|1 - \overline{w}z|^2}$$

and

$$1 - \left|\frac{|w| - |z|}{1 - |w||z|}\right|^2 = \frac{(1 - |z|^2)(1 - |w|^2)}{(1 - |w||z|)^2},$$

along with $0 \leqslant 1 - |w||z| \leqslant |1 - \overline{w}z|$, we deduce that

$$\left|\frac{w - z}{1 - \overline{w}z}\right| \geqslant \frac{\big||w| - |z|\big|}{1 - |wz|}.$$

Hence,

$$\left|\frac{\lambda_k - \lambda_n}{1 - \overline{\lambda}_k\lambda_n}\right| \geqslant \frac{\big||\lambda_k| - |\lambda_n|\big|}{1 - |\lambda_k\lambda_n|} \geqslant \frac{1 - c^{|n-k|}}{1 + c^{|n-k|}}$$

for all $n, k \geqslant 1$, $n \neq k$, so that

$$\prod_{\substack{k\geqslant 1 \\ k\neq n}} \left|\frac{\lambda_k - \lambda_n}{1 - \overline{\lambda}_k\lambda_n}\right| \geqslant \prod_{j\geqslant 1}\left(\frac{1 - c^j}{1 + c^j}\right)^2 > 0. \qquad \square$$

When the sequence $\{\lambda_n\}_{n\geqslant 1}$ lies on a ray, we have the following.

Proposition 12.12 *Suppose $\{\lambda_n\}_{n\geqslant 1} \subset \mathbb{D}$ lies on a ray $[0, e^{i\theta})$. Then $\{\lambda_n\}_{n\geqslant 1}$ is uniformly separated if and only if it is exponential.*

Proof Without loss of generality, we assume that the ray is the interval $[0, 1)$ and that

$$0 \leqslant r_1 < r_2 < \cdots < 1.$$

One direction follows from Proposition 12.11. For the other direction, suppose $\{r_n\}_{n \geqslant 1}$ is uniformly separated, that is,

$$\inf_{n \geqslant 1} \prod_{\substack{k \geqslant 1 \\ k \neq n}} \left| \frac{r_k - r_n}{1 - r_k r_n} \right| > 0.$$

Certainly there is a $\delta > 0$ such that

$$\frac{r_{n+1} - r_n}{1 - r_{n+1} r_n} \geqslant \delta, \quad n \geqslant 1.$$

This inequality is equivalent to

$$1 - r_{n+1} \leqslant \frac{(1 - \delta)(1 - r_n)}{1 + \delta r_n}, \quad n \geqslant 1.$$

Thus,

$$1 - r_{n+1} \leqslant c(1 - r_n), \quad n \geqslant 1$$

with $c = 1 - \delta$. $\qquad\square$

The next result is found in [97, p. 277–278] and connects uniformly separated sequences with Carleson measures (see Chapter 11 for more on Carleson measures).

Proposition 12.13 *Let $\{\lambda_n\}_{n \geqslant 1} \subset \mathbb{D}$ be a uniformly separated sequence with separation constant δ. Then*

$$\sum_{n \geqslant 1} (1 - |\lambda_n|^2) |f(\lambda_n)|^2 \leqslant M \|f\|^2, \qquad f \in H^2, \tag{12.10}$$

where M is a positive constant constant depending only on δ. Furthermore, if $\{\lambda_n\}_{n \geqslant 1} \subset \mathbb{D}$ such that (12.10) holds, then $\{\lambda_n\}_{n \geqslant 1}$ is a finite union of uniformly separated sequences.

Define the discrete measure μ on $\mathbb{D}$ by

$$\mu = \sum_{n \geqslant 1} (1 - |\lambda_n|^2) \delta_{\lambda_n}. \tag{12.11}$$

Writing (12.10) as

$$\int_{\mathbb{D}} |f|^2 d\mu \leqslant M \|f\|^2, \quad f \in H^2,$$

with μ as in (12.11), Proposition 12.13 says that μ is a *Carleson measure* for H^2.

12.4 The mappings Λ, V, and Γ

Let

$$\mathbb{C}^\infty = \{(a_1, a_2, \ldots) : a_j \in \mathbb{C}\}$$

be the set of all sequences of complex numbers. Fix a sequence $\{\mathbf{x}_n\}_{n \geqslant 1} \subset \mathcal{H}$ and define the mapping

$$\Lambda : \mathcal{H} \to \mathbb{C}^\infty, \qquad \Lambda \mathbf{x} = \{\langle \mathbf{x}, \mathbf{x}_n \rangle\}_{n \geqslant 1}. \tag{12.12}$$

In other words, Λ maps $\mathbf{x}$ to its generalized Fourier coefficients with respect to $\{\mathbf{x}_n\}_{n \geqslant 1}$. This linear mapping has several interesting properties. For example, one can see that $\{\mathbf{x}_n\}_{n \geqslant 1}$ is complete in $\mathcal{H}$ if and only if Λ is injective. By Proposition 12.2, we see that $\{\mathbf{x}_n\}_{n \geqslant 1}$ is minimal if and only if the range of Λ contains all of the sequences

$$\mathbf{e}_n = \{\delta_{kn}\}_{k \geqslant 1}, \qquad n \geqslant 1.$$

The range of Λ is contained in $\mathbb{C}^\infty$. We now explore what happens when the range of Λ is contained in the smaller space $\ell^2 := \ell^2(\mathbb{N})$.

Proposition 12.14 *Let $\{\mathbf{x}_n\}_{n \geqslant 1} \subset \mathcal{H}$ be such that* ran $\Lambda \subset \ell^2$. *Then Λ is a bounded linear operator from $\mathcal{H}$ into ℓ^2, and, for each $\mathbf{x} \in \mathcal{H}$,*

$$\langle \mathbf{x}, \mathbf{x}_n \rangle = \langle \Lambda \mathbf{x}, \mathbf{e}_n \rangle_{\ell^2}, \qquad n \geqslant 1. \tag{12.13}$$

Proof Let $\{\mathbf{y}_m\}_{m \geqslant 1} \subset \mathcal{H}$ be such that $\mathbf{y}_m \to \mathbf{y}$ for some $\mathbf{y} \in \mathcal{H}$ and $\Lambda \mathbf{y}_m \to \mathbf{a}$ for some $\mathbf{a} \in \ell^2$. We want to show that $\Lambda \mathbf{y} = \mathbf{a}$.

Indeed, write $\mathbf{a} = \{a_k\}_{k \geqslant 1}$ and fix $n \geqslant 1$. Since

$$|\langle \mathbf{y}_m, \mathbf{x}_n \rangle - a_n|^2 \leqslant \sum_{k \geqslant 1} |\langle \mathbf{y}_m, \mathbf{x}_k \rangle - a_k|^2 = \|\Lambda \mathbf{y}_m - \mathbf{a}\|_{\ell^2}^2,$$

we obtain

$$\lim_{m \to \infty} \langle \mathbf{y}_m, \mathbf{x}_n \rangle = a_n.$$

Because $\mathbf{y}_m \to \mathbf{y}$, we see that

$$\lim_{m \to \infty} \langle \mathbf{y}_m, \mathbf{x}_n \rangle = \langle \mathbf{y}, \mathbf{x}_n \rangle$$

and so $a_n = \langle \mathbf{y}, \mathbf{x}_n \rangle$. Since $\mathbf{a} = \Lambda \mathbf{y}$, the Closed Graph Theorem implies that Λ is a bounded operator from $\mathcal{H}$ into ℓ^2. The formula in (12.13) is obtained by taking the inner product of $\Lambda \mathbf{x}$ and $\mathbf{e}_n$. $\qquad \square$

Note that under the hypotheses of Proposition 12.14, the operator Λ is surjective if and only if $\{x_n\}_{n \geq 1}$ is minimal. Furthermore, Λ is an isometric isomorphism between $\mathcal{H}$ and ℓ^2 if and only if $\{x_n\}_{n \geq 1}$ is complete and minimal.

This next proposition helps us estimate the operator norm of Λ by controlling it on a dense set.

Proposition 12.15 *Let* $\{x_n\}_{n \geq 1} \subset \mathcal{H}$. *Suppose there exists a dense subset* $E \subset \mathcal{H}$ *and a constant* $M > 0$ *such that*

$$\|\Lambda x\|_{\ell^2} \leq M\|x\|, \qquad x \in E. \tag{12.14}$$

Then $\operatorname{ran} \Lambda \subset \ell^2$ *and the preceding estimate holds for all* $x \in \mathcal{H}$.

Proof Fix $y \in \mathcal{H}$. Since E is dense in $\mathcal{H}$, there is a sequence $\{y_m\}_{m \geq 1} \subset E$ such that $y_m \to y$. Since $\{y_m\}_{m \geq 1}$ is a Cauchy sequence in E, we see from (12.14) that $\{\Lambda y_m\}_{m \geq 1}$ is a Cauchy sequence in ℓ^2. Therefore, Λy_m converges to some $a \in \ell^2$.

The hypothesis of the proposition says that for each $y_m \in E$ we have $\Lambda y_m \in \ell^2$. Thus, as seen in the proof of Proposition 12.14, we conclude that

$$\langle y_m, x_n \rangle = \langle \Lambda y_m, e_n \rangle_{\ell^2}, \qquad m, n \geq 1.$$

Let $m \to \infty$ to deduce

$$\langle y, x_n \rangle = \langle a, e_n \rangle_{\ell^2}, \qquad n \geq 1.$$

This means that $\Lambda y = a$, and thus we have $\Lambda y_m \to \Lambda y \in \ell^2$. Finally, if we set $x = y_m$ in (12.14) and then let $m \to \infty$, we see that (12.14) holds for all $y \in \mathcal{H}$. $\qquad \square$

We denote the set of finitely (or compactly) supported sequences in $\mathbb{C}^\infty$ by $\mathbb{C}_c^\infty$. We introduce this class to temporarily avoid dealing with convergence issues in the definitions below. However, under some mild restrictions, all of the corresponding equations remain valid for infinite sequences.

Fix $\{x_n\}_{n \geq 1} \subset \mathcal{H}$ and define

$$V : \mathbb{C}_c^\infty \to \mathcal{H}, \qquad V a = \sum_{n \geq 1} a_n x_n,$$

where $a = \{a_n\}_{n \geq 1} \in \mathbb{C}_c^\infty$. The operator V depends on the sequence $\{x_n\}_{n \geq 1}$. We write $\widehat{V}$ for the mapping corresponding to the biorthogonal sequence $\{\widehat{x}_n\}_{n \geq 1}$ (if it exists). The same convention will apply for the operators U and Λ to be defined in a moment.

If $\mathbf{x}$ is a finite linear combination of elements from $\{\mathbf{x}_n\}_{n \geqslant 1}$ and $\mathbf{a} \in \mathbb{C}_c^\infty$, one can check that the following identities hold:

$$\mathbf{x} = \sum_{n \geqslant 1} \langle \mathbf{x}, \widehat{\mathbf{x}_n} \rangle \mathbf{x}_n, \tag{12.15}$$

$$\mathbf{a} = \sum_{n \geqslant 1} \langle \mathbf{a}, \mathbf{e}_n \rangle_{\ell^2} \mathbf{e}_n, \tag{12.16}$$

$$\Lambda \mathbf{x} = \sum_{n \geqslant 1} \langle \mathbf{x}, \mathbf{x}_n \rangle \mathbf{e}_n, \tag{12.17}$$

$$\widehat{\Lambda} \mathbf{x} = \sum_{n \geqslant 1} \langle \mathbf{x}, \widehat{\mathbf{x}_n} \rangle \mathbf{e}_n, \tag{12.18}$$

$$V \mathbf{a} = \sum_{n \geqslant 1} \langle \mathbf{a}, \mathbf{e}_n \rangle_{\ell^2} \mathbf{x}_n, \tag{12.19}$$

$$\widehat{V} \mathbf{a} = \sum_{n \geqslant 1} \langle \mathbf{a}, \mathbf{e}_n \rangle_{\ell^2} \widehat{\mathbf{x}_n}. \tag{12.20}$$

We will systematically use the identities above in the rest of the chapter. Note that all of the summations in the identities above are actually finite sums. Therefore, under the same assumptions, and by using (12.16), (12.18), and (12.19), we have

$$\widehat{\Lambda} V \mathbf{a} = \mathbf{a} \tag{12.21}$$

and, by (12.15), (12.18), and (12.19),

$$V \widehat{\Lambda} \mathbf{x} = \mathbf{x}. \tag{12.22}$$

Moreover, by (12.15) and (12.19),

$$\begin{aligned}
\langle V \mathbf{a}, \mathbf{y} \rangle &= \Big\langle \sum_{n \geqslant 1} \langle \mathbf{a}, \mathbf{e}_n \rangle_{\ell^2} \mathbf{x}_n, \mathbf{y} \Big\rangle \\
&= \sum_{n \geqslant 1} \langle \mathbf{a}, \mathbf{e}_n \rangle_{\ell^2} \langle \mathbf{x}_n, \mathbf{y} \rangle \\
&= \Big\langle \mathbf{a}, \sum_{n \geqslant 1} \langle \mathbf{y}, \mathbf{x}_n \rangle \mathbf{e}_n \Big\rangle_{\ell^2} \\
&= \langle \mathbf{a}, \Lambda \mathbf{y} \rangle_{\ell^2}. \tag{12.23}
\end{aligned}$$

The *Gram matrix* of the sequence $\{\mathbf{x}_n\}_{n \geqslant 1} \subset \mathcal{H}$ is the infinite matrix

$$\Gamma = [\langle \mathbf{x}_m, \mathbf{x}_n \rangle]_{m,n \geqslant 1}.$$

Note that Γ is a self-adjoint matrix. We let $\widehat{\Gamma}$ denote the Gram matrix for the corresponding biorthogonal sequence $\{\widehat{\mathbf{x}}_n\}_{n \geqslant 1}$ (if it exists).

One of our goals is to explore conditions under which Γ can be interpreted as a bounded operator on ℓ^2. For the time being, we note that for each $\mathbf{a} \in \mathbb{C}_c^\infty$,

$$\|V \mathbf{a}\|^2 = \Big\langle \sum_{m \geqslant 1} a_m \mathbf{x}_m, \sum_{n \geqslant 1} a_n \mathbf{x}_n \Big\rangle = \sum_{m \geqslant 1} \sum_{n \geqslant 1} a_m \overline{a_n} \langle \mathbf{x}_m, \mathbf{x}_n \rangle,$$

whence

$$\|V\mathbf{a}\|^2 = \langle \Gamma \mathbf{a}, \mathbf{a} \rangle_{\ell^2}, \qquad \mathbf{a} \in \mathbb{C}_c^\infty. \tag{12.24}$$

The identity (12.24), along with the identity

$$\|\Gamma\| = \{ |\langle \Gamma \mathbf{a}, \mathbf{a} \rangle_{\ell^2}| : \mathbf{a} \in \mathbb{C}_c^\infty, \|\mathbf{a}\|_{\ell^2} = 1 \}$$

(coming from the fact that Γ is self-adjoint), says that V extends to a bounded operator from ℓ^2 into $\mathcal{H}$ if and only if Γ is a bounded operator on ℓ^2. Similarly, the identity (12.23) says that V extends to a bounded operator from ℓ^2 into $\mathcal{H}$ if and only if Λ is bounded from $\mathcal{H}$ into ℓ^2. We summarize this observation as a proposition.

Proposition 12.16 *If any one of the following three operators*

$$\Lambda : \mathcal{H} \to \ell^2, \qquad V : \ell^2 \to \mathcal{H}, \qquad \Gamma : \ell^2 \to \ell^2,$$

is well-defined and bounded, then so are the other two.

Under the hypotheses of Proposition 12.16, we see from (12.23) and (12.24) that

$$V = \Lambda^* \qquad \text{and} \qquad \Gamma = V^* V. \tag{12.25}$$

Recall from Propositions 12.14 or 12.15 that if $\Lambda \mathcal{H} \subseteq \ell^2$, we may conclude that $\Lambda : \mathcal{H} \to \ell^2$ is a bounded operator.

12.5 Abstract Riesz sequences

Definition 12.17 A sequence $\{\mathbf{x}_n\}_{n \geqslant 1} \subset \mathcal{H}$ is said to be a *Riesz basis* if it is the image of an orthonormal basis under an isomorphism, that is to say, a bounded invertible operator.

As a consequence of the Riesz–Fisher theorem [158, p. 85], this is equivalent to saying that a sequence $\{\mathbf{x}_n\}_{n \geqslant 1}$ is a Riesz basis if and only if there is an isomorphism $U : \mathcal{H} \to \ell^2$ such that

$$U\mathbf{x}_n = \mathbf{e}_n, \qquad n \geqslant 1. \tag{12.26}$$

The operator U is called the *orthogonalizer* of $\{\mathbf{x}_n\}_{n \geqslant 1}$. Since $\{\mathbf{x}_n\}_{n \geqslant 1}$ is complete, U is the unique invertible operator satisfying (12.26). A sequence $\{\mathbf{x}_n\}_{n \geqslant 1}$ is called a *Riesz sequence* in $\mathcal{H}$ if it is a Riesz basis for its closed linear span.

By (12.26),

$$\|\mathbf{x}_n\| = \|U^{-1}\mathbf{e}_n\| \leqslant \|U^{-1}\|, \qquad n \geqslant 1, \tag{12.27}$$

and so a Riesz sequence must be bounded. Moreover, for each $\mathbf{x} \in \mathcal{H}$, we have

$$U\mathbf{x} = \{\langle U\mathbf{x}, \mathbf{e}_n \rangle_{\ell^2}\}_{n \geqslant 1}. \tag{12.28}$$

One has more flexibility with Riesz bases. For example, orthonormal sequences are not stable under small perturbations. However, if $\{\mathbf{y}_n\}_{n \geqslant 1}$ is an orthonormal basis for $\mathcal{H}$ (and thus a Riesz basis) and the sequence $\{\mathbf{x}_n\}_{n \geqslant 1}$ is such that

$$\sum_{n \geqslant 1} \|\mathbf{x}_n - \mathbf{y}_n\|^2 < 1,$$

then $\{\mathbf{x}_n\}_{n \geqslant 1}$ is a Riesz basis for $\mathcal{H}$ [152, Sec. 86]).

Proposition 12.18 *Let $\{\mathbf{x}_n\}_{n \geqslant 1}$ be a Riesz basis for $\mathcal{H}$. Then $\{\mathbf{x}_n\}_{n \geqslant 1}$ is uniformly minimal and the corresponding unique biorthogonal sequence $\{\widehat{\mathbf{x}}_n\}_{n \geqslant 1}$ is given by*

$$\widehat{\mathbf{x}}_n = U^* U \mathbf{x}_n, \qquad n \geqslant 1.$$

Moreover, the sequence $\{\widehat{\mathbf{x}}_n\}_{n \geqslant 1}$ is also a Riesz basis for $\mathcal{H}$ with orthogonalizer $\widehat{U} = U^{*-1}$.

Proof   By the definition of U we have

$$\langle U^* U \mathbf{x}_m, \mathbf{x}_n \rangle = \langle U\mathbf{x}_m, U\mathbf{x}_n \rangle = \langle \mathbf{e}_m, \mathbf{e}_n \rangle_{\ell^2} = \delta_{mn}, \qquad m, n \geqslant 1.$$

Hence by Proposition 12.2, $\{\mathbf{x}_n\}_{n \geqslant 1}$ is minimal and $\{U^* U \mathbf{x}_n\}_{n \geqslant 1}$ is its unique corresponding biorthogonal sequence. Use (12.27), the estimate

$$\|\widehat{\mathbf{x}}_n\| = \|U^* U \mathbf{x}_n\| = \|U^* \mathbf{e}_n\| \leqslant \|U^*\|, \qquad n \geqslant 1,$$

and Corollary 12.7 to conclude that $\{\mathbf{x}_n\}_{n \geqslant 1}$ is uniformly minimal. Finally, the bounded invertible operator $U^{*-1} : \mathcal{H} \to \ell^2$ satisfies

$$U^{*-1} \widehat{\mathbf{x}}_n = \mathbf{e}_n, \qquad n \geqslant 1.$$

Hence, by definition, the sequence $\{\widehat{\mathbf{x}}_n\}_{n \geqslant 1}$ is a Riesz basis for $\mathcal{H}$ with orthogonalizer U^{*-1}. $\square$

The next result establishes the connection between the operators Λ and U.

Proposition 12.19 *Let $\{\mathbf{x}_n\}_{n \geqslant 1}$ be a Riesz basis for $\mathcal{H}$ and let U be its orthogonalizer. Then we have*

$$\Lambda = U^{*-1} \qquad and \qquad \widehat{\Lambda} = U.$$

In particular, Λ and $\widehat{\Lambda}$ are both bounded invertible operators from $\mathcal{H}$ onto ℓ^2.

Proof By Proposition 12.18, $\widehat{\mathbf{x}}_n = U^*U\mathbf{x}_n$, $n \geqslant 1$. By (12.28), we have, for each $\mathbf{x} \in \mathcal{H}$,

$$\begin{aligned}
\widehat{\Lambda}\mathbf{x} &= \{\langle \mathbf{x}, \widehat{\mathbf{x}}_n \rangle\}_{n \geqslant 1} = \{\langle \mathbf{x}, U^*U\mathbf{x}_n \rangle\}_{n \geqslant 1} \\
&= \{\langle U\mathbf{x}, U\mathbf{x}_n \rangle_{\ell^2}\}_{n \geqslant 1} = \{\langle U\mathbf{x}, \mathbf{e}_n \rangle_{\ell^2}\}_{n \geqslant 1} \\
&= U\mathbf{x}.
\end{aligned}$$

Thus, $\widehat{\Lambda} = U$. In particular, $\widehat{\Lambda}$ is a bounded invertible operator from $\mathcal{H}$ onto ℓ^2.

Finally, by Proposition 12.18, $\{\widehat{\mathbf{x}}_n\}_{n \geqslant 1}$ is also a Riesz basis for $\mathcal{H}$ with $\widehat{U} = U^{*-1}$. In the argument above, we can interchange the roles of $\{\mathbf{x}_n\}_{n \geqslant 1}$ and $\{\widehat{\mathbf{x}}_n\}_{n \geqslant 1}$ to conclude that $\Lambda = \widehat{U}$. $\qquad\square$

We now provide several equivalent characterizations of Riesz bases. The following one is the most popular and is considered in several textbooks to be the definition of a Riesz basis.

Proposition 12.20 *Let* $\{\mathbf{x}_n\}_{n \geqslant 1}$ *be a minimal and complete sequence in a Hilbert space* $\mathcal{H}$*. Then* $\{\mathbf{x}_n\}_{n \geqslant 1}$ *is a Riesz basis for* $\mathcal{H}$ *if and only if there is a positive constant* M *such that*

$$\frac{1}{M} \sum_{n \geqslant 1} |a_n|^2 \leqslant \left\| \sum_{n \geqslant 1} a_n \mathbf{x}_n \right\|^2 \leqslant M \sum_{n \geqslant 1} |a_n|^2 \qquad (12.29)$$

for all $\{a_n\}_{n \geqslant 1} \in \mathbb{C}_c^\infty$.

Proof Assume that $\{\mathbf{x}_n\}_{n \geqslant 1}$ is a Riesz basis and let $U : \mathcal{H} \to \ell^2$ denote its orthogonalizer, that is to say, $U\mathbf{x}_n = \mathbf{e}_n$ for $n \geqslant 1$. For each sequence $\{a_n\}_{n \geqslant 1} \in \mathbb{C}_c^\infty$ we have

$$\left\| \sum_{n \geqslant 1} a_n \mathbf{x}_n \right\|^2 = \left\| U^{-1} \left(\sum_{n \geqslant 1} a_n \mathbf{e}_n \right) \right\|^2 \asymp \left\| \sum_{n \geqslant 1} a_n \mathbf{e}_n \right\|_{\ell^2}^2 = \sum_{n \geqslant 1} |a_n|^2$$

which proves (12.29).

Now assume that (12.29) holds. This assumption can be rewritten as

$$\frac{1}{M} \|\mathbf{a}\|_{\ell^2}^2 \leqslant \|V\mathbf{a}\|^2 \leqslant M\|\mathbf{a}\|_{\ell^2}^2, \qquad \mathbf{a} \in \mathbb{C}_c^\infty.$$

Hence V extends to a bounded invertible operator from ℓ^2 onto $\mathcal{H}$. Since $V^{-1}\mathbf{x}_n = \mathbf{e}_n$ for $n \geqslant 1$, we see that $\{\mathbf{x}_n\}_{n \geqslant 1}$ is a Riesz basis for $\mathcal{H}$ with orthogonalizer $U = V^{-1}$. $\qquad\square$

We now provide several other characterizations of Riesz bases.

Theorem 12.21 *Let $\{\mathbf{x}_n\}_{n\geq 1}$ be a minimal and complete sequence in $\mathcal{H}$ and let $\{\widehat{\mathbf{x}}_n\}_{n\geq 1}$ be its unique associated biorthogonal sequence. Then the following statements are equivalent:*

(i) $\{\mathbf{x}_n\}_{n\geq 1}$ *is a Riesz basis for $\mathcal{H}$;*

(ii) *There are two positive constants M and M' such that*

$$\left\|\sum_{n\geq 1} a_n \mathbf{x}_n\right\|^2 \leq M \sum_{n\geq 1} |a_n|^2 \tag{12.30}$$

and

$$\left\|\sum_{n\geq 1} a_n \widehat{\mathbf{x}}_n\right\|^2 \leq M' \sum_{n\geq 1} |a_n|^2 \tag{12.31}$$

for all $\{a_n\}_{n\geq 1} \in \mathbb{C}_c^\infty$;

(iii) $\{\widehat{\mathbf{x}}_n\}_{n\geq 1}$ *is complete and there are two positive constants c and c' such that*

$$\left\|\sum_{n\geq 1} a_n \mathbf{x}_n\right\|^2 \geq c \sum_{n\geq 1} |a_n|^2 \tag{12.32}$$

and

$$\left\|\sum_{n\geq 1} a_n \widehat{\mathbf{x}}_n\right\|^2 \geq c' \sum_{n\geq 1} |a_n|^2 \tag{12.33}$$

for all $\{a_n\}_{n\geq 1} \in \mathbb{C}_c^\infty$;

(iv) $\operatorname{ran}\Lambda \subset \ell^2$ *and* $\operatorname{ran}\widehat{\Lambda} \subset \ell^2$;

(v) *The Gram matrices Γ and $\widehat{\Gamma}$ are bounded operators from ℓ^2 into itself;*

(vi) *The Gram matrix Γ is a bounded invertible operator of ℓ^2 onto itself.*

Proof   The strategy of the proof is illustrated with the following diagram:

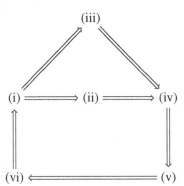

(i) $\Rightarrow$ (iii): By Proposition 12.18, the sequence $\{\mathbf{x}_n\}_{n\geq 1}$ is a Riesz basis for $\mathcal{H}$ if and only if $\{\widehat{\mathbf{x}}_n\}_{n\geq 1}$ is as well. Hence, by Proposition 12.20, we see that

(12.32) and (12.33) are precisely the left-hand side of the inequality (12.29) for the sequences $\{\mathbf{x}_n\}_{n \geqslant 1}$ and $\{\widehat{\mathbf{x}}_n\}_{n \geqslant 1}$.

(iii) $\Rightarrow$ (iv): The inequality (12.32) can be rewritten as

$$c\|\mathbf{a}\|_{\ell^2} \leqslant \|V\mathbf{a}\|, \qquad \mathbf{a} \in \mathbb{C}_c^{\infty}.$$

This inequality and (12.22) imply that

$$c\|\widehat{\Lambda}\mathbf{x}\|_{\ell^2} \leqslant \|V\widehat{\Lambda}\mathbf{x}\| = \|\mathbf{x}\|$$

for all finite linear combinations of elements of the sequence $\{\mathbf{x}_n\}_{n \geqslant 1}$. Therefore, by Proposition 12.15, we deduce that $\widehat{\Lambda}\mathcal{H} \subset \ell^2$. This argument also applies to the biorthogonal sequence $\{\widehat{\mathbf{x}}_n\}_{n \geqslant 1}$ and so $\Lambda\mathcal{H} \subset \ell^2$.

(i) $\Rightarrow$ (ii): By Proposition 12.18, $\{\widehat{\mathbf{x}}_n\}_{n \geqslant 1}$ is a Riesz basis for $\mathcal{H}$. Thus by Proposition 12.20, the inequality (12.29) holds for both $\{\mathbf{x}_n\}_{n \geqslant 1}$ and $\{\widehat{\mathbf{x}}_n\}_{n \geqslant 1}$. In particular, we obtain (12.30) and (12.31).

(ii) $\Rightarrow$ (iv): In the light of (12.19) and (12.20), the assumptions (12.30) and (12.31) can be rewritten as

$$\|V\mathbf{a}\|^2 \leqslant M\|\mathbf{a}\|_{\ell^2}^2 \quad \text{and} \quad \|\widehat{V}\mathbf{a}\|^2 \leqslant M'\|\mathbf{a}\|_{\ell^2}^2$$

for all $\mathbf{a} = \{a_n\}_{n \geqslant 1} \in \mathbb{C}_c^{\infty}$. These inequalities respectively mean that V and $\widehat{V}$ are bounded on $\mathbb{C}_c^{\infty}$. Hence they extend to bounded operators from ℓ^2 into $\mathcal{H}$. By Proposition 12.16, Λ and $\widehat{\Lambda}$ are bounded operators from $\mathcal{H}$ into ℓ^2. In particular, we have $\Lambda\mathcal{H} \subset \ell^2$ and $\widehat{\Lambda}\mathcal{H} \subset \ell^2$.

(iv) $\Rightarrow$ (v): According to Proposition 12.15, the operators Λ and $\widehat{\Lambda}$ are bounded from $\mathcal{H}$ into ℓ^2. Therefore, by Proposition 12.16, Γ and $\widehat{\Gamma}$ extend to be bounded operators on ℓ^2.

(v) $\Rightarrow$ (vi): Since Γ and $\widehat{\Gamma}$ are bounded operators on ℓ^2, again Proposition 12.16 implies that $\widehat{\Lambda} : \mathcal{H} \rightarrow \ell^2$ and $V : \ell^2 \rightarrow \mathcal{H}$ are bounded operators. The identities (12.21) and (12.22) now show that the operator V is actually a bounded invertible operator from ℓ^2 onto $\mathcal{H}$ with $V^{-1} = \widehat{\Lambda}$. Therefore, by (12.25), $\Gamma = V^*V$ is a bounded invertible operator on ℓ^2.

(vi) $\Rightarrow$ (i): The relation (12.24) shows that Γ is a positive operator. Thus, by considering the positive invertible operator $\Gamma^{1/2}$, we see that there is a constant $M > 0$ such that

$$\frac{1}{M^2}\|\mathbf{a}\|_{\ell^2}^2 \leqslant \langle\Gamma\mathbf{a}, \mathbf{a}\rangle_{\ell^2} \leqslant M^2\|\mathbf{a}\|_{\ell^2}^2, \qquad \mathbf{a} \in \ell^2.$$

Hence, again by (12.24), we can say

$$\frac{1}{M}\|\mathbf{a}\|_{\ell^2} \leqslant \|V\mathbf{a}\| \leqslant M\|\mathbf{a}\|_{\ell^2}, \qquad \mathbf{a} \in \ell^2.$$

These estimates implicitly mean that V is a bounded invertible operator from ℓ^2 onto its range. Moreover, $V\mathbf{e}_n = \mathbf{x}_n$ for $n \geqslant 1$, and the completeness of $\{\mathbf{x}_n\}_{n\geqslant 1}$ forces $V\ell^2 = \mathcal{H}$. Therefore V is a bounded invertible operator from ℓ^2 onto $\mathcal{H}$. At the same time, the relation $V\mathbf{e}_n = \mathbf{x}_n$ implies that $\{\mathbf{x}_n\}_{n\geqslant 1}$ is a Riesz basis for $\mathcal{H}$ with orthogonalizer V^{-1}. This completes the proof.   □

12.6 Riesz sequences in $\mathcal{K}_B$

Theorem 12.3 says that if a sequence $\{c_{\lambda_n}\}_{n\geqslant 1}$ of Cauchy kernels is minimal, then $\{\lambda_n\}_{n\geqslant 1}$ must be a Blaschke sequence. Denoting the corresponding Blaschke product by B, we have $c_{\lambda_n} = k_{\lambda_n}$. In particular, since $|\lambda_n| \to 1$ as $n \to \infty$, we have

$$\|k_{\lambda_n}\| = \frac{1}{(1 - |\lambda_n|^2)^{1/2}} \to \infty.$$

By (12.27), this means (since a Riesz sequence must be bounded) that the sequence $\{k_{\lambda_n}\}_{n\geqslant 1}$ is never a Riesz sequence in $\mathcal{K}_B$. Instead, we can consider the *normalized* reproducing kernels, that is,

$$\kappa_\lambda = \frac{k_\lambda}{\|k_\lambda\|} = \frac{(1 - |\lambda|^2)^{1/2}}{1 - \bar{\lambda}z},$$

and then ask if $\{\kappa_{\lambda_n}\}_{n\geqslant 1}$ forms a Riesz sequence.

By Theorem 12.4, the unique biorthogonal sequence associated with $\{\kappa_{\lambda_n}\}_{n\geqslant 1}$ in $\mathcal{K}_B$ is

$$\widehat{\kappa_{\lambda_n}} = \|k_{\lambda_n}\|\widehat{k_{\lambda_n}} = \frac{(1 - |\lambda_n|^2)^{1/2}}{B_n(\lambda_n)} \frac{\lambda_n}{|\lambda_n|} C k_{\lambda_n}, \tag{12.34}$$

where C denotes the conjugation (8.6) on $\mathcal{K}_B$.

By (12.12), the operator $\Lambda : \mathcal{K}_B \to \mathbb{C}^\infty$ corresponding to $\{\kappa_{\lambda_n}\}_{n\geqslant 1}$ is

$$\begin{aligned}
\Lambda f &= \{\langle f, \kappa_{\lambda_n}\rangle\}_{n\geqslant 1} \\
&= \{\langle f, k_{\lambda_n}/\|k_{\lambda_n}\|\rangle\}_{n\geqslant 1} \\
&= \left\{(1 - |\lambda_n|^2)^{1/2}\langle f, k_{\lambda_n}\rangle\right\}_{n\geqslant 1} \\
&= \left\{(1 - |\lambda_n|^2)^{1/2} f(\lambda_n)\right\}_{n\geqslant 1}.
\end{aligned} \tag{12.35}$$

Similarly, by (12.34), the operator $\widehat{\Lambda} : \mathcal{K}_B \to \mathbb{C}^\infty$ corresponding to $\{\widehat{\kappa_{\lambda_n}}\}_{n\geqslant 1}$ is given by

$$\begin{aligned}
\widehat{\Lambda} f &= \{\langle f, \widehat{\kappa_{\lambda_n}}\rangle\}_{n\geqslant 1} \\
&= \left\{\frac{(1 - |\lambda_n|^2)^{1/2}}{\overline{B_n(\lambda_n)}} \frac{\bar{\lambda}_n}{|\lambda_n|} \langle f, C k_{\lambda_n}\rangle\right\}_{n\geqslant 1}
\end{aligned}$$

$$= \left\{ \frac{(1 - |\lambda_n|^2)^{1/2}}{\overline{B_n(\lambda_n)}} \frac{\overline{\lambda_n}}{|\lambda_n|} \langle k_{\lambda_n}, Cf \rangle \right\}_{n \geqslant 1}$$

$$= \left\{ \frac{(1 - |\lambda_n|^2)^{1/2}}{\overline{B_n(\lambda_n)}} \frac{\overline{\lambda_n}}{|\lambda_n|} \overline{\langle Cf, k_{\lambda_n} \rangle} \right\}_{n \geqslant 1}$$

$$= \left\{ \frac{(1 - |\lambda_n|^2)^{1/2}}{\overline{B_n(\lambda_n)}} \frac{\overline{\lambda_n}}{|\lambda_n|} \overline{(Cf)(\lambda_n)} \right\}_{n \geqslant 1}. \tag{12.36}$$

Theorem 12.22 *Let $\{\lambda_n\}_{n \geqslant 1}$ be a Blaschke sequence of distinct points in $\mathbb{D}$, let B be the corresponding Blaschke product, and let $\{\kappa_{\lambda_n}\}_{n \geqslant 1}$ be the associated sequence of normalized reproducing kernels. Then the following statements are equivalent:*

(i) $\{\kappa_{\lambda_n}\}_{n \geqslant 1}$ is a Riesz basis for $\mathcal{K}_B$;
(ii) $\{\kappa_{\lambda_n}\}_{n \geqslant 1}$ is uniformly minimal;
(iii) $\{\lambda_n\}_{n \geqslant 1}$ is a uniformly separated sequence.

Proof   (i) $\implies$ (ii): Theorem 12.18 says that any Riesz basis is uniformly minimal.

(ii) $\implies$ (iii): This is precisely the content of Theorem 12.9.

(iii) $\implies$ (i): By Proposition 12.13 and the fact that

$$\delta = \inf_{n \geqslant 1} |B_n(\lambda_n)| > 0,$$

we have

$$\sum_{n \geqslant 1} (1 - |\lambda_n|^2) |f(\lambda_n)|^2 \leqslant M \|f\|^2 < \infty$$

and

$$\sum_{n \geqslant 1} \frac{1 - |\lambda_n|^2}{|B_n(\lambda_n)|^2} |(Cf)(\lambda_n)|^2 \leqslant \frac{M}{\delta^2} \|f\|^2 < \infty \quad \forall f \in \mathcal{K}_B.$$

In light of (12.35) and (12.36), the preceding two inequalities imply that Λ and $\widehat{\Lambda}$ map $\mathcal{K}_B$ into ℓ^2. Since $\{\kappa_{\lambda_n}\}_{n \geqslant 1}$ is minimal and complete in $\mathcal{K}_B$, Theorem 12.21 now implies that it is a Riesz basis for $\mathcal{K}_B$. $\square$

12.7 Completeness problems

For an inner function u and Λ a sequence of *distinct* points in $\mathbb{D}$, when does

$$K_\Lambda := \bigvee \{k_\lambda : \lambda \in \Lambda\} = \mathcal{K}_u?$$

Such problems are known as *completeness problems*.

By considering orthogonal complements and the reproducing property of k_λ, we see that $K_\Lambda = \mathcal{K}_u$ precisely when there are no functions from $\mathcal{K}_u \setminus \{0\}$ that vanish on Λ. We have already seen that $K_\Lambda = \mathcal{K}_u$ if Λ is not a Blaschke sequence or if Λ has an accumulation point in $\mathbb{D}$ (Proposition 5.20). A little more thought shows that the same holds when Λ has an accumulation point in $\mathbb{T} \setminus \sigma(u)$ since otherwise there would exist a non-zero function in $\mathcal{K}_u$ whose analytic continuation vanishes on a set having a limit point in its domain (Proposition 7.21). For general Blaschke sequences, the problem becomes more delicate and we can rephrase the completeness problem in terms of kernels of Toeplitz operators on H^2.

Proposition 12.23 *Suppose u is an inner function, Λ is a Blaschke sequence of distinct points in $\mathbb{D}$, and B is the Blaschke product corresponding to Λ. Then*

$$K_\Lambda \neq \mathcal{K}_u \quad \Longleftrightarrow \quad \ker T_{\bar{u}B} \neq \{0\}.$$

Proof Note that if $K_\Lambda \neq \mathcal{K}_u$, then there is an $f \in \mathcal{K}_u \setminus \{0\}$ that vanishes on Λ. This happens precisely when $f = gB$ for some $g \in H^2 \setminus \{0\}$. Since $f \in \mathcal{K}_u$, we know that $\bar{u}f \in \overline{zH^2}$. This means that $\bar{u}Bg \in \overline{zH^2}$. Now apply the Riesz projection P to $\bar{u}Bg$ to see that $T_{\bar{u}B}(g) = P(\bar{u}Bg) = 0$. For the other direction, simply reverse the argument. □

Example 12.24 If u divides $B = B_\Lambda$ (a Blaschke product with simple zeros), then $\bar{u}B \in H^\infty$ and the kernel of the analytic Toeplitz operator $T_{\bar{u}B}$ is indeed zero since it is just a multiplication operator on H^2. So in this case, $K_\Lambda = \mathcal{K}_u$.

Example 12.25 On the other extreme, if B divides u and u does not divide B, then $u/B \in H^\infty$ with $S^*(u/B) \neq 0$ and so

$$T_{\bar{u}B}\left(S^*\frac{u}{B}\right) = T_{\bar{u}B}T_{\bar{z}}\frac{u}{B} = T_{\bar{u}B\bar{z}}\frac{u}{B} = P\left(\bar{u}B\bar{z}\frac{u}{B}\right) = P(\bar{z}) = 0.$$

Thus $K_\Lambda \neq \mathcal{K}_u$.

12.8 Notes

12.8.1 References

The material in this section is available in the book of Singer [181]. The Gram matrices and other operators introduced in this chapter are well known in the geometric studies of Hilbert spaces. More on this can be found in Akhiezer and Glazman [6]. The important classification Theorem 12.21 is due to Bari [26] and partly to Boas [32]. A good general source for bases and frames is [40].

12.8.2 Kernels of Toeplitz operators

Kernels of Toeplitz operators have been studied before and we refer the reader to the papers [66, 86, 164] for further details.

12.8.3 Completeness problems and Clark theory

The original paper of Clark [46] examined kernel function completeness problems by using the Paley–Wiener theory, comparing a known orthonormal basis of eigenvectors for a Clark unitary operator with a given sequence of kernel functions. This idea was also explored by Sarason in [161].

12.8.4 Completeness problems for differential operators

One can relate kernel completeness problems for model spaces with completeness theorems for solutions to differential operators such as the Schrödinger and Sturm–Liouville operators [130, 188].

12.8.5 Interpolating sequences

Uniformly separated sequences are equivalently known as "interpolating sequences" for H^2 [63, 97].

12.8.6 Eigenvectors of complex symmetric operators

Recall from Chapter 8 that a conjugation on a complex Hilbert space $\mathcal{H}$ is a conjugate-linear function $C : \mathcal{H} \to \mathcal{H}$ that is involutive and isometric (Definition 8.1). An operator $T \in \mathcal{B}(\mathcal{H})$ is called C-symmetric if $T = CT^*C$. We say that T is a *complex symmetric operator* if there exists a conjugation C with respect to which T is C-symmetric. For example, Proposition 9.8 tells us that the compressed shift S_u is a complex symmetric operator, since $S_u = CS_u^*C$, where C denotes the conjugation $Cf = \overline{f}zu$ on $\mathcal{K}_u$. Similarly, $S_u^* = S^*|_{\mathcal{K}_u}$ is C-symmetric as well.

Associated to each conjugation C on $\mathcal{H}$ is the bilinear form

$$[\mathbf{x}, \mathbf{y}] = \langle \mathbf{x}, C\mathbf{y} \rangle. \tag{12.37}$$

Indeed, since the standard sesquilinear form $\langle \cdot, \cdot \rangle$ is conjugate-linear in the second position, it follows from the fact that C is conjugate-linear that $[\cdot, \cdot]$ is linear in both positions. Although we have the Cauchy–Schwarz inequality

$$|[\mathbf{x}, \mathbf{y}]| \leqslant \|\mathbf{x}\|\,\|\mathbf{y}\|,$$

it turns out that $[\cdot, \cdot]$ is not a true inner product since it is possible for $[\mathbf{x}, \mathbf{x}] = 0$ to hold even when $\mathbf{x} \neq \mathbf{0}$. Two vectors $\mathbf{x}$ and $\mathbf{y}$ are C-*orthogonal* if $[\mathbf{x}, \mathbf{y}] = 0$ (denoted by $\mathbf{x} \perp_C \mathbf{y}$). We say that two subspaces $\mathcal{E}_1$ and $\mathcal{E}_2$ are C-orthogonal (denoted $\mathcal{E}_1 \perp_C \mathcal{E}_2$) if $[\mathbf{x}_1, \mathbf{x}_2] = 0$ for every $\mathbf{x}_1$ in $\mathcal{E}_1$ and $\mathbf{x}_2$ in $\mathcal{E}_2$.

With respect to the bilinear form $[\cdot, \cdot]$, it turns out that C-symmetric operators superficially resemble self-adjoint operators. For instance, an operator T is C-symmetric if and only if $[T\mathbf{x}, \mathbf{y}] = [\mathbf{x}, T\mathbf{y}]$ for all $\mathbf{x}, \mathbf{y}$ in $\mathcal{H}$. As another example, the eigenvectors of a C-symmetric operator corresponding to distinct eigenvalues are orthogonal with respect to $[\cdot, \cdot]$, even though they are not necessarily orthogonal with respect to the original sesquilinear form $\langle \cdot, \cdot \rangle$.

Suppose that $\{\mathbf{x}_n\}_{n \geqslant 1}$ is a complete system of C-orthonormal vectors in $\mathcal{H}$:

$$[\mathbf{x}_i, \mathbf{x}_j] = \delta_{ij}. \tag{12.38}$$

In other words, suppose that $\{\mathbf{x}_n\}_{n \geqslant 1}$ and $\{C\mathbf{x}_n\}_{n \geqslant 1}$ are complete biorthogonal sequences in $\mathcal{H}$. In light of (12.38), a finite linear combination $\mathbf{x} = \sum_{n=1}^N c_n \mathbf{x}_n$ has coefficients given by $c_n = [\mathbf{x}, \mathbf{x}_n]$ and hence each such $\mathbf{x}$ can be recovered via the skew Fourier expansion

$$\mathbf{x} = \sum_{1 \leqslant n \leqslant N} [\mathbf{x}, \mathbf{x}_n]\mathbf{x}_n,$$

where N depends on $\mathbf{x}$.

As an example, suppose that u is a Blaschke product whose zeros $\{\lambda_n\}_{n \geqslant 1}$ are all simple. The preceding lemma tells us that the eigenvectors k_{λ_n} of the C-symmetric operator S_u^* satisfy $[k_{\lambda_i}, k_{\lambda_j}] = 0$ if $i \neq j$. This can be verified directly:

$$[k_{\lambda_i}, k_{\lambda_j}] = \langle k_{\lambda_i}, Ck_{\lambda_j} \rangle = \langle k_{\lambda_i}, \overline{Ck_{\lambda_j}} \rangle = \overline{\langle Ck_{\lambda_j}, k_{\lambda_i} \rangle} = \overline{Ck_{\lambda_j}(\lambda_i)}$$

$$= \begin{cases} 0 & \text{if } i \neq j, \\ \overline{u'(\lambda_i)} & \text{if } i = j. \end{cases}$$

In particular, for any determination of $(\overline{u'(\lambda_n)})^{1/2}$ it follows that

$$\mathbf{x}_n = \frac{k_{\lambda_n}}{(\overline{u'(\lambda_n)})^{1/2}}$$

is a complete C-orthonormal system in $\mathcal{K}_u$.

A key object in the study of the C-orthonormal systems is the densely defined operator

$$A_0\Big(\sum_{1 \leqslant n \leqslant N} c_n \mathbf{x}_n \Big) = \sum_{1 \leqslant n \leqslant N} c_n C \mathbf{x}_n.$$

The following general theorem tells us precisely when a C-orthonormal system forms a Riesz basis for $\mathcal{H}$ [84, 88].

Theorem 12.26 *If $\{\mathbf{x}_n\}_{n \geqslant 1}$ is a complete C-orthonormal system in $\mathcal{H}$, then the following are equivalent:*

(i) *$\{\mathbf{x}_n\}_{n \geqslant 1}$ is a Bessel sequence with Bessel bound M;*
(ii) *$\{\mathbf{x}_n\}_{n \geqslant 1}$ is a Riesz basis with lower and upper bounds M^{-1} and M;*
(iii) *A_0 extends to a bounded linear operator on H satisfying $\|A_0\| \leqslant M$;*
(iv) *There exists an $M > 0$ satisfying*

$$\Big\| \sum_{1 \leqslant n \leqslant N} \overline{c_n} \mathbf{x}_n \Big\| \leqslant M \Big\| \sum_{1 \leqslant n \leqslant N} c_n \mathbf{x}_n \Big\|,$$

for every finite sequence $c_1, c_2, \ldots, c_N$;
(v) *The Gram matrix $(\langle \mathbf{x}_j, \mathbf{x}_k \rangle)_{j,k \geqslant 1}$ dominates its transpose:*

$$(M^2 \langle \mathbf{x}_j, \mathbf{x}_k \rangle - \langle \mathbf{x}_k, \mathbf{x}_j \rangle)_{j,k \geqslant 1} \geqslant 0$$

for some $M > 0$;
(vi) *The Gram matrix $G = (\langle \mathbf{x}_j, \mathbf{x}_k \rangle)_{j,k \geqslant 1}$ is bounded on $\ell^2(\mathbb{N})$ and orthogonal ($G^T G = I$ as matrices). Furthermore, $\|G\| \leqslant M$;*
(vii) *The skew Fourier expansion*

$$\sum_{n \geqslant 1} [\mathbf{x}, \mathbf{x}_n] \mathbf{x}_n$$

converges in norm for each $\mathbf{x} \in \mathcal{H}$ and

$$\frac{1}{M} \|\mathbf{x}\|^2 \leqslant \sum_{n \geqslant 1} |[\mathbf{x}, \mathbf{x}_n]|^2 \leqslant M \|\mathbf{x}\|^2.$$

In all cases, the infimum over all such M equals the norm of A_0.

A non-trivial application of the above result to free interpolation in the Hardy space of the disk is described in [84]. Further information about complex symmetric operators in general can be found in [55, 80–83, 88].

13

Truncated Toeplitz operators

We have already seen truncated Toeplitz operators in Theorem 10.9 when we identified the commutant $\{S_u\}'$ of the compressed shift S_u. This led us to the operators $A_\varphi^u f = P_u(\varphi f)$, where $\varphi \in H^\infty$. We also encountered them in Theorem 9.28 when we identified the C^*-algebra generated by S_u, which led us to the operators A_φ^u, where $\varphi \in C(\mathbb{T})$.

In this chapter, we consider the class of truncated Toeplitz operators A_φ^u with symbols $\varphi \in L^2$. For these operators, several interesting technical difficulties arise. First, the symbol φ for A_φ^u is never unique. Indeed, there are many symbols that represent A_φ^u. Second, there are bounded operators A_φ^u for $\varphi \in L^2$ for which there is no bounded symbol ψ for which $A_\varphi^u = A_\psi^u$. Though initially disheartening, there will be a wonderful structure to this class of operators and, since the field is relatively new, there is a lot of work to be done.

To give these operators some context, the reader might want to review our discussion of Toeplitz operators from Chapter 4. Finally, we point out that many of the proofs in the early part of this chapter are based on Sarason's original paper [166].

13.1 The basics

Definition 13.1 A *truncated Toeplitz operator* A_φ^u with *symbol* $\varphi \in L^2$ on the model space $\mathcal{K}_u$ is the operator initially defined on the dense linear manifold $\mathcal{K}_u \cap H^\infty$ (Proposition 5.21) by the formula

$$A_\varphi^u f := P_u(\varphi f), \tag{13.1}$$

where P_u is the orthogonal projection of L^2 onto $\mathcal{K}_u$ (Proposition 5.13).

In certain situations, one can enlarge the initial domain of definition $\mathcal{K}_u \cap H^\infty$. For example, when $\varphi \in L^\infty$ we can define A^u_φ on all of $\mathcal{K}_u$ since $\varphi f \in L^2$ for all $f \in \mathcal{K}_u$ and so $P_u(\varphi f) \in \mathcal{K}_u$. Note that $A^u_z = S_u$ is the compressed shift operator from Chapter 9 while $A^u_{\bar{z}} = S^*|_{\mathcal{K}_u}$ is the restriction of the backward shift to $\mathcal{K}_u$ (equivalently S^*_u).

Definition 13.2 Let $\mathcal{T}_u$ denote the set of all truncated Toeplitz operators, initially defined on $\mathcal{K}_u \cap H^\infty$, that extend to be *bounded* operators on $\mathcal{K}_u$.

The operator A^u_φ can alternatively be understood as follows. By Proposition 5.13, the orthogonal projection $P_u g$ of $g \in L^2$ onto $\mathcal{K}_u$ is given by the formula

$$(P_u g)(\lambda) = \int_{\mathbb{T}} g \overline{k^u_\lambda} \, dm, \qquad \lambda \in \mathbb{D}.$$

For each $\lambda \in \mathbb{D}$, we have $k^u_\lambda \in H^\infty$. For fixed $\zeta \in \mathbb{T}$, the function $\lambda \mapsto k^u_\lambda(\zeta)$ is a conjugate-analytic function on $\mathbb{D}$. Thus the preceding integral formula still makes sense, and defines an analytic function on $\mathbb{D}$, even when g belongs to the larger space L^1. Hence for any $\varphi \in L^2$, we have $\varphi f \in L^1$ for all $f \in \mathcal{K}_u$. Thus we can define the linear transformation

$$A^u_\varphi : \mathcal{K}_u \to \mathcal{O}(\mathbb{D})$$

(here $\mathcal{O}(\mathbb{D})$ denotes the analytic functions on $\mathbb{D}$) by the integral formula

$$(A^u_\varphi f)(\lambda) := \int_{\mathbb{T}} \varphi f \overline{k^u_\lambda} \, dm, \qquad f \in \mathcal{K}_u, \, \lambda \in \mathbb{D}. \tag{13.2}$$

The Cauchy–Schwarz Inequality yields

$$|(A^u_\varphi f)(\lambda)| \leqslant \|f\| \, \|\varphi\| \, \|k^u_\lambda\|_\infty, \qquad \lambda \in \mathbb{D}.$$

Proposition 13.3 *Let u be inner and $\varphi \in L^2$. If $A^u_\varphi f \in \mathcal{K}_u$ for every $f \in \mathcal{K}_u$, then $A^u_\varphi \in \mathcal{T}_u$. Furthermore, when $A^u_\varphi \in \mathcal{T}_u$, the definitions in (13.1) and (13.2) for A^u_φ coincide.*

Proof To show that the operator defined by (13.2) is bounded, we will use the Closed Graph Theorem (Theorem 1.28). Suppose $\{f_n\}_{n \geqslant 1} \subset \mathcal{K}_u$ with $f_n \to f$ in $\mathcal{K}_u$ and $A^u_\varphi f_n \to g$ in $\mathcal{K}_u$. Then $f_n \to f$ weakly in L^2 and so for each $\lambda \in \mathbb{D}$, we use the fact that $k^u_\lambda \in L^\infty$ to get

$$(A^u_\varphi f_n)(\lambda) = \langle \varphi f_n, k^u_\lambda \rangle = \langle f_n, \overline{\varphi} k^u_\lambda \rangle \to \langle f, \overline{\varphi} k^u_\lambda \rangle = \langle \varphi f, k_\lambda \rangle = (A^u_\varphi f)(\lambda).$$

Since $A^u_\varphi f_n \to g$ in $\mathcal{K}_u$ we conclude that $A^u_\varphi f_n \to g$ pointwise on $\mathbb{D}$ (Proposition 3.4). Thus $A^u_\varphi f = g$. By the Closed Graph Theorem, $A^u_\varphi \in \mathcal{T}_u$.

For the proof of the second part of the theorem, observe that the (bounded) operators defined by the formulas in (13.1) and (13.2) agree on the dense set $\mathcal{K}_u \cap H^\infty$ and therefore must be equal.   □

At this point, the reader might be confused as to why we bother defining A_φ^u for $\varphi \in L^2$ (and possibly unbounded) when one can define A_φ^u everywhere on $\mathcal{K}_u$ if $\varphi \in L^\infty$. We did not go through all this trouble when defining Toeplitz operators on H^2.

As we will see in a moment, there are some substantial differences between Toeplitz operators on H^2 and truncated Toeplitz operators on $\mathcal{K}_u$. For Toeplitz operators, the symbol is unique in that $T_{\varphi_1} = T_{\varphi_2}$ if and only if $\varphi_1 = \varphi_2$ almost everywhere. Furthermore, one can show that for a symbol $\varphi \in L^2$, the densely defined operator $T_\varphi f = P(\varphi f)$ on H^∞ has a bounded extension to H^2 if and only if $\varphi \in L^\infty$. For truncated Toeplitz operators, the symbol is never unique (Theorem 13.6). Moreover, there are A_φ^u with $\varphi \in L^2$ that extend to be bounded operators on $\mathcal{K}_u$ but for which there is no bounded symbol that represents A_φ^u.

The skeptical reader might still question the need to consider truncated Toeplitz operators with unbounded symbols. From Corollary 4.18, the set of bounded Toeplitz operators on H^2 forms a weakly closed linear space in $\mathcal{B}(H^2)$. To have the same result for truncated Toeplitz operators (Theorem 13.11), we need to include such unusual operators. Finally, we are deliberately using the term *truncated* Toeplitz operator and not *compressed* Toeplitz operator since, though the formula $A_\varphi^u f = P_u(\varphi f)$ suggests a compression of the Toeplitz operator T_φ to $\mathcal{K}_u$, we are not assuming the extra condition $(A_\varphi^u)^n f = P_u(\varphi^n f)$ for all $n \in \mathbb{N}$ needed to make A_φ^u a true *compression* of T_φ to $\mathcal{K}_u$ (see Remark 9.2). For certain φ, for example if $\varphi \in H^\infty$, then A_φ^u is indeed a compression of T_φ to $\mathcal{K}_u$. However, for general φ, this is not always the case since the product of two truncated Toeplitz operators is not always a truncated Toeplitz operator.

Example 13.4   When $u = z^n$, the set $\{1, z, \ldots, z^{n-1}\}$ is an orthonormal basis for $\mathcal{K}_u$. Moreover, any operator in $\mathcal{T}_u$, when represented with respect to this basis, yields a Toeplitz matrix. Conversely, any $n \times n$ Toeplitz matrix gives rise to a truncated Toeplitz operator on $\mathcal{K}_u$. Indeed, for $0 \leqslant j, k \leqslant n - 1$,

$$\langle A_\varphi^u z^j, z^k \rangle = \langle P_u(\varphi z^j), z^k \rangle$$
$$= \langle \varphi z^j, P_u z^k \rangle$$
$$= \langle \varphi z^j, z^k \rangle$$
$$= \langle \varphi, z^{k-j} \rangle$$
$$= \widehat{\varphi}(k - j)$$

and so the matrix representation $[A_\varphi^u]$ of A_φ^u with respect to the basis $\{1, z, \ldots, z^{n-1}\}$ is the Toeplitz matrix

$$\begin{bmatrix} \widehat{\varphi}(0) & \widehat{\varphi}(-1) & \widehat{\varphi}(-2) & \cdots & \cdots & \widehat{\varphi}(-n+1) \\ \widehat{\varphi}(1) & \widehat{\varphi}(0) & \widehat{\varphi}(-1) & \ddots & & \vdots \\ \widehat{\varphi}(2) & \widehat{\varphi}(1) & \ddots & \ddots & \ddots & \vdots \\ \vdots & \ddots & \ddots & \ddots & \widehat{\varphi}(-1) & \widehat{\varphi}(-2) \\ \vdots & & \ddots & \widehat{\varphi}(1) & \widehat{\varphi}(0) & \widehat{\varphi}(-1) \\ \widehat{\varphi}(n-1) & \cdots & \cdots & \widehat{\varphi}(2) & \widehat{\varphi}(1) & \widehat{\varphi}(0) \end{bmatrix}. \quad (13.3)$$

It is also worth pointing out in this particular case that since only a finite number of Fourier coefficients are involved, we have $A_\varphi^u \in \mathscr{T}_u$ for *all* $\varphi \in L^2$.

There are some similarities between truncated Toeplitz operators on $\mathcal{K}_u$ and Toeplitz operators on H^2. Here is one example (see Proposition 4.14 for comparison).

Proposition 13.5 *If $A_\varphi^u \in \mathscr{T}_u$, then $A_{\overline{\varphi}}^u \in \mathscr{T}_u$ and $(A_\varphi^u)^* = A_{\overline{\varphi}}^u$.*

Proof For $f, g \in \mathcal{K}_u \cap H^\infty$, we have

$$\begin{aligned} \langle A_\varphi^u f, g \rangle = \langle P_u(\varphi f), g \rangle &= \langle \varphi f, g \rangle \\ &= \langle f, \overline{\varphi} g \rangle = \langle P_u f, \overline{\varphi} g \rangle = \langle f, P_u(\overline{\varphi} g) \rangle \\ &= \langle f, A_{\overline{\varphi}}^u g \rangle. \end{aligned}$$

This proves that $A_{\overline{\varphi}}^u \in \mathscr{T}_u$ and $(A_\varphi^u)^* = A_{\overline{\varphi}}^u$. □

Another similarity is that $\mathscr{T}_u$ forms a linear space since

$$\alpha A_\varphi^u + \beta A_\psi^u = A_{\alpha\varphi+\beta\psi}^u, \quad \alpha, \beta \in \mathbb{C}, \ \varphi, \psi \in L^2.$$

There are, however, many differences between Toeplitz operators and truncated Toeplitz operators. For instance, the symbol φ of a Toeplitz operator T_φ is unique (Corollary 4.13). For truncated Toeplitz operators, the story is much different.

Theorem 13.6 (Sarason) *A truncated Toeplitz operator A_φ^u is identically zero if and only if $\varphi \in uH^2 + \overline{uH^2}$. Consequently, $A_{\varphi_1}^u = A_{\varphi_2}^u$ if and only if $\varphi_1 - \varphi_2 \in uH^2 + \overline{uH^2}$.*

Before getting to the proof, let us decompose L^2 as

$$L^2 = H^2 \oplus \overline{\zeta H^2} = uH^2 \oplus \mathcal{K}_u \oplus \overline{\zeta H^2}$$

and observe that

$$\bar{u}\mathcal{K}_u \subset \overline{\zeta H^2}. \tag{13.4}$$

Indeed, for any $f \in \mathcal{K}_u$, we have $0 = \langle f, uh \rangle = \langle \bar{u}f, h \rangle$ for all $h \in H^2$. Thus $\bar{u}f \in (H^2)^{\perp} = \overline{\zeta H^2}$.

Proof of Theorem 13.6   Suppose that $\varphi = uh_1 + \overline{uh_2}$, where $h_1, h_2 \in H^2$. Then, for $f \in \mathcal{K}_u \cap H^{\infty}$,

$$\varphi f = uh_1 f + \overline{uh_2} f.$$

The first term belongs to uH^2 and so $P_u(uh_1 f) = 0$. By (13.4), the second term belongs to $\overline{zH^2}$ and thus $P_u(\overline{uh_2} f) = 0$. Hence

$$A^u_{\varphi} f = P_u(\varphi f) = P_u(uh_1 f) + P_u(\overline{uh_2} f) = 0.$$

For the other direction, we need to recall two facts. The first is the pair of operator identities

$$I - S_u S_u^* = k_0 \otimes k_0, \quad I - S_u^* S_u = Ck_0 \otimes Ck_0$$

from Lemma 9.9. The second is the fact that if $\psi \in H^2$, one can prove a version of (9.12) to show that A^u_{ψ} commutes with S_u on $H^{\infty} \cap \mathcal{K}_u$. This will be true whether or not A^u_{ψ} extends to be bounded on $\mathcal{K}_u$. A similar argument will show that A^u_{ψ} commutes with S_u^* on $\mathcal{K}_u \cap H^{\infty}$.

Write the symbol $\varphi \in L^2$ for $A^u_{\varphi} \in \mathcal{T}_u$ as $\varphi = \psi + \bar{\chi}$, where $\varphi, \chi \in H^2$. Note that this decomposition is not unique since there is an additive constant involved. Since we are assuming that A^u_{φ} is the zero operator, this means that $A^u_{\psi} = -A^u_{\bar{\chi}}$. Using the fact, from the discussion in the previous paragraph, that

$$A^u_{\psi} S_u = S_u A^u_{\psi}, \tag{13.5}$$

we also see, from the identities $A^u_{\psi} = -A^u_{\bar{\chi}}$ and

$$A^u_{\bar{\chi}} S_u^* = S_u^* A^u_{\bar{\chi}}, \tag{13.6}$$

that A^u_{ψ} and $A^u_{\bar{\chi}}$ commute with both S_u and S_u^*. Thus

$$A^u_{\psi}(k_0 \otimes k_0) = A^u_{\psi}(I - S_u S_u^*) = (I - S_u S_u^*)A^u_{\psi} = (k_0 \otimes k_0)A^u_{\psi}.$$

Evaluating the operator identity $A^u_{\psi}(k_0 \otimes k_0) = (k_0 \otimes k_0)A^u_{\psi}$ at k_0 yields

$$\|k_0\|^2 A^u_{\psi} k_0 = \langle A^u_{\psi} k_0, k_0 \rangle k_0$$

and so

$$A_\psi^u k_0 = c k_0$$

for some constant $c \in \mathbb{C}$. From here we have

$$
\begin{aligned}
0 &= (A_\psi^u - cI)k_0 \\
&= P_u((\psi - c)(1 - \overline{u(0)}u)) \\
&= P_u(\psi - c) - \overline{u(0)}P_u((\psi - c)u) \\
&= P_u(\psi - c) - 0.
\end{aligned}
$$

Thus $\psi - c \in uH^2$ and so $A_\psi^u = cI$, which also implies that $A_{\overline{\chi}}^u = -cI$. Now repeat a version of the same argument to see that $\chi + \overline{c} \in uH^2$ and so $\overline{\chi} + c \in \overline{uH^2}$. Therefore

$$\varphi = (\psi - c) + (\overline{\chi} + c) \in uH^2 + \overline{uH^2},$$

which completes the proof.   □

Corollary 13.7   *If $A \in \mathcal{T}_u$, there are $\psi, \chi \in \mathcal{K}_u$ such that $A = A_{\psi + \overline{\chi}}^u$.*

Proof   Write the symbol $\varphi \in L^2$ for $A = A_\varphi^u \in \mathcal{T}_u$ as $\varphi = \varphi_1 + \overline{\varphi_2}$, where $\varphi_1, \varphi_2 \in H^2$. Now define $\psi = P_u\varphi_1$ and $\chi = P_u\varphi_2$. Then $\varphi_1 - \psi$ and $\varphi_2 - \chi$ both belong to uH^2. Furthermore,

$$\varphi = \psi + \overline{\chi} + (\varphi_1 - \psi) + (\overline{\varphi_2 - \chi}).$$

Theorem 13.6 ensures that $A_{\varphi_1 - \psi}^u = A_{\overline{\varphi_2 - \chi}}^u = 0$ and so $A_\varphi^u = A_{\psi + \overline{\chi}}^u$.   □

Theorem 13.6 implies that the symbol φ for A_φ^u actually belongs to a set of symbols: $\varphi + uH^2 + \overline{uH^2}$. Thus, even when $\varphi \in L^\infty$, there are unbounded symbols ψ for which $A_\varphi^u = A_\psi^u$. This leads to a natural question: if A_φ^u is bounded, does there exist a $\psi \in L^\infty$ for which $A_\varphi^u = A_\psi^u$? We will address this issue in the end notes of this chapter.

13.2 A characterization

The Brown–Halmos Theorem (Theorem 4.16) says that $T \in \mathcal{B}(H^2)$ is a Toeplitz operator if and only if $T = STS^*$, where S is the unilateral shift on H^2. The following is the truncated Toeplitz operator analogue of this result, where the compressed shift S_u plays the role of S.

Theorem 13.8 (Sarason) *A bounded operator A on $\mathcal{K}_u$ belongs to $\mathcal{T}_u$ if and only if there are $\psi, \chi \in \mathcal{K}_u$ such that*

$$A = S_u A S_u^* + \psi \otimes k_0 + k_0 \otimes \chi,$$

in which case, $A = A_{\psi + \bar{\chi}}^u$.

Proof First let us prove the identity

$$A_{\psi + \bar{\chi}}^u - S_u A_{\psi + \bar{\chi}}^u S_u^* = \psi \otimes k_0 + k_0 \otimes \chi, \qquad \psi, \chi \in \mathcal{K}_u. \tag{13.7}$$

Since $\psi \in \mathcal{K}_u$, we have

$$A_\psi^u k_0 = \psi. \tag{13.8}$$

Indeed,

$$A_\psi^u k_0 = P_u((1 - \overline{u(0)}u)\psi) = P_u(\psi) - \overline{u(0)}P_u(u\psi) = \psi.$$

We now compute $A_{\psi + \bar{\chi}}^u - S_u A_{\psi + \bar{\chi}}^u S_u^*$ as follows:

$$
\begin{aligned}
A_{\psi + \bar{\chi}}^u &- S_u A_{\psi + \bar{\chi}}^u S_u^* \\
&= (A_\psi^u + A_{\bar{\chi}}^u) - S_u(A_\psi^u + A_{\bar{\chi}}^u)S_u^* \\
&= (A_\psi - S_u A_\psi S_u^*) + (A_{\bar{\chi}} - S_u A_{\bar{\chi}} S_u^*) \\
&= (A_\psi^u - A_\psi^u S_u S_u^*) + (A_{\bar{\chi}}^u - S_u S_u^* A_{\bar{\chi}}^u) && \text{(by (13.5), (13.6))} \\
&= A_\psi^u(I - S_u S_u^*) + (I - S_u S_u^*)A_{\bar{\chi}}^u \\
&= A_\psi^u(k_0 \otimes k_0) + (k_0 \otimes k_0)A_{\bar{\chi}}^u && \text{(Lemma 9.9)} \\
&= (A_\psi^u k_0) \otimes k_0 + k_0 \otimes (A_\chi^u k_0) && \text{(Proposition 1.32)} \\
&= \psi \otimes k_0 + k_0 \otimes \chi. && \text{(by (13.8))}
\end{aligned}
$$

This verifies (13.7).

Next we want to prove, for fixed $\psi, \chi \in \mathcal{K}_u$, that

$$\langle A_{\psi + \bar{\chi}}^u f, g \rangle = \sum_{n \geqslant 0} \left(\langle f, S_u^n k_0 \rangle \langle S_u^n \psi, g \rangle + \langle f, S_u^n \chi \rangle \langle S_u^n k_0, g \rangle \right) \tag{13.9}$$

for every $f, g \in \mathcal{K}_u \cap H^\infty$. To see this, use (13.7) along with Proposition 1.32 to obtain the identities

$$S_u^n A_{\psi + \bar{\chi}}^u S_u^{*n} - S_u^{n+1} A_{\psi + \bar{\chi}}^u S_u^{*(n+1)} = (S_u^n \psi \otimes S_u^n k_0) + (S_u^n k_0 \otimes S_u^n \chi), \qquad n \geqslant 0.$$

Now sum both sides of the equation above from $n = 0$ to $n = N$ (and use a telescoping series argument) to get

$$A_{\psi + \bar{\chi}}^u = \sum_{0 \leqslant n \leqslant N} ((S_u^n \psi \otimes S_u^n k_0) + (S_u^n k_0 \otimes S_u^n \chi)) + S_u^{N+1} A_{\psi + \bar{\chi}}^u S_u^{*(N+1)}. \tag{13.10}$$

For any $f, g \in \mathcal{K}_u \cap H^\infty$ we obtain, using (13.5) and (13.6),

$$\langle S_u^N A_{\psi+\bar{\chi}}^u S_u^{*N} f, g \rangle = \langle S_u^{*N} f, S_u^{*N} A_{\bar{\psi}}^u g, \rangle + \langle S_u^{*N} A_{\bar{\chi}}^u f, S_u^{*N} g \rangle.$$

Since $S^{*N} \to 0$ in the strong operator topology (Lemma 9.12), both of the terms on the right-hand side of the preceding identity approach zero as $N \to \infty$. The identity in (13.9) now follows by first taking inner products in (13.10) and then taking limits as $N \to \infty$.

We are now ready for the main body of the proof. Indeed, one direction follows directly from the identity in (13.7). For the other direction, assume that

$$A = S_u A S_u^* + \psi \otimes k_0 + k_0 \otimes \chi$$

for some $\psi, \chi \in \mathcal{K}_u$. We then apply the proof of (13.9) to obtain

$$A = \sum_{0 \leqslant n \leqslant N} ((S_u^n \psi \otimes S_u^n k_0) + (S_u^n k_0 \otimes S_u^n \chi)) + S_u^{N+1} A S_u^{*(N+1)}.$$

Again use the fact that $S_u^{*N} \to 0$ in the strong operator topology to see that

$$A = \sum_{n \geqslant 0} \left((S_u^n \psi \otimes S_u^n k_0) + (S_u^n k_0 \otimes S_u^n \chi) \right), \tag{13.11}$$

where the series in (13.11) converges in the strong operator topology. To finish, use (13.9) to see that the right-hand side of (13.11) is equal to $A_{\psi+\bar{\chi}}^u$. □

Remark 13.9 Note that

$$A_{\psi-ck_0}^u = A_\psi^u - c A_{1-\overline{u(0)}u}^u = A_\psi^u - cI - \overline{u(0)} A_u^u = A_\psi^u - cI$$

and

$$A_{\bar{\chi}+\overline{ck_0}}^u = A_{\bar{\chi}}^u + c A_{1-u(0)\bar{u}}^u = A_{\bar{\chi}}^u + cI.$$

Thus

$$A_{\psi+\bar{\chi}}^u = A_{(\psi-ck_0)+(\bar{\chi}+\overline{ck_0})}^u,$$

which means, if desired, we can adjust the value of either ψ or χ at the origin.

Example 13.10 Let $u = z^3$ and note that $\mathcal{K}_u = \bigvee\{1, z, z^2\}$ (Proposition 5.16). From (13.3), the truncated Toeplitz operators can be viewed as Toeplitz matrices. Suppose that A is a generic 3×3 Toeplitz matrix

$$\begin{bmatrix} a & d & e \\ b & a & d \\ c & b & a \end{bmatrix}.$$

The compressed shift S_u has the matrix representation

$$S_u = \begin{bmatrix} 0 & 0 & 0 \\ 1 & 0 & 0 \\ 0 & 1 & 0 \end{bmatrix}$$

and a computation will show that

$$S_u A S_u^* - A = \begin{bmatrix} -a & -d & -e \\ -b & 0 & 0 \\ -c & 0 & 0 \end{bmatrix}.$$

With respect to the basis $\{1, z, z^2\}$, the function $k_0 = 1$ can be written as the vector $(1, 0, 0)$. Furthermore, one can check the identities

$$(z_1, z_2, z_3) \otimes (1, 0, 0) = \begin{bmatrix} z_1 & 0 & 0 \\ z_2 & 0 & 0 \\ z_3 & 0 & 0 \end{bmatrix}, \quad (1, 0, 0) \otimes (z_1, z_2, z_3) = \begin{bmatrix} z_1 & z_2 & z_3 \\ 0 & 0 & 0 \\ 0 & 0 & 0 \end{bmatrix}.$$

Hence

$$S_u A S_u^* - A = (-\tfrac{a}{2}, -b, -c) \otimes (1, 0, 0) + (1, 0, 0) \otimes (-\tfrac{a}{2}, -d, -e).$$

Describing this in terms of elements of $\mathcal{K}_u$, we have the formula

$$S_u A S_u^* = A + (-\tfrac{a}{2} - bz - cz^2) \otimes 1 + 1 \otimes (-\tfrac{a}{2} - dz - ez^2).$$

Notice the rank-two perturbation of A in the formula above.

Recall from Theorem 4.18 that the set of Toeplitz operators forms a linear space that is closed in the weak operator topology. The same is true for the truncated Toeplitz operators.

Theorem 13.11 (Sarason) $\mathcal{T}_u$ *is closed in the weak operator topology.*

Proof   Suppose that a net $\{A_\alpha^u\}_{\alpha \in I} \subset \mathcal{T}_u$ converges in the weak operator topology to $A \in \mathcal{B}(\mathcal{K}_u)$. By Theorem 13.8, there are $\psi_\alpha, \chi_\alpha \in \mathcal{K}_u$ for which

$$A_\alpha^u - S_u A_\alpha^u S_u^* = (\psi_\alpha \otimes k_0) + (k_0 \otimes \chi_\alpha). \tag{13.12}$$

By Remark 13.9, we can assume that $\chi_\alpha(0) = 0$ for every $\alpha \in I$. Hence inserting k_0 into (13.12), we see that

$$A_\alpha^u k_0 - S_u A_\alpha^u S_u^* k_0 = \|k_0\|^2 \psi_\alpha.$$

This implies that the net ψ_α converges weakly to

$$\psi = \frac{1}{\|k_0\|} (A k_0 - S_u A S_u^* k_0) \in \mathcal{K}_u.$$

Furthermore, the net of rank-one operators $\psi_\alpha \otimes k_0$ converges in the weak operator topology to $\psi \otimes k_0$. Using this fact and (13.12), we see that χ_α converges weakly to some $\chi \in \mathcal{K}_u$ and hence the net $k_0 \otimes \chi_\alpha$ converges in the weak operator topology to $k_0 \otimes \chi$. Putting this all together, we obtain

$$A - S_u A S_u^* = (\psi \otimes k_0) + (k_0 \otimes \chi),$$

whence, by Theorem 13.8, $A \in \mathcal{T}_u$. □

13.3 C-symmetric operators

Recall from Chapter 8 that a conjugation on a complex Hilbert space $\mathcal{H}$ is a map $C : \mathcal{H} \to \mathcal{H}$ that is conjugate-linear ($C(a\mathbf{x} + \mathbf{y}) = \overline{a}C\mathbf{x} + C\mathbf{y}$), involutive ($C^2 = I$), and isometric ($\langle C\mathbf{x}, C\mathbf{y} \rangle = \langle \mathbf{y}, \mathbf{x} \rangle$). We say $T \in \mathcal{B}(\mathcal{H})$ is C-symmetric if $T = CT^*C$ and complex symmetric if there exists a conjugation C with respect to which T is C-symmetric.

For example, the conjugation

$$(Cf)(x) = \overline{f(1 - x)}$$

on $L^2[0, 1]$ from Example 8.4 satisfies $CVC = V^*$, where

$$V : L^2[0, 1] \to L^2[0, 1], \quad (Vf)(x) = \int_0^x f(t)dt,$$

is the classical Volterra operator. Thus V is a complex symmetric operator. With the Toeplitz conjugation C on $\mathbb{C}^n$ defined by

$$C(z_1, \ldots, z_n) = (\overline{z_n}, \ldots, \overline{z_1})$$

from Example 8.3, one can show that $T = CT^*C$ for any Toeplitz matrix T. Thus any Toeplitz matrix defines a complex symmetric operator on $\mathbb{C}^n$.

Let C denote the conjugation (8.6) on $\mathcal{K}_u$. The next result from [80] says that truncated Toeplitz operators are complex symmetric operators. We have already seen a version of this for the special truncated Toeplitz operator $A_{\overline{z}}^u = S_u$, the compressed shift (Proposition 9.8).

Theorem 13.12 *For any $A \in \mathcal{T}_u$, we have $A = CA^*C$.*

Proof Recall that $Cf = \overline{\zeta f} u$ on $\mathbb{T}$ and so for $f, g \in \mathcal{K}_u \cap H^\infty$ we obtain

$$\langle CA_\varphi^u Cf, g \rangle = \langle Cg, A_\varphi^u Cf \rangle = \langle Cg, P_u(\varphi Cf) \rangle = \langle Cg, \varphi Cf \rangle$$
$$= \langle \overline{\zeta g}u, \varphi \overline{\zeta f}u \rangle = \langle \overline{g}, \varphi \overline{f} \rangle = \langle \overline{\varphi}f, g \rangle$$

$$= \langle \overline{\varphi} f, P_u g \rangle = \langle P_u(\overline{\varphi} f), g \rangle$$
$$= \langle A_{\overline{\varphi}}^u f, g \rangle = \langle (A_\psi^u)^* f, g \rangle \qquad \square$$

It is suspected that the truncated Toeplitz operators might serve as some sort of model operator for various classes of complex symmetric operators. See the end notes of this chapter for further references. For the moment, let us note that the matrix representation of a truncated Toeplitz operator A_φ^u with respect to a modified Clark basis (see Example 8.15) is complex symmetric (self-transpose). This was first observed in [80] and developed further in [78].

13.4 The spectrum of A_φ^u

The Livšic–Möller Theorem (Theorem 9.22) characterized the spectrum of the compressed shift $S_u = A_z^u$ as the spectrum (see Proposition 7.19)

$$\sigma(u) = \left\{ \lambda \in \mathbb{D}^- : \varliminf_{z \to \lambda} |u(z)| = 0 \right\}$$

of the inner function u. The following result of P. Fuhrmann [76, 77] generalizes this for the analytic truncated Toeplitz operators $\{A_\varphi^u : \varphi \in H^\infty\}$.

Theorem 13.13 *Let u be inner and $\varphi \in H^\infty$. Then*

$$\sigma(A_\varphi^u) = \left\{ \lambda \in \mathbb{C} : \inf_{z \in \mathbb{D}} \left(|u(z)| + |\varphi(z) - \lambda| \right) = 0 \right\}.$$

Proof Let $\psi \in H^\infty$ and observe, by the H^∞-functional calculus for the compressed shift (Theorem 9.20), that

$$(A_\psi^u)^* = T_{\overline{\psi}}|_{\mathcal{K}_u}.$$

For each $w \in \mathbb{D}$, we can write the reproducing kernel k_w for $\mathcal{K}_u$ as

$$k_w(z) = \frac{1 - \overline{u(w)}u(z)}{1 - \overline{w}z} = c_w(z) + \overline{u(w)}u(z)c_w(z).$$

Combine these two facts to get

$$\begin{aligned} (A_\psi^u)^* k_w = A_{\overline{\psi}}^u k_w &= P(\overline{\psi} k_w) \\ &= P(\overline{\psi} c_w) - \overline{u(w)} P(\overline{\psi} u c_w) \\ &= T_{\overline{\psi}} c_w - \overline{u(w)} P(\overline{\psi} u c_w) \\ &= \overline{\psi(w)} c_w - \overline{u(w)} P(\overline{\psi} u c_w). \qquad \text{(by Proposition 4.19)} \end{aligned}$$

Thus

$$\|(A_\psi^u)^* k_w\| \leqslant |\psi(w)| \|c_w\| + |u(w)| \|\psi\|_\infty \|c_w\|$$

$$\leqslant (|\psi(w)| + |u(w)|)(1 + \|\psi\|_\infty)\|c_w\|.$$

Hence, with $\kappa_w = k_w / \|k_w\|$, we get

$$\|(A_\psi^u)^* \kappa_w\| \leqslant \frac{1 + \|\psi\|_\infty}{\sqrt{1 - |u(w)|^2}} (|\psi(w)| + |u(w)|). \tag{13.13}$$

Let $\lambda \in \mathbb{C}$ satisfy

$$\inf_{z \in \mathbb{D}} (|u(z)| + |\varphi(z) - \lambda|) = 0.$$

Then there is a sequence $\{z_n\}_{n \geqslant 1} \subset \mathbb{D}$ such that

$$u(z_n) \to 0 \quad \text{and} \quad \varphi(z_n) \to \lambda.$$

Applying (13.13) to $\psi = \varphi - \lambda$, we have

$$\|(A_\varphi^u - \overline{\lambda}I)\kappa_w\| = \|A_\psi^u \kappa_w\| = \|(A_\psi^u)^* \kappa_w\|$$

$$\leqslant \frac{(1 + |\lambda|) + \|\varphi\|_\infty}{\sqrt{1 - |u(w)|^2}} (|\varphi(w) - \lambda| + |u(w)|).$$

Thus if we let $w = z_n$ and $n \to \infty$, we obtain

$$\inf \left\{ \|(A_\varphi^u - \overline{\lambda}I)f\| : f \in \mathcal{K}_u, \|f\| = 1 \right\} = 0. \tag{13.14}$$

The previous identity says that the operator $A_\varphi^u - \overline{\lambda}I$ is not bounded below and hence is not invertible. Thus

$$\overline{\lambda} \in \sigma(A_\varphi^u) = \sigma((A_\varphi^u)^*) = \overline{\sigma(A_\varphi^u)},$$

and so

$$\left\{ \lambda \in \mathbb{C} : \inf_{z \in \mathbb{D}} (|u(z)| + |\varphi(z) - \lambda|) = 0 \right\} \subset \sigma(A_\varphi^u).$$

To prove the reverse inclusion, we will now show that

$$\inf_{z \in \mathbb{D}} (|u(z)| + |\varphi(z) - \lambda|) > 0 \implies \lambda \notin \sigma(A_\varphi^u).$$

To do this, we appeal to the Corona Theorem [97]. This deep result of Carleson ensures that when

$$\inf_{z \in \mathbb{D}} (|u(z)| + |\varphi(z) - \lambda|) > 0,$$

there are $f, g \in H^\infty$ such that

$$f(z)u(z) + g(z)(\varphi(z) - \lambda) = 1, \qquad z \in \mathbb{D}.$$

By Theorem 9.19 (the H^∞-functional calculus for S_u), the preceding function identity implies the operator identity

$$A_f^u A_u^u + A_g^u (A_\varphi^u - \lambda I) = I.$$

Since $A_u^u = 0$ (Theorem 9.20), the operator identity above reduces to

$$A_g^u (A_\varphi^u - \lambda I) = I.$$

But A_g^u and $A_\varphi^u - \lambda I = A_{\varphi - \lambda}^u$ are analytic truncated Toeplitz operators and hence they commute (see (9.12)). Thus $(A_\varphi^u - \lambda I) A_g^u = I$ and so $\lambda \notin \sigma(A_\varphi^u)$. □

We now give an alternate description of $\sigma(A_\varphi^u)$ in terms of cluster sets for φ. In the course of the proof of Theorem 13.13, we used the fact that

$$\inf_{z \in \mathbb{D}} (|u(z)| + |\varphi(z) - \lambda|) = 0$$

if and only if there is a sequence $\{z_n\}_{n \geqslant 1} \subset \mathbb{D}$ such that

$$u(z_n) \to 0 \quad \text{and} \quad \varphi(z_n) \to \lambda.$$

By the Bolzano–Weierstrass Theorem, we may pass to a subsequence, if necessary, and assume that $\{z_n\}_{n \geqslant 1}$ converges to some point $z_0 \in \mathbb{D}^-$. By Proposition 7.19, we know that $z_0 \in \sigma(u)$. Certainly if $z_0 \in \mathbb{D}$, the continuity of φ near z_0 will imply that $\lambda = \varphi(z_0)$. However, the situation becomes more complicated if $z_0 \in \mathbb{T}$ since $\varphi \in H^\infty$ and is not necessarily continuous near the boundary point z_0. This naturally leads us to the notion of cluster sets. The text [50] is a good source for further information about this.

Let $B(z, r) = \{w : |w - z| < r\}$. For $f \in H^\infty$ and $\zeta \in \mathbb{T}$, the *cluster set* $C(f, \zeta)$ and the *range* $\mathcal{R}(f, \zeta)$ of f at ζ are defined as

$$C(f, \zeta) = \bigcap_{r > 0} f (\mathbb{D} \cap B(\zeta, r))^-$$

and

$$\mathcal{R}(f, \zeta) = \bigcap_{r > 0} f (\mathbb{D} \cap B(\zeta, r)).$$

One can see that $C(f, \zeta)$ is a non-empty, compact, connected subset of $\mathbb{C}$. Moreover, f is continuous at ζ if and only if $C(f, \zeta)$ is a singleton. On the other hand, by the Open Mapping Theorem, $\mathcal{R}(f, \zeta)$ is a connected G_δ subset of $\mathbb{C}$ (possibly empty). Observe that $w \in C(f, \zeta)$ if and only if there is a sequence $\{z_n\}_{n \geqslant 1} \subset \mathbb{D}$ such that

$$\lim_{n \to \infty} z_n = \zeta \quad \text{and} \quad \lim_{n \to \infty} f(z_n) = w,$$

while $w \in \mathcal{R}(f, \zeta)$ if and only if there is a sequence $\{z_n\}_{n \geqslant 1} \subset \mathbb{D}$ such that

$$\lim_{n \to \infty} z_n = \zeta \quad \text{and} \quad f(z_n) \equiv w.$$

This means that $\mathcal{R}(f, \zeta)$ is the set of values assumed by f infinitely many times when we approach ζ from within $\mathbb{D}$, while $C(f, \zeta)$ is the set of values that can be approximated by f as we get closer to ζ. Using this interpretation, one can show that if f is not a constant function and f has an analytic extension to a neighborhood of ζ, then

$$C(f, \zeta) = \{f(\zeta)\} \quad \text{and} \quad \mathcal{R}(f, \zeta) = \varnothing.$$

On the other extreme we have the following.

Proposition 13.14 *If u is a non-constant inner function and $\zeta \in \sigma(u) \cap \mathbb{T}$, then*

$$C(u, \zeta) = \mathbb{D}^- \quad \text{and} \quad \mathbb{D} \setminus \mathcal{E} \subset \mathcal{R}(f, \zeta) \subset \mathbb{D}$$

where

$$\mathcal{E} = \left\{ w \in \mathbb{D} : \frac{w - u}{1 - \overline{w}u} \text{ is not a Blaschke product} \right\}.$$

We remark that $\mathcal{E}$ is the exceptional set for u. It is a set of logarithmic capacity zero (Remark 2.23).

Proof First note that the containment $\mathcal{R}(u, \zeta) \subset \mathbb{D}$ follows from the fact that $u(\mathbb{D}) \subset \mathbb{D}$ since u is non-constant and inner. By Theorem 2.22, there is an exceptional set $\mathcal{E} \subset \mathbb{D}$ such that for any $w \in \mathbb{D} \setminus \mathcal{E}$, the Frostman shift

$$u_w = \tau_{1,w} \circ u = \frac{w - u}{1 - \overline{w}u}$$

is a Blaschke product. We also remind the reader that the singular points of a Blaschke product (that is, those points ξ on $\mathbb{T}$ where the Blaschke product does not analytically continue to a neighborhood of ξ) are precisely the accumulation points of its zero set on $\mathbb{T}$ (Theorem 7.18 and Proposition 7.20). Since we are assuming that $\zeta \in \sigma(u) \cap \mathbb{T}$, the inner function u cannot be analytically continued to a neighborhood of ζ (Proposition 7.20). Hence, for any $w \in \mathbb{D}$, the function u_w also fails to have an analytic continuation to a neighborhood of ζ, since otherwise, $u = \tau_{1,w} \circ u_w$ would have an analytic continuation in a neighborhood of ζ. This means that ζ is also a singular point of u_w and thus, for $w \in \mathbb{D} \setminus \mathcal{E}$, must be an accumulation point of the zeros of u_w. Hence there is a sequence $\{z_n\}_{n \geqslant 1} \subset \mathbb{D}$ such that $z_n \to \zeta$ and

$$u_w(z_n) = \frac{w - u(z_n)}{1 - \overline{w}u(z_n)} = 0.$$

Therefore, $w \in \mathcal{R}(u, \zeta)$, which establishes $\mathbb{D} \setminus \mathcal{E} \subset \mathcal{R}(u, \zeta)$.

Since $\mathcal{E}$ has no interior (Theorem 2.22) and $C(u, \zeta)$ is a compact subset of $\mathbb{D}^-$, the first assertion follows from the containment

$$\mathbb{D} \setminus \mathcal{E} \subset \mathcal{R}(u, \zeta) \subset C(u, \zeta) \subset \mathbb{D}^-. \qquad \square$$

The preceding result shows that a non-constant inner function assumes all values in $\mathbb{D}$, except possibly those in a *small* exceptional subset, infinitely many times in each neighborhood of any of its singular points. Consider a Blaschke product whose zeros accumulate at all points of $\mathbb{T}$, or perhaps a singular inner function with corresponding singular measure whose support is all of $\mathbb{T}$. Such a function exhibits bizarre behavior at all points of $\mathbb{T}$. On one hand, according to Fatou's Theorem, this function has unimodular *non-tangential* limits almost everywhere. On the other hand, this function assumes nearly every value in $\mathbb{D}$ infinitely many times in the vicinity of any boundary point.

The above discussion shows that Theorem 13.13 can be rewritten as

$$\sigma(A_\varphi^u) = \{\varphi(z) : z \in \sigma(u) \cap \mathbb{D}\} \cup \{C(\varphi, \zeta) : \zeta \in \sigma(u) \cap \mathbb{T}\}. \qquad (13.15)$$

For certain φ, we have the following *spectral mapping* result. Recall from Definition 5.23 that the disk algebra $\mathcal{A}$ is the set of all analytic functions on $\mathbb{D}$ which extend to be continuous on $\mathbb{D}^-$.

Corollary 13.15 *Let u be inner and let $\varphi \in \mathcal{A}$. Then*

$$\sigma(A_\varphi^u) = \varphi(\sigma(A_z^u)) = \varphi(\sigma(u)).$$

Proof   Since φ is continuous on $\mathbb{D}^-$, we have $C(\varphi, \zeta) = \{\varphi(\zeta)\}$. The result now follows from (13.15). $\qquad \square$

Recall Theorem 9.28, where a similar result holds for the essential spectrum of A_φ^u. The following result complements Theorem 13.13 by giving a complete description of the point spectrum of A_φ^u.

Theorem 13.16 *Let u be inner and $\varphi \in H^\infty$. Fix $\lambda \in \mathbb{C}$ and set*

$$v = \gcd\left(((\varphi - \lambda)_i, u\right),$$

where $(\varphi - \lambda)_i$ is the inner part of $\varphi - \lambda$. Then

$$\ker(A_\varphi^u - \lambda I) = \frac{u}{v} \mathcal{K}_v \qquad (13.16)$$

and

$$\ker(A_{\bar\varphi}^u - \bar\lambda I) = \mathcal{K}_v. \qquad (13.17)$$

In particular, the following are equivalent:

(i) $\lambda \in \sigma_p(A_\varphi^u)$;

(ii) $\overline{\lambda} \in \sigma_p(A_{\overline{\varphi}}^u)$;

(iii) $v - \gcd((\varphi - \lambda)_i, u)$ *is not constant.*

Proof Let $\widetilde{u} = u/v$. For $f \in \mathcal{K}_u$ we have

$$f \in \ker(A_\varphi^u - \lambda I) \iff f \in \ker A_{\varphi-\lambda}^u$$
$$\iff P_u((\varphi - \lambda)f) = 0$$
$$\iff u|(\varphi - \lambda)f$$
$$\iff \widetilde{u}|f.$$

Therefore,

$$\ker(A_\varphi^u - \lambda I) = \mathcal{K}_u \cap \widetilde{u}H^2$$
$$= (\mathcal{K}_{\widetilde{u}} \oplus \widetilde{u}\mathcal{K}_v) \cap \widetilde{u}H^2 \qquad \text{(Lemma 5.10)}$$
$$= \widetilde{u}\mathcal{K}_v,$$

which proves (13.16).

To prove (13.17), we use Theorem 13.12 to see that if $Cf = \overline{zfu}$ is the conjugation operator on $\mathcal{K}_u$ we have

$$C A_\varphi^u C = A_{\overline{\varphi}}^u.$$

From here it follows that

$$\ker(A_{\overline{\varphi}}^u - \overline{\lambda}I) = C \ker(A_\varphi^u - \lambda I)$$
$$= C(\frac{u}{v}\mathcal{K}_v)$$
$$= \{C(\frac{u}{v}f) : f \in \mathcal{K}_v\}$$
$$= \{C(\frac{u}{v}\overline{zg}v) : g \in \mathcal{K}_v\} \qquad \text{(by Proposition 5.4)}$$
$$= \{\frac{\overline{u}}{\overline{v}}zg\overline{v}u\overline{z} : g \in \mathcal{K}_v\}$$
$$= \mathcal{K}_v.$$

The equivalence of statements (i), (ii), and (iii) now follow. □

Corollary 13.17 *Let u be inner and let $z_0 \in \mathbb{D}$ be such that $u(z_0) = 0$. For any $\varphi \in H^\infty$ we have*

$$\varphi(z_0) \in \sigma_p(A_\varphi^u) \quad \text{and} \quad \overline{\varphi(z_0)} \in \sigma_p(A_{\overline{\varphi}}^u). \qquad (13.18)$$

If φ is univalent, then

$$\ker(A_\varphi^u - \varphi(z_0)I) = \mathbb{C}Q_{z_0}u \quad and \quad \ker(A_\varphi^u - \overline{\varphi(z_0)}I) = \mathbb{C}c_{z_0}. \qquad (13.19)$$

Proof   The proof of (13.18) follows directly from Theorem 13.16. To prove the first identity in (13.19), observe that since φ is univalent, we get

$$(\varphi - \varphi(z_0))_i = \frac{z - z_0}{1 - \overline{z_0}z},$$

Furthermore, the single Blaschke factor on the right-hand side of the above is an inner factor of u and so

$$v = \gcd((\varphi - \varphi(z_0))_i, u) = \frac{z - z_0}{1 - \overline{z_0}z}.$$

Finally, from Proposition 5.16 we have

$$\mathcal{K}_v = \mathbb{C}\frac{1}{1 - \overline{z_0}z}$$

and so, by (13.16), we see that

$$\ker(A_\varphi^u - \varphi(z_0)I) = u\frac{1 - \overline{z_0}z}{z - z_0}\mathbb{C}\frac{1}{1 - \overline{z_0}z} = \mathbb{C}Q_{z_0}u.$$

One can prove the other identity in (13.19) by noting that $Q_{z_0}u = Ck_{z_0}$ and $k_{z_0} = c_{z_0}$ (since $u(z_0) = 0$), together with the identity

$$\ker(A_\varphi^u - \overline{\varphi(z_0)}I) = C\ker(A_\varphi^u - \varphi(z_0)I). \qquad \square$$

For Blaschke products, we can be even more specific. We write B_Λ for the Blaschke product corresponding to the Blaschke sequence Λ.

Corollary 13.18   *Let $\varphi \in H^\infty$ and let $B = B_\Lambda$ be a Blaschke product with simple zeros. Then*

$$\lambda \in \sigma_p(A_\varphi^B) \quad \Longleftrightarrow \quad \Lambda' = \varphi^{-1}(\{\lambda\}) \cap \Lambda \neq \varnothing.$$

Moreover, for each $\lambda \in \sigma_p(A_\varphi^B)$,

$$\ker(A_\varphi^B - \lambda I) = B_{\Lambda''}\mathcal{K}_{B_{\Lambda'}}, \qquad (13.20)$$

and

$$\ker(A_\varphi^B - \overline{\lambda}I) = \mathcal{K}_{B_{\Lambda'}}, \qquad (13.21)$$

where $\Lambda'' = \Lambda \setminus \Lambda'$. In particular, for each $z_0 \in \Lambda$,

$$A_\varphi^B\left(\frac{B}{z - z_0}\right) = \varphi(z_0)\frac{B}{z - z_0} \qquad (13.22)$$

and

$$A_{\bar\varphi}^B c_{z_0} = \overline{\varphi(z_0)} c_{z_0}. \tag{13.23}$$

Proof The first part follows from Theorem 13.16 and the fact that the only divisors of a Blaschke product are its partial products. The equations (13.20) and (13.21) follow respectively from equations (13.16) and (13.17). Finally, for each $z_0 \in \Lambda$ and $\lambda := \varphi(z_0)$, we have $z_0 \in \Lambda'$. Thus, both functions $k_{z_0} = c_{z_0}$ and $\frac{B}{z - z_0}$ belong to $\mathcal{K}_{B_{\Lambda'}}$. This observation implies (13.22) and (13.23). □

13.5 An operator disintegration formula

Here is a fascinating operator integral formula that relates truncated Toeplitz operators, the Clark unitary operators U_α (see (11.6)), and the Aleksandrov disintegration formula (Theorem 11.16). The Spectral Theorem ensures that if φ is a bounded Borel function on $\mathbb{T}$ then $\varphi(U_\alpha)$ is a well-defined normal operator.

Theorem 13.19 *If φ is a bounded Borel function on $\mathbb{T}$ and u is inner, then A_φ^u can be written as*

$$A_\varphi^u = \int_{\mathbb{T}} \varphi(U_\alpha)\, dm(\alpha).$$

By this we mean that for any $f, g \in \mathcal{K}_u$

$$\langle A_\varphi^u f, g \rangle = \int_{\mathbb{T}} \langle \varphi(U_\alpha) f, g \rangle\, dm(\alpha).$$

Proof Let $f, g \in \mathcal{K}_u$. Then

$$\langle A_\varphi^u f, g \rangle = \langle P_u(\varphi f), g \rangle = \langle \varphi f, P_u g \rangle = \langle \varphi f, g \rangle$$

$$= \int_{\mathbb{T}} \varphi f \bar{g}\, dm$$

$$= \int_{\mathbb{T}} \left(\int_{\mathbb{T}} \varphi f \bar{g}\, d\sigma_\alpha \right) dm(\alpha) \qquad \text{(Remark 11.17)}$$

$$= \int_{\mathbb{T}} \left(\int_{\mathbb{T}} \varphi V_\alpha f \overline{V_\alpha g}\, d\sigma_\alpha \right) dm(\alpha) \qquad \text{(Theorem 11.14)}$$

$$= \int_{\mathbb{T}} \langle M_\varphi V_\alpha f, V_\alpha g \rangle_{L^2(\sigma_\alpha)}\, dm(\alpha) \qquad \text{(Theorem 11.4)}$$

$$= \int_{\mathbb{T}} \langle V_\alpha^* M_\varphi V_\alpha f, V_\alpha^* V_\alpha g \rangle_{H^2}\, dm(\alpha) \qquad \text{(Theorem 11.6)}$$

$$= \int_{\mathbb{T}} \langle \varphi(U_\alpha) f, g \rangle_{H^2}\, dm(\alpha). \qquad \square$$

There are problems that arise when φ is unbounded. For example, how does one define $\varphi(U_\alpha)$ in this case? The paper [166] carefully documents some of the technical difficulties that arise when ψ is unbounded.

13.6 Norm of a truncated Toeplitz operator

For Toeplitz operators, recall that $\|T_\varphi\| = \|\varphi\|_\infty$ for each $\varphi \in L^\infty$. In contrast to this, we can say little more than

$$0 \leqslant \|A_\varphi^u\| \leqslant \|\varphi\|_\infty \qquad (13.24)$$

for general truncated Toeplitz operators with bounded symbols. In fact, computing, or at least estimating, the norm of a truncated Toeplitz operator is a difficult problem. In a way, the problem seems unfair since, as we have seen in Theorem 13.6, there are many symbols that represent the same truncated Toeplitz operator. This makes it difficult to pose theorems about norms in terms of the symbol as one did so neatly for Toeplitz operators (and composition operators).

However, it is possible to obtain lower estimates of $\|A_\varphi^u\|$ for general φ in L^2. This can be helpful, for instance, in determining whether or not a given truncated Toeplitz operator is unbounded. Although a variety of lower bounds on $\|A_\varphi^u\|$ are provided in [92], we focus here on perhaps the most useful of these. Fatou's Theorem (Theorem 1.10) says that $\lim_{r\to 1^-} \mathscr{P}(\varphi)(r\xi) = \varphi(\xi)$ whenever φ is continuous at ξ, or, more generally, whenever ξ is a Lebesgue point of φ. Here ξ is a *Lebesgue point* of φ if

$$\lim_{\delta \to 0^+} \frac{1}{m(I_{\delta,\xi})} \int_{I_{\delta,\xi}} |\varphi(\zeta) - \varphi(\xi)| \, dm(\zeta) = 0,$$

where $I_{\delta,\xi}$ is the arc of $\mathbb{T}$ subtended by the points $e^{-i\delta}\xi$ and $e^{i\delta}\xi$. The Lebesgue Differentiation Theorem says that almost every point of $\mathbb{T}$ is a Lebesgue point for φ [158, p. 165].

Theorem 13.20   *If $\varphi \in L^2$ and u is inner, then*

$$\|A_\varphi^u\| \geqslant \sup\{|\mathscr{P}(\varphi)(\lambda)| : \lambda \in \mathbb{D} : u(\lambda) = 0\},$$

where the supremum above is regarded as 0 if u never vanishes on $\mathbb{D}$.

Proof   The proof is similar to the one used in Proposition 4.12 to compute the norm of a Toeplitz operator. When $u(\lambda) = 0$, the corresponding kernel function k_λ turns out to be c_λ. Now follow the rest of the proof of Proposition 4.12.   □

Corollary 13.21 *If $\varphi \in C(\mathbb{T})$ and u is an inner function whose zeros accumulate almost everywhere on $\mathbb{T}$, then $\|A_\varphi^u\| = \|\varphi\|_\infty$.*

At the expense of wordiness, the hypothesis of the previous corollary can be considerably weakened. We only need $\zeta \in \mathbb{T}$ to be a limit point of the zeros of u, the symbol $\varphi \in L^\infty$ to be continuous on an open arc containing ζ, and $|\varphi(\zeta)| = \|\varphi\|_\infty$.

13.7 Notes

Although we covered the basics of truncated Toeplitz operators, the subject is quickly growing [93]. Below are some topics of interest in the literature.

13.7.1 The bounded symbol problem

From the estimate $\|A_\varphi^u\| \leqslant \|\varphi\|_\infty$ we see that $A_\varphi^u \in \mathscr{T}_u$ whenever $\varphi \in L^\infty$. However, one can take any unbounded $\psi \in uH^2 + \overline{uH^2}$ and observe that $A_\varphi^u = A_{\varphi+\psi}^u$. Thus bounded truncated Toeplitz operators always have unbounded symbols. If a truncated Toeplitz operator is a bounded operator, does it have a bounded symbol? The answer, in general, is no and a specific example was provided in [24]. There are, however, technical conditions one can impose on the inner function u to guarantee that every operator from $\mathscr{T}_u$ has a bounded symbol [25].

13.7.2 Finite-rank operators

By Theorem 4.15, there are no non-zero compact Toeplitz operators on H^2. In contrast to this, there are many examples of finite-rank, hence compact, truncated Toeplitz operators. In fact, the rank-one truncated Toeplitz operators were first identified by Sarason [166, Thm. 5.1] who proved that

$$k_\lambda \otimes Ck_\lambda = A_{\frac{\overline{u}}{\overline{z}-\overline{\lambda}}}^u,$$

$$Ck_\lambda \otimes k_\lambda = A_{\frac{u}{z-\lambda}}^u,$$

$$k_\zeta \otimes k_\zeta = A_{k_\zeta + \overline{k_\zeta} - 1}^u,$$

where $\lambda \in \mathbb{D}$ and u has a finite angular derivative, in the sense of Carathéodory, at $\zeta \in \mathbb{T}$. Furthermore, any truncated Toeplitz operator of rank-one is a scalar multiple of one of the above operators.

We should also mention the somewhat more involved results of Sarason [166, Thms. 6.1 & 6.2] which identify a variety of natural finite-rank truncated Toeplitz operators. The full classification of the finite-rank truncated Toeplitz operators was given by Bessonov in [29].

13.7.3 Topologies on $\mathscr{T}_u$

Baranov, Bessonov, and Kapustin identified the predual of $\mathscr{T}_u$ and discussed the weak-$*$ topology on $\mathscr{T}_u$ [25]. Consider the space

$$\mathcal{X}_u := \Big\{ F = \sum_{n \geqslant 1} f_n \overline{g_n} \; : \; f_n, g_n \in \mathcal{K}_u, \; \sum_{n \geqslant 1} \|f_n\| \, \|g_n\| < \infty \Big\}$$

with norm

$$\|F\|_{\mathcal{X}_u} := \inf \Big\{ \sum_{n \geqslant 1} \|f_n\| \, \|g_n\| : F = \sum_{n \geqslant 1} f_n \overline{g_n} \Big\}.$$

It turns out that $\mathcal{X}_u$ is a Banach space of functions on $\mathbb{D}$ and, in terms of non-tangential boundary values,

$$\mathcal{X}_u \subset \overline{u} z H^1 \cap \overline{u z H^1}.$$

Furthermore, each element of $\mathcal{X}_u$ can be written as a linear combination of four elements of the form $f\overline{g}$, where $f, g \in \mathcal{K}_u$. It turns out that $\mathcal{X}_u^*$, the dual space of $\mathcal{X}_u$, is isometrically isomorphic to $\mathscr{T}_u$ via the dual pairing

$$(F, A) := \sum_{n \geqslant 1} \langle A f_n, g_n \rangle, \quad F = \sum_{n \geqslant 1} f_n \overline{g_n} \in \mathcal{X}_u, \; A \in \mathscr{T}_u.$$

Moreover, if $\mathscr{T}_u^c$ denotes the compact truncated Toeplitz operators, then $(\mathscr{T}_u^c)^*$, the dual of $\mathscr{T}_u^c$, is isometrically isomorphic to $\mathcal{X}_u$.

These results go on to say that the weak topology and the weak-$*$ topology on $\mathscr{T}_u$ are the same. Moreover, the norm closed linear span of the rank-one truncated Toeplitz operators is $\mathscr{T}_u^c$ and $\mathscr{T}_u^c$ is weakly dense in $\mathscr{T}_u$. Finally, the truncated Toeplitz operators with bounded symbols are weakly dense in $\mathscr{T}_u$.

13.7.4 The spatial isomorphism problem

For two inner functions u_1 and u_2, when is $\mathscr{T}_{u_1}$ *spatially isomorphic* to $\mathscr{T}_{u_2}$? In other words, when does there exist a unitary operator $U : \mathcal{K}_{u_1} \to \mathcal{K}_{u_2}$

such that $U \mathscr{T}_{u_1} U^* = \mathscr{T}_{u_2}$? This is evidently a stronger condition than isometric isomorphism since one insists that the isometric isomorphism is implemented in a particularly restrictive manner.

A concrete solution to the spatial isomorphism problem posed above was given in [45]. To describe its solution, let us consider several special cases.

(i) Theorem 6.8 says that if ψ is a disk automorphism, then the weighted composition operator

$$U_\psi : \mathcal{K}_u \to \mathcal{K}_{u \circ \psi}, \qquad U_\psi f = \sqrt{\psi'}(f \circ \psi),$$

is unitary. Moreover, the map

$$\Lambda_\psi : \mathscr{T}_u \to \mathscr{T}_{u \circ \psi}, \qquad \Lambda_\psi(A) = U_\psi A U_\psi^*,$$

satisfies $\Lambda_\psi(A_\varphi^u) = A_{\varphi \circ \psi}^{u \circ \psi}$ and thus implements a spatial isomorphism between $\mathscr{T}_u$ and $\mathscr{T}_{u \circ \psi}$.

(ii) From Theorem 6.7, recall that for each $a \in \mathbb{D}$ and disk automorphism $\tau_{1,a}$, one can verify that the Crofoot transform

$$J_a : \mathcal{K}_u \to \mathcal{K}_{\tau_{1,a} \circ u}, \qquad J_a f = \frac{\sqrt{1 - |a|^2}}{1 - \bar{a}u} f,$$

is unitary. Moreover, the map

$$\Lambda_a : \mathscr{T}_u \to \mathscr{T}_{\tau_{1,a} \circ u}, \qquad \Lambda_a(A) = J_a A J_a^*,$$

implements a spatial isomorphism between $\mathscr{T}_u$ and $\mathscr{T}_{\tau_{1,a} \circ u}$.

(iii) From Theorem 6.9, recall the unitary operator $U_\# : \mathcal{K}_u \to \mathcal{K}_{u^\#}$ defined by

$$[U_\# f](\zeta) = \bar{\zeta} f(\bar{\zeta}) u^\#(\zeta), \qquad u^\#(z) := \overline{u(\bar{z})}.$$

The map

$$\Lambda_\# : \mathscr{T}_u \to \mathscr{T}_{u^\#}, \qquad \Lambda_\#(A) = U_\# A U_\#^*,$$

satisfies $\Lambda_\#(A_\varphi^u) = A_{\varphi^\#}^{u^\#}$ whence $\mathscr{T}_u$ is spatially isomorphic to $\mathscr{T}_{u^\#}$.

It turns out that any spatial isomorphism between truncated Toeplitz operator spaces can be written in terms of the three basic types described above. Indeed, for any two inner functions u_1 and u_2, the spaces $\mathscr{T}_{u_1}$ and $\mathscr{T}_{u_2}$ are spatially isomorphic if and only if either $u_1 = \psi \circ u_2 \circ \varphi$ or $u_1 = \psi \circ u_2^\# \circ \varphi$ for some disk automorphisms φ, ψ. Moreover, any spatial isomorphism $\Lambda : \mathscr{T}_{u_1} \to \mathscr{T}_{u_2}$ can be written as $\Lambda = \Lambda_a \Lambda_\psi$ or $\Lambda_a \Lambda_\# \Lambda_\psi$, where we allow $a = 0$ or $\psi(z) = z$.

13.7.5 Algebras of truncated Toeplitz operators

Recall that $\mathscr{T}_u$ is a weakly closed subspace of $\mathcal{B}(\mathcal{K}_u)$ (Theorem 13.11). Although $\mathscr{T}_u$ is not an algebra, there are many interesting (weakly closed) algebras contained within $\mathscr{T}_u$. In fact, the thesis [169] and subsequent paper [170] of Sedlock describes them all. We discuss the properties of these so-called *Sedlock algebras* below, along with several further results from [96]. For the sake of comparison, recall Theorem 4.22 which describes when the product of two Toeplitz operators is another Toeplitz operator. What is the corresponding result for $\mathscr{T}_u$?

To begin, we require the following generalization (see [166, Sect. 10]) of the Clark unitary operators defined earlier in (11.6):

$$S_u^a = S_u + \frac{a}{1 - \overline{u(0)}a} k_0 \otimes C k_0, \quad a \in \mathbb{D}^-. \tag{13.25}$$

These operators turn out to be fundamental to the study of Sedlock algebras.

Before proceeding, let us recall a few facts from Chapter 1. For $A \in \mathscr{T}_u$, the commutant $\{A\}'$ of A is defined to be the set of all $B \in \mathcal{B}(\mathcal{K}_u)$ such that $AB = BA$. The (WOT) closed linear span of $\{A^n : n \geqslant 0\}$ will be denoted by $\mathcal{W}(A)$. Note that $\mathcal{W}(A) \subset \{A\}'$ and that $\{A\}'$ is a (WOT) closed subset of $\mathcal{B}(\mathcal{K}_u)$. The relevance of these concepts lies in the following two results from [166, p. 515] and [96], respectively. Namely that for each $a \in \mathbb{D}^-$,

$$\{S_u^a\}' \subset \mathscr{T}_u \quad \text{and} \quad \{S_u^a\}' = \mathcal{W}(S_u^a).$$

The preceding results tell us that $\mathcal{W}(S_u^a)$ and $\mathcal{W}((S_u^b)^*)$, where a, b belong to $\mathbb{D}^-$, are weakly closed commutative algebras contained in $\mathscr{T}_u$. We adopt the following notation introduced by Sedlock [170]:

$$\mathscr{B}_u^a := \begin{cases} \mathcal{W}(S_u^a) & \text{if } a \in \mathbb{D}^-, \\ \mathcal{W}((S_u^{1/\overline{a}})^*) & \text{if } a \in \widehat{\mathbb{C}} \setminus \mathbb{D}^-. \end{cases}$$

Note that $\mathscr{B}_u^0$ is the algebra of analytic truncated Toeplitz operators ($\mathscr{B}_u^0 = \mathcal{W}(A_z^u) = \{A_z^u\}'$) while $\mathscr{B}_u^\infty$ is the algebra of conjugate analytic truncated Toeplitz operators ($\mathscr{B}_u^\infty = \mathcal{W}(A_{\overline{z}}^u) = \{A_{\overline{z}}^u\}'$). Sedlock's result says that the algebras $\mathscr{B}_u^a$ for $a \in \widehat{\mathbb{C}}$ are the only maximal abelian algebras in $\mathscr{T}_u$.

13.7.6 Norms of truncated Toeplitz operators

The norm of A_φ^u is related to the norm of a Hankel operator as well as to certain classical extremal problems. We refer the reader to the papers [38, 91] for more on this.

13.8 For further exploration

13.8.1 Models for complex symmetric operators

The class of complex symmetric operators has undergone much study recently [39, 55, 78–83, 88–90, 98, 115, 116, 123, 191, 192, 194–196] and is deserving of further study.

We have seen that every truncated Toeplitz operator is complex symmetric (Theorem 13.12). An increasingly long list of complex symmetric operators have been proven to be unitarily equivalent to truncated Toeplitz operators [93]. Is every complex symmetric operator unitarily equivalent to a truncated Toeplitz operator? If not, which ones are? One can also explore when a bounded operator is similar to a truncated Toeplitz operator. In finite dimensions, every operator is similar to a truncated Toeplitz operator [45]. What happens in infinite dimensions?

13.8.2 Operator theory for truncated Toeplitz operators

Some natural operator theoretic questions about truncated Toeplitz operators (when are they normal, self-adjoint, unitary?) are beginning to be answered [37]. Others, such as a description of the spectrum and the norm, remain open.

13.8.3 Unitary equivalence

When are A_φ^u and A_ψ^v unitarily equivalent? This problem was solved for compressed shifts $S_u = A_z^u$ and $S_v = A_z^v$ (see the end notes for Chapter 9). What happens in general? Be aware that problems like this could be tricky to answer due to the fact that there is an entire family of symbols that represent the operator (Theorem 13.6). In either case, the answer to these types of operator theoretic questions about truncated Toeplitz operators should depend on the relationship between the inner function (which defines the space) and the family of symbols (which define the operator).

13.8.4 Products of truncated Toeplitz operators

As discussed earlier in the end notes for this chapter, $\mathscr{T}_u$ is not an algebra. This leads us to the question as to what types of operators in $\mathcal{B}(\mathcal{K}_u)$ can be written as a finite product of elements from $\mathscr{T}_u$. One could also take infinite products but, as expected, there are convergence issues to overcome. When $u = z^n$, $\mathscr{T}_u$ can be identified with the $n \times n$ Toeplitz matrices (see (13.3)) and it is known that every matrix can be written as a product of Toeplitz matrices [124]. There are even bounds on how many Toeplitz matrices are needed to do this. It seems reasonable that the same holds (every operator on $\mathcal{K}_u$ can be written as a finite product of elements from $\mathcal{T}_u$) when u is a finite Blaschke product. For general inner functions u, how far can one go with this?

13.8.5 Weak factorization

The authors in [25] considered the Banach space

$$\mathcal{K}_u \odot \mathcal{K}_u := \Big\{ \sum_{n \geqslant 1} f_n g_n : f_n, g_n \in \mathcal{K}_u, \sum_{n \geqslant 1} \|f_n\| \|g_n\| < \infty \Big\}$$

of analytic functions on $\mathbb{D}$ with the norm

$$\|f\|_{\mathcal{K}_u \odot \mathcal{K}_u} := \inf \Big\{ \sum_{n \geqslant 1} \|f_n\| \|g_n\| : f = \sum_{n \geqslant 1} f_n g_n \Big\}.$$

For certain inner functions u, there is a description of $\mathcal{K}_u \odot \mathcal{K}_u$ that relates to the problem of whether or not every operator in $\mathcal{T}_u$ has a bounded symbol. For general u, the space $\mathcal{K}_u \odot \mathcal{K}_u$ is not completely understood. Does anything interesting come out of exploring $\mathcal{K}_u \odot \mathcal{K}_v$ for two inner functions u and v?

References

[1] Abakumov, E. V. 1995. Cyclicity and approximation by lacunary power series. *Michigan Math. J.*, **42**(2), 277–299.

[2] Agler, J. and McCarthy, J. E. 2002. *Pick Interpolation and Hilbert Function Spaces*. Graduate Studies in Mathematics, vol. 44. Providence, RI: American Mathematical Society.

[3] Ahern, P. R. and Clark, D. N. 1970a. On functions orthogonal to invariant subspaces. *Acta Math.*, **124**, 191–204.

[4] Ahern, P. R. and Clark, D. N. 1970b. Radial limits and invariant subspaces. *Amer. J. Math.*, **92**, 332–342.

[5] Ahern, P. R. and Clark, D. N. 1971. Radial nth derivatives of Blaschke products. *Math. Scand.*, **28**, 189–201.

[6] Akhiezer, N. I. and Glazman, I. M. 1993. *Theory of Linear Operators in Hilbert Space*. New York: Dover Publications. Translated from the Russian and with a preface by Merlynd Nestell, reprint of the 1961 and 1963 translations, two volumes bound as one.

[7] Aleksandrov, A. B. 1987. Multiplicity of boundary values of inner functions. *Izv. Akad. Nauk Armyan. SSR Ser. Mat.*, **22**(5), 490–503, 515.

[8] Aleksandrov, A. B. 1989. Inner functions and related spaces of pseudocontinuable functions. *Zap. Nauchn. Sem. Leningrad. Otdel. Mat. Inst. Steklov. (LOMI)*, **170**(17), 7–33, 321.

[9] Aleksandrov, A. B. 1995. On the existence of angular boundary values of pseudocontinuable functions. *Zap. Nauchn. Sem. S.-Peterburg. Otdel. Mat. Inst. Steklov. (POMI)*, **222**(23), 5–17, 307.

[10] Aleksandrov, A. B. 1996. Isometric embeddings of co-invariant subspaces of the shift operator. *Zap. Nauchn. Sem. S.-Peterburg. Otdel. Mat. Inst. Steklov. (POMI)*, **232**(24), 5–15, 213.

[11] Aleksandrov, A. B. 1997. Gap series and pseudocontinuations: an arithmetic approach. *Algebra i Analiz*, **9**(1), 3–31.

[12] Aleksandrov, A. B. 1999. Embedding theorems for coinvariant subspaces of the shift operator. II. *Zap. Nauchn. Sem. S.-Peterburg. Otdel. Mat. Inst. Steklov. (POMI)*, **262**(27), 5–48, 231.

[13] Aleman, A. and Richter, S. 1996. Simply invariant subspaces of H^2 of some multiply connected regions. *Integral Equations Operator Theory*, **24**(2), 127–155.

[14] Aleman, A. and Ross, W. T. 1996. The backward shift on weighted Bergman spaces. *Michigan Math. J.*, **43**(2), 291–319.

[15] Aleman, A., Richter, S. and Ross, W. T. 1998. Pseudocontinuations and the backward shift. *Indiana Univ. Math. J.*, **47**(1), 223–276.

[16] Aleman, A., Feldman, N. S., and Ross, W. T. 2009. *The Hardy Space of a Slit Domain*. Frontiers in Mathematics. Basel: Birkhäuser Verlag.

[17] Arias de Reyna, J. 2002. *Pointwise Convergence of Fourier Series*. Lecture Notes in Mathematics, vol. 1785. Berlin: Springer-Verlag.

[18] Aronszajn, N. 1950. Theory of reproducing kernels. *Trans. Amer. Math. Soc.*, **68**, 337–404.

[19] Arov, D. Z. 1971. Darlington's method in the study of dissipative systems. *Dokl. Akad. Nauk SSSR*, **201**(3), 559–562.

[20] Arveson, W. 2002. *A Short Course on Spectral Theory*. Graduate Texts in Mathematics, vol. 209. New York: Springer-Verlag.

[21] Bagemihl, F. and Seidel, W. 1954. Some boundary properties of analytic functions. *Math. Z.*, **61**, 186–199.

[22] Ball, J. 1973. Unitary perturbations of contractions. Ph.D. thesis, University of Virginia.

[23] Ball, J. A. and Lubin, A. 1976. On a class of contractive perturbations of restricted shifts. *Pacific J. Math.*, **63**(2), 309–323.

[24] Baranov, A. I., Chalendar, E., Fricain, E., Mashreghi, J., and Timotin, D. 2010. Bounded symbols and reproducing kernel thesis for truncated Toeplitz operators. *J. Funct. Anal.*, **259**(10), 2673–2701.

[25] Baranov, A., Bessonov, R. and Kapustin, V. 2011. Symbols of truncated Toeplitz operators. *J. Funct. Anal.*, **261**(12), 3437–3456.

[26] Bari, N. K. 1951. Biorthogonal systems and bases in Hilbert space. *Moskov. Gos. Univ. Učenye Zapiski Matematika*, **148**(4), 69–107.

[27] Bercovici, H. 1988. *Operator Theory and Arithmetic in H^∞*. Mathematical Surveys and Monographs, vol. 26. Providence, RI: American Mathematical Society.

[28] Berman, R. D. and Cohn, W. S. 1987. Tangential limits of Blaschke products and functions of bounded mean oscillation. *Illinois J. Math.*, **31**(2), 218–239.

[29] Bessonov, R. V. 2014. Truncated Toeplitz operators of finite rank. *Proc. Amer. Math. Soc.*, **142**(4), 1301–1313.

[30] Beurling, A. 1948. On two problems concerning linear transformations in Hilbert space. *Acta Math.*, **81**, 17.

[31] Blandignères, A., Fricain, E., Gaunard, F., Hartmann, A. and Ross, W. 2013. Reverse Carleson embeddings for model spaces. *J. Lond. Math. Soc. (2)*, **88**(2), 437–464.

[32] Boas, Jr., R. P. 1941. A general moment problem. *Amer. J. Math.*, **63**, 361–370.

[33] Böttcher, A. and Silbermann, B. 2006. *Analysis of Toeplitz Operators*, 2nd edn. Springer Monographs in Mathematics. Berlin: Springer-Verlag. Prepared jointly with Alexei Karlovich.

[34] Brown, A. and Halmos, P. R. 1963/1964. Algebraic properties of Toeplitz operators. *J. Reine Angew. Math.*, **213**, 89–102.

[35] Brown, L., and Shields, A. L. 1984. Cyclic vectors in the Dirichlet space. *Trans. Amer. Math. Soc.*, **285**(1), 269–303.

[36] Cargo, G. T. 1962. Angular and tangential limits of Blaschke products and their successive derivatives. *Canad. J. Math.*, **14**, 334–348.

[37] Chalendar, I. and Timotin, D. Commutation relations for truncated Toeplitz operators. arXiv:1305.6739.

[38] Chalendar, I., Fricain, E., and Timotin, D. 2009. On an extremal problem of Garcia and Ross. *Oper. Matrices*, **3**(4), 541–546.

[39] Chevrot, N., Fricain, E., and Timotin, D. 2007. The characteristic function of a complex symmetric contraction. *Proc. Amer. Math. Soc.*, **135**(9), 2877–2886 (electronic).

[40] Christensen, O. 2003. *An Introduction to Frames and Riesz Bases*. Applied and Numerical Harmonic Analysis. Boston, MA: Birkhäuser Boston Inc.

[41] Cima, J. A. and Ross, W. T. 2000. *The Backward Shift on the Hardy Space*. Mathematical Surveys and Monographs, vol. 79. Providence, RI: American Mathematical Society.

[42] Cima, J. A., Matheson, A. L., and Ross, W. T. 2006. *The Cauchy Transform*. Mathematical Surveys and Monographs, vol. 125. Providence, RI: American Mathematical Society.

[43] Cima, J. A., Ross, W. T., and Wogen, W. R. 2008. Truncated Toeplitz operators on finite dimensional spaces. *Oper. Matrices*, **2**(3), 357–369.

[44] Cima, J. A. and Matheson, A. L. 1997. Essential norms of composition operators and Aleksandrov measures. *Pacific J. Math.*, **179**(1), 59–64.

[45] Cima, J. A., Garcia, S. R., Ross, W. T., and Wogen, W. R. 2010. Truncated Toeplitz operators: spatial isomorphism, unitary equivalence, and similarity. *Indiana Univ. Math. J.*, **59**(2), 595–620.

[46] Clark, D. N. 1972. One dimensional perturbations of restricted shifts. *J. Analyse Math.*, **25**, 169–191.

[47] Coburn, L. A. 1967. The C^*-algebra generated by an isometry. *Bull. Amer. Math. Soc.*, **73**, 722–726.

[48] Coburn, L. A. 1969. The C^*-algebra generated by an isometry. II. *Trans. Amer. Math. Soc.*, **137**, 211–217.

[49] Cohn, B. 1982. Carleson measures for functions orthogonal to invariant subspaces. *Pacific J. Math.*, **103**(2), 347–364.

[50] Collingwood, E. F. and Lohwater, A. J. 1966. *The Theory of Cluster Sets*. Cambridge Tracts in Mathematics and Mathematical Physics, No. 56. Cambridge: Cambridge University Press.

[51] Conway, J. B. 1990. *A Course in Functional Analysis*, 2nd edn. Graduate Texts in Mathematics, vol. 96. New York: Springer-Verlag.

[52] Conway, J. B. 2000. *A Course in Operator Theory*. Graduate Studies in Mathematics, vol. 21. Providence, RI: American Mathematical Society.

[53] Cowen, C. and MacCluer, B. D. 1995. *Composition Operators on Spaces of Analytic Functions*. Studies in Advanced Mathematics. Boca Raton, FL: CRC Press.

[54] Crofoot, R. B. 1994. Multipliers between invariant subspaces of the backward shift. *Pacific J. Math.*, **166**(2), 225–246.

[55] Danielger, J., Garcia, S. R., and Putinar, M. 2008. Variational principles for symmetric bilinear forms. *Math. Nachr.*, **281**(6), 786–802.

[56] Davidson, K. R. 1996. *C*-algebras by example*. Fields Institute Monographs, vol. 6. Providence, RI: American Mathematical Society.

[57] de Branges, L. 1968. *Hilbert Spaces of Entire Functions*. Englewood Cliffs, NJ: Prentice-Hall.

[58] de Branges, L. and Rovnyak, J. 1966. *Square Summable Power Series*. London: Holt, Rinehart and Winston.

[59] Douglas, R. G. 1998. *Banach Algebra Techniques in Operator Theory*, 2nd edn. Graduate Texts in Mathematics, vol. 179. New York: Springer-Verlag.

[60] Douglas, R. G., and Helton, J. W. 1973. Inner dilations of analytic matrix functions and Darlington synthesis. *Acta Sci. Math. (Szeged)*, **34**, 61–67.

[61] Douglas, R. G., Shapiro, H. S., and Shields, A. L. 1970. Cyclic vectors and invariant subspaces for the backward shift operator. *Ann. Inst. Fourier (Grenoble)*, **20**(1), 37–76.

[62] Douglas, R. G. 1972. *Banach Algebra Techniques in Operator Theory*. Pure and Applied Mathematics, vol. 49. New York: Academic Press.

[63] Duren, P. L. 1970. *Theory of H^p Spaces*. New York: Academic Press.

[64] Duren, P. and Schuster, A. 2004. *Bergman Spaces*. Mathematical Surveys and Monographs, vol. 100. Providence, RI: American Mathematical Society.

[65] Dyakonov, K. and Khavinson, D. 2006. Smooth functions in star-invariant subspaces. In *Recent Advances in Operator-Related Function Theory*. Contemporary Mathematics, vol. 393. Providence, RI: American Mathematical Society, pp. 59–66.

[66] Dyakonov, K. M. 2000. Kernels of Toeplitz operators via Bourgain's factorization theorem. *J. Funct. Anal.*, **170**(1), 93–106.

[67] Dyakonov, K. M. 2012. Zeros of analytic functions, with or without multiplicities. *Math. Ann.*, **352**(3), 625–641.

[68] Dyakonov, K. M. 2014. Two problems on coinvariant subspaces of the shift operator. *Integral Equations Operator Theory*, **78**(2), 151–154.

[69] Ebbinghaus, H.-D., Hermes, H., Hirzebruch, F., Koecher, M., Mainzer, K., Neukirch, J., Prestel, A., and Remmert, R. 1991. *Numbers*. Graduate Texts in Mathematics, vol. 123. New York: Springer-Verlag. Translated from the second 1988 German edition by H. L. S. Orde.

[70] El-Fallah, O., Kellay, K., Mashreghi, J., and Ransford, T. 2014. *A Primer on the Dirichlet Space*. Cambridge Tracts in Mathematics, vol. 203. Cambridge: Cambridge University Press.

[71] Erdélyi, I. 1966. On partial isometries in finite-dimensional Euclidean spaces. *SIAM J. Appl. Math.*, **14**, 453–467.

[72] Fabry, E. 1898–1899. Sur les séries de Taylor qui ont une infinité de points singuliers. *Acta Math.*, **22**, 65–88.

[73] Fricain, E., and Mashreghi, J. 2014. *The Theory of H(b) spaces, Volume 1*. New Mathematical Monographs, vol. 20. Cambridge: Cambridge University Press.

[74] Fricain, E., and Mashreghi, J. 2015. *The Theory of H(b) Spaces, Volume 2*. New Mathematical Monographs, vol. 21. Cambridge: Cambridge University Press.

[75] Frostman, O. 1942. Sur les produits de Blaschke. *Kungl. Fysiografiska Sällskapets i Lund Förhandlingar [Proc. Roy. Physiog. Soc. Lund]*, **12**(15), 169–182.

[76] Fuhrmann, P. A. 1968. On the corona theorem and its application to spectral problems in Hilbert space. *Trans. Amer. Math. Soc.*, **132**, 55–66.

[77] Fuhrmann, P. A. 1976. Exact controllability and observability and realization theory in Hilbert space. *J. Math. Anal. Appl.*, **53**(2), 377–392.

[78] Garcia, S. R. 2006. Conjugation and Clark operators. In *Recent Advances in Operator-related Function Theory*. Contemporary Mathematics, vol. 393. Providence, RI: American Mathematical Society, pp. 67–111.

[79] Garcia, S. R. 2008a. Aluthge transforms of complex symmetric operators. *Integral Equations Operator Theory*, **60**(3), 357–367.

[80] Garcia, S. R. and Putinar, M. 2006. Complex symmetric operators and applications. *Trans. Amer. Math. Soc.*, **358**(3), 1285–1315 (electronic).

[81] Garcia, S. R. and Putinar, M. 2007. Complex symmetric operators and applications. II. *Trans. Amer. Math. Soc.*, **359**(8), 3913–3931 (electronic).

[82] Garcia, S. R. and Wogen, W. R. 2009. Complex symmetric partial isometries. *J. Funct. Anal.*, **257**(4), 1251–1260.

[83] Garcia, S. R. and Wogen, W. R. 2010. Some new classes of complex symmetric operators. *Trans. Amer. Math. Soc.*, **362**(11), 6065–6077.

[84] Garcia, S. R. and Putinar, M. 2008. Interpolation and complex symmetry. *Tohoku Math. J. (2)*, **60**(3), 423–440.

[85] Garcia, S. R. 2005a. A *-closed subalgebra of the Smirnov class. *Proc. Amer. Math. Soc.*, **133**(7), 2051–2059 (electronic).

[86] Garcia, S. R. 2005b. Conjugation, the backward shift, and Toeplitz kernels. *J. Operator Theory*, **54**(2), 239–250.

[87] Garcia, S. R. 2005c. Inner matrices and Darlington synthesis. *Methods Funct. Anal. Topology*, **11**(1), 37–47.

[88] Garcia, S. R. 2008b. The eigenstructure of complex symmetric operators. Pages 169–183 of: *Recent advances in matrix and operator theory*. Oper. Theory Adv. Appl., vol. 179. Basel: Birkhäuser.

[89] Garcia, S. R. and Poore, D. E. 2013a. On the closure of the complex symmetric operators: compact operators and weighted shifts. *J. Funct. Anal.*, **264**(3), 691–712.

[90] Garcia, S. R. and Poore, D. E. 2013b. On the norm closure problem for complex symmetric operators. *Proc. Amer. Math. Soc.*, **141**(2), 549.

[91] Garcia, S. R. and Ross, W. T. 2009. A non-linear extremal problem on the Hardy space. *Comput. Methods Funct. Theory*, **9**(2), 485–524.

[92] Garcia, S. R. and Ross, W. T. 2010. The norm of a truncated Toeplitz operator. In *Hilbert Spaces of Analytic Functions*. CRM Proc. Lecture Notes, vol. 51. Providence, RI: American Mathematical Society, pp. 59–64.

[93] Garcia, S. R. and Ross, W. T. 2013. Recent progress on truncated Toeplitz operators. In *Blaschke Products and their Applications*. Fields Institute Communications, vol. 65. New York: Springer, pp. 275–319.

[94] Garcia, S. R. and Sarason, D. 2003. Real outer functions. *Indiana Univ. Math. J.*, **52**(6), 1397–1412.

[95] Garcia, S. R., Ross, W. T., and Wogen, W. R. 2010a. Spatial isomorphisms of algebras of truncated Toeplitz operators. *Indiana Univ. Math. J.*, **59**(6), 1971–2000.

[96] Garcia, S. R., Ross, W. T., and Wogen, W. R. 2010b. Spatial isomorphisms of algebras of truncated Toeplitz operators. *Indiana Univ. Math. J.*, **59**(6), 1971–2000.

[97] Garnett, J. 2007. *Bounded Analytic Functions*. Graduate Texts in Mathematics, vol. 236. New York: Springer.

[98] Gilbreath, T. M. and Wogen, W. R. 2007. Remarks on the structure of complex symmetric operators. *Integral Equations Operator Theory*, **59**(4), 585–590.

[99] Gohberg, I. C. and Krupnik, N. Ja. 1969. The algebra generated by the Toeplitz matrices. *Funkcional. Anal. i Priložen.*, **3**(2), 46–56.

[100] Gorbachuk, M. L. and Gorbachuk, V. I. 1997. *M. G. Krein's Lectures on Entire Operators*. Operator Theory: Advances and Applications, vol. 97. Basel: Birkhäuser Verlag.

[101] Hadamard, J. 1892. Essai sur l'étude des fonctions données par leur développement de Taylor. *J. Math.*, **8**, 101–186.

[102] Hartmann, A. and Ross, W. T. 2012a. Bad boundary behavior in star-invariant subspaces I. *Arkiv for Matematik*, 1–22.

[103] Hartmann, A. and Ross, W. T. 2012b. Bad boundary behavior in star invariant subspaces II. *Ann. Acad. Sci. Fenn. Math.*, **37**(2), 467–478.

[104] Hartmann, A. and Ross, W. T. 2013. Truncated Toeplitz operators and boundary values in nearly invariant subspaces. *Complex Anal. Oper. Theory*, **7**(1), 261–273.

[105] Havin, V. and Mashreghi, J. 2003a. Admissible majorants for model subspaces of H^2. I. Slow winding of the generating inner function. *Canad. J. Math.*, **55**(6), 1231–1263.

[106] Havin, V. and Mashreghi, J. 2003b. Admissible majorants for model subspaces of H^2. II. Fast winding of the generating inner function. *Canad. J. Math.*, **55**(6), 1264–1301.

[107] Hayashi, E. 1986. The kernel of a Toeplitz operator. *Integral Equations Operator Theory*, **9**(4), 588–591.

[108] Hayashi, E. 1990. Classification of nearly invariant subspaces of the backward shift. *Proc. Amer. Math. Soc.*, **110**(2), 441–448.

[109] Hedenmalm, H., Korenblum, B., and Zhu, K. 2000. *Theory of Bergman spaces*. Graduate Texts in Mathematics, vol. 199. New York: Springer-Verlag.

[110] Helson, H. 1990. Large analytic functions. II. In *Analysis and partial differential equations*. Lecture Notes in Pure and Applied Mathematics, vol. 122. New York: Dekker, pp. 217–220.

[111] Hitt, D. 1988. Invariant subspaces of H^2 of an annulus. *Pacific J. Math.*, **134**(1), 101–120.

[112] Hoffman, K. 1962. *Banach Spaces of Analytic Functions*. Prentice-Hall Series in Modern Analysis. Englewood Cliffs, NJ: Prentice-Hall Inc.

[113] Hruščev, S. V., and Vinogradov, S. A. 1981. Inner functions and multipliers of Cauchy type integrals. *Ark. Mat.*, **19**(1), 23–42.

[114] Inoue, J. 1994. An example of a nonexposed extreme function in the unit ball of H^1. *Proc. Edinburgh Math. Soc. (2)*, **37**(1), 47–51.

[115] Jung, S., Ko, E., and Lee, J. On scalar extensions and spectral decompositions of complex symmetric operators. *J. Math. Anal. Appl.* preprint.

[116] Jung, S., Ko, E., Lee, M., and Lee, J. 2011. On local spectral properties of complex symmetric operators. *J. Math. Anal. Appl.*, **379**, 325–333.

[117] Kelley, J. L. 1975. *General topology*. New York-Berlin: Springer-Verlag. Reprint of the 1955 edition [Van Nostrand, Toronto, Ont.], Graduate Texts in Mathematics, No. 27.

[118] Koosis, P. 1998. *Introduction to H_p Spaces*, 2nd edn. Cambridge Tracts in Mathematics, vol. 115. Cambridge: Cambridge University Press.

[119] Kriete, III, T. L. 1970/71. On the Fourier coefficients of outer functions. *Indiana Univ. Math. J.*, **20**, 147–155.

[120] Kriete, III, T. L. 1971. A generalized Paley-Wiener theorem. *J. Math. Anal. Appl.*, **36**, 529–555.

[121] Lacey, M. T., Sawyer, E. T., Shen, C.-Y., Uriarte-Tuero, I., and Wick, B. D. Two weight inequalities for the Cauchy transform from $\mathbb{R}$ to $\mathbb{C}_+$. ArXiv:1310 .4820v2.

[122] Łanucha, B. 2014. Matrix representations of truncated Toeplitz operators. *J. Math. Anal. Appl.*, **413**(1), 430–437.

[123] Li, C. G., Zhu, S., and Zhou, T. Foguel operators with complex symmetry. preprint.

[124] Lim, L.-H., and Ye, K. Every matrix is a product of Toeplitz matrices. arXiv:1307.5132v2.

[125] Livšic, M. 1946. On a class of linear operators in Hilbert space. *Mat. Sb.*, **19**, 239–262.

[126] Livšic, M. S. 1960. Isometric operators with equal deficiency indices, quasi-unitary operators. *Amer. Math. Soc. Transl. (2)*, **13**, 85–103.

[127] Lohwater, A. J. and Piranian, G. 1957. The boundary behavior of functions analytic in a disk. *Ann. Acad. Sci. Fenn. Ser. A. I.*, **1957**(239), 17.

[128] Lotto, B. A. and McCarthy, J. E. 1993. Composition preserves rigidity. *Bull. London Math. Soc.*, **25**(6), 573–576.

[129] Lusin, N. and Priwaloff, J. 1925. Sur l'unicité et la multiplicité des fonctions analytiques. *Ann. Sci. École Norm. Sup. (3)*, **42**, 143–191.

[130] Makarov, N. and Poltoratski, A. 2005. Meromorphic inner functions, Toeplitz kernels and the uncertainty principle. In *Perspectives in Analysis*. Mathematical Physics Studies, vol. 27. Berlin: Springer, pp. 185–252.

[131] Martin, R. T. W. 2010. Symmetric operators and reproducing kernel Hilbert spaces. *Complex Anal. Oper. Theory*, **4**(4), 845–880.

[132] Martin, R. T. W. 2011. Representation of simple symmetric operators with deficiency indices $(1, 1)$ in de Branges space. *Complex Anal. Oper. Theory*, **5**(2), 545–577.

[133] Martin, R. T. W. 2013. Unitary perturbations of compressed n-dimensional shifts. *Complex Anal. Oper. Theory*, **7**(4), 765–799.

[134] Martínez-Avendaño, R. A. and Rosenthal, P. 2007. *An Introduction to Operators on the Hardy-Hilbert Space*. Graduate Texts in Mathematics, vol. 237. New York: Springer.

[135] Mashreghi, J. and Shabankhah, M. 2014. Composition of inner functions. *Canad. J. Math.*, **66**(2), 387–399.

[136] Mashreghi, J. 2009. *Representation Theorems in Hardy Spaces*. London Mathematical Society Student Texts, vol. 74. Cambridge: Cambridge University Press.

[137] Mashreghi, J. 2013. *Derivatives of Inner Functions*. Fields Institute Monographs, vol. 31. New York: Springer.

[138] Mashreghi, J., and Shabankhah, M. 2013. Composition operators on finite rank model subspaces. *Glasg. Math. J.*, **55**(1), 69–83.

[139] Mason, R. C. 1984. *Diophantine equations over function fields*. London Mathematical Society Lecture Note Series, vol. 96. Cambridge: Cambridge University Press.

[140] Moeller, J. W. 1962. On the spectra of some translation invariant spaces. *J. Math. Anal. Appl.*, **4**, 276–296.

[141] Nikolski, N. 1986. *Treatise on the Shift Operator*. Berlin: Springer-Verlag.

[142] Nikolski, N. 2002a. *Operators, Functions, and Systems: An Easy Reading. Vol. 1*. Mathematical Surveys and Monographs, vol. 92. Providence, RI: American Mathematical Society. Translated from the French by Andreas Hartmann.

[143] Nikolski, N. 2002b. *Operators, Functions, and Systems: An Easy Reading. Vol. 2*. Mathematical Surveys and Monographs, vol. 93. Providence, RI: American Mathematical Society. Translated from the French by Andreas Hartmann and revised by the author.

[144] Paulsen, V. 2009. *An Introduction to the Theory of Reproducing Kernel Hilbert Spaces*. www.math.uh.edu/vern/rkhs.pdf.

[145] Peller, V. V. 1993. Invariant subspaces of Toeplitz operators with piecewise continuous symbols. *Proc. Amer. Math. Soc.*, **119**(1), 171–178.

[146] Poltoratski, A. and Sarason, D. 2006. Aleksandrov–Clark measures. In *Recent Advances in Operator-related Function Theory*. Contemporary Mathematics, vol. 393. Providence, RI: American Mathematical Society, pp. 1–14.

[147] Poltoratski, A. G. 2001. Properties of exposed points in the unit ball of H^1. *Indiana Univ. Math. J.*, **50**(4), 1789–1806.

[148] Poltoratskiĭ, A. G. 1993. Boundary behavior of pseudocontinuable functions. *Algebra i Analiz*, **5**(2), 189–210.

[149] Privalov, I. I. 1956. *Randeigenschaften Analytischer Funktionen*. Berlin: VEB Deutscher Verlag der Wissenschaften.

[150] Radjavi, H. and Rosenthal, P. 1973. *Invariant Subspaces*. Heidelberg: Springer-Verlag.

[151] Richter, S. 1988. Invariant subspaces of the Dirichlet shift. *J. Reine Angew. Math.*, **386**, 205–220.

[152] Riesz, F. and Sz.-Nagy, B. 1955. *Functional Analysis*. New York: Frederick Ungar. Translated by Leo F. Boron.

[153] Rosenblum, M. and Rovnyak, J. 1985. *Hardy Classes and Operator Theory*. Oxford Mathematical Monographs. New York: The Clarendon Press/Oxford University Press.

[154] Ross, W. T., and Shapiro, H. S. 2002. *Generalized Analytic Continuation*. University Lecture Series, vol. 25. Providence, RI: American Mathematical Society.

[155] Ross, W. T. 2008. Indestructible Blaschke products. In *Banach Spaces of Analytic Functions*. Contemporary Mathematics, vol. 454. Providence, RI: American Mathematical Society, pp. 119–134.

[156] Ross, W. T. and Shapiro, H. S. 2003. Prolongations and cyclic vectors. *Comput. Methods Funct. Theory*, 3(1–2), 453–483.

[157] Royden, H. L. 1988. *Real Analysis*, 3rd edn. New York: Macmillan.

[158] Rudin, W. 1987. *Real and Complex Analysis*, 3rd edn. New York: McGraw-Hill.

[159] Rudin, W. 1991. *Functional Analysis*, 2nd edn. New York: McGraw-Hill.

[160] Saksman, E. 2007. An elementary introduction to Clark measures. *Topics in Complex Analysis and Operator Theory*. Málaga: University of Málaga, pp. 85–136.

[161] Sarason, D. 1965a. A remark on the Volterra operator. *J. Math. Anal. Appl.*, **12**, 244–246.

[162] Sarason, D. 1967. Generalized interpolation in H^∞. *Trans. Amer. Math. Soc.*, **127**, 179–203.

[163] Sarason, D. 1988. Nearly invariant subspaces of the backward shift. In *Contributions to Operator Theory and its Applications (Mesa, AZ, 1987)*. Operations Theory and Advanced Applications, vol. 35. Basel: Birkhäuser, pp. 481–493.

[164] Sarason, D. 1994a. Kernels of Toeplitz operators. In *Toeplitz Operators and Related Topics (Santa Cruz, CA, 1992)*. Operations Theory Advanced Applications, vol. 71. Basel: Birkhäuser, pp. 153–164.

[165] Sarason, D. 1994b. *Sub-Hardy Hilbert Spaces in the Unit Disk*. University of Arkansas Lecture Notes in the Mathematical Sciences, 10. New York: John Wiley & Sons.

[166] Sarason, D. 2007. Algebraic properties of truncated Toeplitz operators. *Oper. Matrices*, **1**(4), 491–526.

[167] Sarason, D. 1965b. On spectral sets having connected complement. *Acta Sci. Math. (Szeged)*, **26**, 289–299.

[168] Sarason, D. 1989. Exposed points in H^1. I. In *The Gohberg Anniversary Collection, Vol. II (Calgary, AB, 1988)*. Operations Theory Advanced Applications, vol. 41. Birkhäuser, Basel, pp. 485–496.

[169] Sedlock, N. 2010. Properties of truncated Toeplitz operators. Ph.D. thesis. ProQuest LLC, Ann Arbor, MI; Washington University in St. Louis.

[170] Sedlock, N. 2011. Algebras of truncated Toeplitz operators. *Oper. Matrices*, **5**(2), 309–326.

[171] Shapiro, H. S. 1964. Weakly invertible elements in certain function spaces, and generators in ℓ_1. *Michigan Math. J.*, **11**, 161–165.

[172] Shapiro, H. S. 1968. Generalized analytic continuation. In *Symposia on Theoretical Physics and Mathematics, Vol. 8 (Symposium, Madras, 1967)*. New York: Plenum, pp. 151–163.

[173] Shapiro, H. S. 1968/1969. Functions nowhere continuable in a generalized sense. *Publ. Ramanujan Inst. No.*, **1**, 179–182.

[174] Shapiro, J. H. 1993. *Composition Operators and Classical Function Theory*. Universitext: Tracts in Mathematics. New York: Springer-Verlag.

[175] Shimorin, S. 2001. Wold-type decompositions and wandering subspaces for operators close to isometries. *J. Reine Angew. Math.*, **531**, 147–189.

[176] Shirokov, N. A. 1978. Ideals and factorization in algebras of analytic functions that are smooth up to the boundary. *Trudy Mat. Inst. Steklov.*, **130**, 196–223.

[177] Shirokov, N. A. 1981. Division and multiplication by inner functions in spaces of analytic functions smooth up to the boundary. In *Complex Analysis and Spectral Theory (Leningrad, 1979/1980)*. Lecture Notes in Mathematics, vol. 864. Berlin: Springer, pp. 413–439.

[178] Simon, B. 1995. Spectral analysis of rank one perturbations and applications. In *Mathematical Quantum Theory. II. Schrödinger Operators (Vancouver, BC, 1993)*. CRM Proceedings Lecture Notes, vol. 8. Providence, RI: American Mathematical Society, pp. 109–149.

[179] Simon, B. 2005. *Orthogonal Polynomials on the Unit Circle. Part 1*. American Mathematical Society Colloquium Publications, vol. 54. Providence, RI: American Mathematical Society.

[180] Simon, B. and Wolff, T. 1986. Singular continuous spectrum under rank one perturbations and localization for random Hamiltonians. *Comm. Pure Appl. Math.*, **39**(1), 75–90.

[181] Singer, I. 1970. *Bases in Banach Spaces. I*. New York: Springer-Verlag.

[182] Stegenga, D. A. 1980. Multipliers of the Dirichlet space. *Illinois J. Math.*, **24**(1), 113–139.

[183] Stothers, W. W. 1981. Polynomial identities and Hauptmoduln. *Quart. J. Math. Oxford Ser. (2)*, **32**(127), 349–370.

[184] Sz.-Nagy, B. 1953. Sur les contractions de l'espace de Hilbert. *Acta Sci. Math. Szeged*, **15**, 87–92.

[185] Sz.-Nagy, B. and Foiaş, C. 1968. Dilatation des commutants d'opérateurs. *C. R. Acad. Sci. Paris Sér. A-B*, **266**, A493–A495.

[186] Sz.-Nagy, B., Foias, C., Bercovici, H., and Kérchy, L. 2010. *Harmonic Analysis of Operators on Hilbert Space*, 2nd edn. Universitext. New York: Springer.

[187] Takenaka, S. 1925. On the orthonormal functions and a new formula of interpolation. *Jap. J. Math.*, **2**, 129 – 145.

[188] Teschl, G. 2009. *Mathematical methods in quantum mechanics*. Graduate Studies in Mathematics, vol. 99. Providence, RI: American Mathematical Society.

[189] Vol'berg, A. L. 1981. Thin and thick families of rational fractions. In *Complex Analysis and Spectral Theory (Leningrad, 1979/1980)*. Lecture Notes in Mathematics, vol. 864. Berlin: Springer, pp. 440–480.

[190] Vol'berg, A. L., and Treil', S. R. 1986. Embedding theorems for invariant subspaces of the inverse shift operator. *Zap. Nauchn. Sem. Leningrad. Otdel. Mat. Inst. Steklov. (LOMI)*, **149**(XV), 38–51, 186–187.

[191] Wang, X. and Gao, Z. 2009. A note on Aluthge transforms of complex symmetric operators and applications. *Integral Equations Operator Theory*, **65**(4), 573–580.

[192] Wang, X. and Gao, Z. 2010. Some equivalence properties of complex symmetric operators. *Math. Pract. Theory*, **40**(8), 233–236.

[193] Yabuta, K. 1971. Remarks on extremum problems in H^1. *Tôhoku Math. J. (2)*, **23**, 129–137.

[194] Zagorodnyuk, S. M. 2010. On a J-polar decomposition of a bounded operator and matrix representations of J-symmetric, J-skew-symmetric operators. *Banach J. Math. Anal.*, **4**(2), 11–36.

[195] Zhu, S. and Li, C. G. 2013. Complex symmetric weighted shifts. *Trans. Amer. Math. Soc.*, **365**(1), 511–530.

[196] Zhu, S., Li, C. G., and Ji, Y. Q. 2012. The class of complex symmetric operators is not norm closed. *Proc. Amer. Math. Soc.*, **140**(5), 1705–1708.

Index

318

Printed in the United States
by Baker & Taylor Publisher Services